Achieving sustainable production of poultry meat

Volume 3: Animal health and welfare

It is widely recognised that agriculture is a significant contributor to global warming and climate change. Agriculture needs to reduce its environmental impact and adapt to current climate change whilst still feeding a growing population, i.e. become more 'climate-smart'. Burleigh Dodds Science Publishing is playing its part in achieving this by bringing together key research on making the production of the world's most important crops and livestock products more sustainable.

Based on extensive research, our publications specifically target the challenge of climate-smart agriculture. In this way we are using 'smart publishing' to help achieve climate-smart agriculture.

Burleigh Dodds Science Publishing is an independent and innovative publisher delivering high quality customer-focused agricultural science content in both print and online formats for the academic and research communities. Our aim is to build a foundation of knowledge on which researchers can build to meet the challenge of climate-smart agriculture.

For more information about Burleigh Dodds Science Publishing simply call us on +44 (0) 1223 839365, email info@bdspublishing.com or alternatively please visit our website at www.bdspublishing.com.

Related titles:

Achieving sustainable production of poultry meat Volume 1: Safety, quality and sustainability
Print (ISBN 978-1-78676-064-7); Online (ISBN 978-1-78676-066-1, 978-1-78676-067-8)

Achieving sustainable production of poultry meat Volume 2: Breeding and nutrition
Print (ISBN 978-1-78676-068-5); Online (ISBN 978-1-78676-070-8, 978-1-78676-071-5)

Achieving sustainable production of eggs Volume 1: Safety and quality
Print (ISBN 978-1-78676-076-0); Online (ISBN 978-1-78676-078-4, 978-1-78676-079-1)

Achieving sustainable production of eggs Volume 2: Animal welfare and sustainability
Print (ISBN 978-1-78676-080-7); Online (ISBN 978-1-78676-082-1, 978-1-78676-083-8)

Chapters are available individually from our online bookshop: https://shop.bdspublishing.com

climate•SMART•publishing
IN AGRICULTURAL SCIENCE

BURLEIGH DODDS SERIES IN AGRICULTURAL SCIENCE

NUMBER 15

Achieving sustainable production of poultry meat

Volume 3: Animal health and welfare

Edited by Professor Todd Applegate, University of Georgia, USA

burleigh dodds
SCIENCE PUBLISHING

Published by Burleigh Dodds Science Publishing Limited
82 High Street, Sawston, Cambridge CB22 3HJ, UK
www.bdspublishing.com

Burleigh Dodds Science Publishing, 1518 Walnut Street, Suite 900, Philadelphia, PA 19102-3406, USA

First published 2017 by Burleigh Dodds Science Publishing Limited

Library of Congress Control Number: 2016956458

British Library Cataloguing in Publication Data
A catalogue record for this book is available from the British Library

ISBN 978-1-78676-072-2 (print)
ISBN 978-1-78676-074-6 (online)
ISBN 978-1-78676-075-3 (online)
ISSN 2059-6936 (print)
ISSN 2059-6944 (online)

Typeset by Deanta Global Publishing Services, Chennai, India
Printed by Lightning Source

Contents

Series list

Title	Series number
Achieving sustainable cultivation of maize - Vol 1 From improved varieties to local applications Edited by: Dr Dave Watson, CGIAR Maize Research Program Manager, CIMMYT, Mexico	001
Achieving sustainable cultivation of maize - Vol 2 Cultivation techniques, pest and disease control Edited by: Dr Dave Watson, CGIAR Maize Research Program Manager, CIMMYT, Mexico	002
Achieving sustainable cultivation of rice - Vol 1 Breeding for higher yield and quality Edited by: Prof. Takuji Sasaki, Tokyo University of Agriculture, Japan	003
Achieving sustainable cultivation of rice - Vol 2 Cultivation, pest and disease management Edited by: Prof. Takuji Sasaki, Tokyo University of Agriculture, Japan	004
Achieving sustainable cultivation of wheat - Vol 1 Breeding, quality traits, pests and diseases Edited by: Prof. Peter Langridge, The University of Adelaide, Australia	005
Achieving sustainable cultivation of wheat - Vol 2 Cultivation techniques Edited by: Prof. Peter Langridge, The University of Adelaide, Australia	006
Achieving sustainable cultivation of tomatoes Edited by: Dr Autar Mattoo, USDA-ARS, USA & Prof. Avtar Handa, Purdue University, USA	007
Achieving sustainable production of milk - Vol 1 Milk composition, genetics and breeding Edited by: Dr Nico van Belzen, International Dairy Federation (IDF), Belgium	008
Achieving sustainable production of milk - Vol 2 Safety, quality and sustainability Edited by: Dr Nico van Belzen, International Dairy Federation (IDF), Belgium	009
Achieving sustainable production of milk - Vol 3 Dairy herd management and welfare Edited by: Prof. John Webster, University of Bristol, UK	010
Ensuring safety and quality in the production of beef - Vol 1 Safety Edited by: Prof. Gary Acuff, Texas A&M University, USA & Prof.James Dickson, Iowa State University, USA	011
Ensuring safety and quality in the production of beef - Vol 2 Quality Edited by: Prof. Michael Dikeman, Kansas State University, USA	012
Achieving sustainable production of poultry meat - Vol 1 Safety, quality and sustainability Edited by: Prof. Steven C. Ricke, University of Arkansas, USA	013
Achieving sustainable production of poultry meat - Vol 2 Breeding and nutrition Edited by: Prof. Todd Applegate, University of Georgia, USA	014
Achieving sustainable production of poultry meat - Vol 3 Health and welfare Edited by: Prof. Todd Applegate, University of Georgia, USA	015

Introduction

Poultry production faces a range of challenges. These are addressed in the three volumes of *Achieving sustainable production of poultry meat*. The three volumes are:

- Volume 1: Safety, quality and sustainability
- Volume 2: Breeding and nutrition
- Volume 3: Health and welfare

Volume 3 discusses recent research on improving poultry health and welfare.

Part 1 Health

Chapters in Part 1 look at key aspects of poultry health, disease prevention and management. Poultry are infected by numerous pathogens that cause a wide variety of diseases. Effective disease surveillance is critical to identify and combat both existing and emerging diseases. Disease surveillance is a staple program in most developed poultry systems and is a cooperative effort between individual poultry companies, regional and national disease control agencies and programs, public and private diagnostic laboratories and other partners.

As Chapter 1 points out, surveillance can be active in the form of sacrificing birds for clinical necropsy examination or submitting samples for diagnosis during a disease outbreak. It can also be passive in the form of routine diagnostic assays designed to provide a historical picture of a flock or region. The chapter reviews key challenges and best practice for both types of surveillance, covering diagnostic laboratories, diagnostic assays, high quality sampling and submission. The data that these submissions and clinical cases provide is invaluable both in highlighting particular problems and in identifying epidemiological patterns of disease. These patterns can help to develop models that can predict the spread of disease and make management of disease outbreaks timely and effective.

Most pathogens of poultry and other vertebrates enter the body via a mucosal portal of entry, and for most birds, this means either the respiratory or gastrointestinal tract (GIT). Chapter 2 surveys the main bacterial pathogens found in the GIT of poultry such as *Salmonella*, *E.coli* and *Clostridium*. The chapter also discusses what we know about the intestinal microflora of chickens, notably phyla such as Firmicutes and Bacteroidetes, and how the assembly of gut microflora is regulated by complex host–microbial and microbial–microbial interactions. The chapter then goes on to review what causes dysbiosis, the shift in the microbiome which results in an imbalance between beneficial and harmful bacteria. This imbalance has a negative impact in the GIT, including reduced nutrient digestibility, increased gut permeability and upregulated inflammatory responses. The eubiosis/dysbiosis status of the microbiome is thus a major factor in poultry health or disease. As the chapter points out, the fragile balance can be altered by any kind of stress that induces inflammation, including environmental factors, diet or pathogens. The chapter also examines the effects of inflammation of the GIT on anaerobiosis and assesses the role of mucins, glycoproteins that cover epithelial surfaces of the intestine and form a mucus layer to protect epithelial cells from infection. After exploring the effect of gut infections in disrupting dietary function, it assesses the impact of airborne infections of enteric bacteria on poultry.

As suggested in Chapter 2, infectious diseases, particularly those caused by viral pathogens, pose one of the most serious threats to the sustainability of the global poultry industry. The industry has relied heavily on vaccination-based control strategies, but there are major drawbacks in this approach, such as the immunosuppressive effects of some viral pathogens and the increasing diversity of pathogens. This chapter reviews viral diseases affecting poultry, including the most significant viral diseases: Newcastle disease, infectious bronchitis, infectious bursal disease, Marek's disease, infectious laryngotracheitis and avian influenza. It discusses each in terms of their distribution, effect on avian health, and role in producing economic losses. For each disease, current findings on symptoms, diagnosis and control measures are discussed. The major threats to vaccination-based strategies are reviewed together with potential solutions.

Many of the avian pathogens continue to emerge and re-emerge, showing huge diversity which poses immense challenges to traditional vaccination-based methods of control. These challenges are evident in diseases such as Marek's disease, where there is something of a biological arms race between vaccines and pathogens able to evolve resistance. Pathogen diversity also poses problems in diagnosis, although the advent of new generation sequencing and PCR-based diagnostic tests are helping to tackle the issues. In this context, development of DIVA (differentiating infected from vaccinated animals) tests and rapid diagnostic tools that could be deployed in the field are vital for curbing outbreaks. Finally, it is important to examine the heavy reliability on vaccination strategy for sustainable control. The recent success of recombinant vectored vaccines against a number of avian diseases is a welcome change as some of the more traditional vaccines are becoming less effective in inducing protection.

Like viruses, parasitic infections and infestations are a major challenge to poultry production worldwide. Chapter 4 provides a comprehensive survey of parasites affecting poultry, divided into the broad categories of protozoa, flagellates and worms. It also discusses arthropods including lice, bedbugs, flies, beetles, mites, fleas and ticks. The chapter examines diagnosis, control, management and disinfection for each of the major parasites associated with poultry farming. It also addresses the pathology, epidemiology and molecular biology of the major parasitic infections.

As the chapter points out, the most severe parasitic diseases are caused by protozoans and include coccidiosis and histomoniasis which are discussed in detail. The chapter also discusses blood-borne protozoan parasites, roundworms, tapeworms and trematodes (flukes) in poultry. As the chapter highlights, future control of parasites in poultry will have to be maintained with fewer, less-effective vaccines. Current avenues of research include selection of poultry for improved disease resistance, and the use of natural products to stimulate innate and acquired resistance to infection. With the increasing importance of free-range rearing of poultry, more research is needed on roundworms, tapeworms and methods for their prevention. The interaction of nematode parasites and protozoan parasites, is of major importance, as well as the interaction of parasites with bacterial diseases such as necrotic enteritis. The chapter concludes by noting the need to find new ways to control parasites if we are to maintain production to meet the demand for high quality food by a growing world population.

As described in previous chapters, diseases in intensive poultry flocks may occur due to viral, bacterial, fungal, protozoal, metazoan or arthropod infections or infestations. Chapter 5 describes current disease preventative measures, health monitoring and disease investigation techniques, and the management of infected poultry flocks. The chapter discusses both routine monitoring and more targeted investigations in the face of a particular outbreak. It also highlights the value of systematic investigation of outbreaks to both improve understanding of a disease and improve techniques for its prevention and management. This systematic approach involves: systematic epidemiological investigation to identify potential risk factors; investigation of risk factors through controlled experiments; and confirmation of the findings in field situations.

Chapter 5 also provides a comprehensive review of biosecurity measures to prevent disease and minimise its spread. Key measures include an 'all-in all-out' approach which, for example, avoids multi-age flocks where older, infected flocks spread disease to young birds. The chapter looks at appropriate quarantine and hygiene procedures, including disinfection regimes and litter management, including composting. The chapter also summarises best practice for dealing with an outbreak of disease in the flock, including hygiene procedures in these situations and methods of disposing of dead birds.

In the perpetual battle against common poultry pathogens and other environmental antigens, the role of the immune system is an important and ongoing topic of research. Chapter 6 reviews the various elements of the host defense system in poultry, including innate and adaptive immune systems. By boosting local immune function at mucosal sites (digestive and respiratory), subclinical infections could be averted or suppressed, thus enhancing growth and general health. The chapter discusses examples of how dietary supplements can modulate the immune system and its response to common diseases. Among the most commonly used modulators are feed and/or water supplements such as probiotics and prebiotics which have been the subject of extensive research.

The chapter focuses particularly on the challenge of coccidiosis and the role of probiotics. As the chapter explains, prior to the activation of an adaptive immune response, the innate immune system will attempt to inhibit the infection through various pathways such as competitive exclusion by commensal microbiota, gastric secretions, phagocytosis and other components. The adaptive immune system uses three main mechanisms: competitive exclusion, bacterial antagonism (for example bacteriocins) and stimulation of the immune system. Research shows the role of probiotics in supporting these mechanisms in the immune system, reducing the rate and severity of enteric infections while maintaining optimal performance in poultry. It also demonstrates the potential of combining probiotics with prebiotics as well as the role of pre-hatch immune interventions. The chapter concludes by highlighting the need for more research on the immune system and its response to nutritional and other interventions.

As mentioned in Chapter 6, competitive exclusion (CE) describes the protective effect of the intestinal microflora against colonization of pathogenic bacteria. Building on Chapter 6, Chapter 7 discusses the current use of treatments to boost CE to prevent *Salmonella* infections in poultry, focussing on the administration of cultures of intestinal origin to day-old chickens. The chapter reviews research on the effectiveness of these treatments in preventing the risk for *Salmonella* infections in newly hatched chickens from being further spread horizontally in a flock or vertically in the case of breeding animals. The chapter reviews the development of undefined and defined CE cultures together with methods of evaluation and administration. It also looks at pathogen and host specificity of CE treatments and factors affecting their efficacy. As the chapter points out, the administrated cultures compensate for the delayed development of the protective intestinal microflora in chickens hatched under hygienic conditions in modern poultry production.

As well as bacterial, viral and other infections, poultry are susceptible to a range of other health problems. Bacterial chondronecrosis with osteomyelitis (BCO) is widely recognized as one of the most common causes of lameness in broilers. Chapter 8 starts by summarizing the pathogenesis of BCO. Experimental models that successfully trigger BCO are then reviewed. Sources of bacteria that infect the bones are discussed, with the primary emphasis on bacterial translocation across the gastrointestinal epithelium. Finally, the chapter reviews the efficacy of probiotics as a prophylactic treatment for BCO.

Part 2 Welfare

In the context of poultry production, behavioral studies have been particularly useful in optimizing management strategies to promote the welfare of commercially raised poultry. Chapter 9 provides an overview of the contribution that the study of animal behavior has made to the debate about poultry welfare. It also discusses the design, execution and interpretation of behavioral research. The chapter begins with an overview of commonly used research methods and methodological challenges that should be considered before behavioral data is collected. These include issues such as sample size and sampling strategy. These issues are explored in detail through a case study looking at the importance and consequences (positive and negative) of improving perch access for poultry.

As noted earlier, genetic selection for highly efficient, fast-growing broilers has had negative effects on broiler welfare. The high stocking densities used in some broiler production systems has also been an issue. These factors mean that activity levels of fast-growing broilers can be low, especially during the second half of their life. This can lead to leg health problems such as lameness, footpad dermatitis and hock burns It has also been found that fast-growing broilers are more sensitive to heart and circulation problems compared with slower growing genotypes, especially when placed in a suboptimal environment. Chapter 10 examines the welfare problems arising from intensive poultry systems, and reviews how both factors have been addressed within the European Union.

These issues are discussed in more detail through two case studies. The first examines the welfare effects of hatching environment and early feeding, whilst the second looks at the benefits of alternative and free-range broiler production systems. Both case studies illustrate how research can contribute to better standards of welfare and more sustainable broiler production. As an example, allowing chicks access to feed and water directly after hatching has the major advantage that chicks that hatch early do not have to wait for access to feed and water until all chicks have hatched. This long fasting period is increasingly criticised by animal protection organisations. Furthermore, early access to feed seems to benefit early muscle development as well as an early colonisation of the gut, boosting gut health and immunity. This may also be the reason why broilers from this type of systems seem to require fewer antibiotic treatments. The chapter also highlights the potential benefits of poultry operations that use slower growing genotypes in lower stocking densities. Evidence suggests fewer health and welfare problems and lower mortality rates in these alternative systems; however, as time to market weight is delayed, it can increase the risk of exposure to behavioural and disease maladies. In some cases these 'middle-market' systems have been shown to outperform conventional and organic systems both in welfare and cost efficiency, but remain to be demonstrated across bird final body weights and market segments.

Broiler breeders are the parent stock of broiler chickens. Appropriate management in rearing and in the production period is essential to produce a high number of fertile eggs and high quality broiler chickens. However, research has shown that current management practices may affect the welfare state of broiler breeders, both during rearing and in the production period. Chapter 11 describes the housing and management of broiler breeders during the rearing and the production periods. It then reviews major welfare issues relating to the management of broiler breeders, including feed and water restrictions, still one of the most important welfare issues in broiler breeders. It suggests that using feeds with higher insoluble fibre contents or lower protein content in the rearing period, as compared to standard commercial diets, may help to alleviate the effects of feed restriction. Another alternative is the use of slower growing or dwarf female broiler breeders that do not need to be feed restricted. The chapter also review research on excisions as well as on environmental enrichment and facilitating mating behaviour. Finally, it discusses the potential for management strategies to have transgenerational effects on the health and quality of offspring.

Chapter 12 examines the specific effects of exogenous and endogenous heat on embryonic growth during different stages of incubation, with an emphasis on control of incubation temperature. The chapter explores the challenges inherent in establishing a temperature recommendation for incubation of poultry eggs, and considers forms of pre-incubation holding as well as the outcomes of both early and late incubation temperatures. Finally, the chapter addresses the connection between incubation temperatures and poultry immunology, as well as thermal manipulations and the impact of temperature on post-hatch stress in chicks.

Environmental enrichment comprises stimuli added to the poultry housing environment to enhance the biological adaptation of the birds and improve their welfare. By promoting species-typical behavior, it has the potential to reduce the risk of harmful behaviors and health conditions, and to guide birds to use the available resources more uniformly. Environmental enrichment may also promote positive emotional states and enhance the birds' ability to cope with unpredictable environmental changes. Chapter 13 reviews environmental enrichment strategies that are relevant to commercial meat poultry production. These include cover panels, perches, barriers and ramps as well as lighting. The chapter also looks at ways of enriching foraging and grooming behaviour. The chapter discusses the benefits of such strategies in terms of the behavior and welfare of the birds and their potential for contributing to more efficient and sustainable poultry production. Potential risks or problems that may arise when environmental enrichment is not appropriately implemented are also discussed.

As discussed in other chapters, maintaining an environment that minimizes stress on poultry is a significant part of obtaining a healthy bird that reaches its full genetic potential. One challenge that producers face is preventing heat stress and maintaining performance in hot weather conditions. By keeping bird body temperature at the right level, birds will continue to grow, develop and efficiently utilize feed. Chapter 14 focuses on the principles of poultry management that farmers can implement in their poultry houses during hot weather to prevent heat stress and maintain normal bird body temperature. The chapter reviews ventilation, evaporative cooling and management practices that influence the ability of the farmer to keep birds cool.

A variety of factors before and during transport affect the welfare of poultry. Chapter 15 begins by considering the pre-loading factors which can affect the birds' welfare during transportation. It then reviews the physiological, behavioural and psychological responses which birds may have during transport. The chapter goes on to discuss a wide range of factors which may affect welfare during transport, including the design of containers and trailers, journey duration, noise, vibrations and stocking density. The chapter also considers the effects of lairage as well as the transport of hens other than adult broilers.

Stunning of animals before slaughter is widely recognised as an essential element in humane slaughter. Chapter 16 reviews current research on the physiological basis of stunning. It then assesses the relative merits of differing stunning systems, starting with electrical stunning methods and gas stunning systems. It also discusses other systems such as maceration, manual neck dislocation, concussive stunning and low atmosphere pressure systems.

Part 1

Animal health

Chapter 1

Monitoring trends in diseases of poultry

Brian Jordan, University of Georgia, USA

1 Introduction

Poultry, encompassing meat and eggs from chickens, turkeys, game birds and waterfowl, has become a staple protein source around the world. Poultry meat accounts for 28% of total meat produced in the world (http://www.fao.org/docrep/005/y4252e/y4252e05b.htm). Poultry meat consumption has risen from 3.2 kg per capita in 1966 to 13.8 kg per capita in 2015 in developed countries, and is expected to increase to 17.2 kg per capita by 2030 (http://www.fao.org/docrep/005/y4252e/y4252e05b.htm). In developing countries, poultry meat consumption has risen by nearly the same amount but on a slightly smaller scale (1.2 kg per capita in 1966 to 10.5 kg per capita in 2015). By far, the largest production sector in poultry is chickens, and the United States leads the world in chicken production. As the demand for poultry meat continues to grow, production of chickens will have to match pace. To do this, advances in husbandry, genetics and nutrition are essential. These advances are all for naught, however, if the main driver of loss in poultry production is not controlled: disease.

Commercial poultry are susceptible to many diseases, and often show more clinical pathology associated with these diseases than other bird types (Banat et al., 2013). Some poultry diseases cause high mortality (highly pathogenic avian influenza (HPAI), velogenic Newcastle disease), but most cause mild to moderate clinical pathology and mainly effect body weight gain, feed conversion or egg production. Other diseases do not affect production parameters dramatically, but can lead to condemnation of meat at processing (e.g. infectious bronchitis virus). Poultry producers have traditionally used vaccines, chemicals, antibiotics and so on to combat the most economically significant poultry diseases, but understanding what pathogens are affecting a flock is essential in making

http://dx.doi.org/10.19103/AS.2016.0011.15

decisions on which type of treatment to use. Farm history and regional disease prevalence can guide veterinarians and producers towards a diagnosis when clinical presentation is typical, but gross and microscopic pathology found at necropsy of diseased poultry and routine and diagnostic disease surveillance must be used when disease presentation is atypical or new to a farm or region. Furthermore, these tools must be used to positively identify a pathogen so that the correct course of action can be taken to treat a flock and take preventative action on susceptible flocks in the proximity. For these reasons, disease surveillance is critical for poultry producers.

Disease surveillance is a staple programme in most developed poultry systems and is a cooperative effort between individual poultry companies, regional and national disease control agencies and programmes, public and private diagnostic laboratories, research universities, wildlife agencies and allied animal health industries. Surveillance can be active in the form of sacrificing birds for clinical necropsy examination or submitting samples for diagnosis during a disease outbreak (including surveillance of wild migratory birds for detecting potential outbreaks of highly pathogenic viruses that would devastate poultry production if introduced) or passive in the form of routine diagnostic assays designed to provide a historical picture of a flock or region (discussed more below). Non-laboratory-based surveillance techniques are also included in this effort, incorporating surveys, telephone and personal contact, and direct disease reporting hotlines to name a few. Both types of surveillance have advantages and drawbacks, but are critical for understanding the aetiological agents causing disease in poultry. As pathogens continue to emerge and evolve, and as treatments are being removed from use in poultry, disease surveillance becomes even more important for maintaining healthy poultry.

2 Diagnostic laboratories

Diagnostic sampling for disease surveillance is a critical component of a poultry health plan, and diagnostic laboratories are the source point for testing and analysis of the surveillance samples. Proper sample collection is imperative to the success of a disease surveillance programme, and there are resources available that describe sample collection technique in detail (https://www.youtube.com/playlist?list=PLzf8tGKj10zzs2GQ70unp0d6oUUMNrG9, also see below). These samples must then be sent to a laboratory possessing the knowledge and technical expertise to properly evaluate the submission, perform the correct diagnostic assay and interpret the results so that the information can be used by the company veterinarian or producer when deciding on the most appropriate intervention. There are many accredited diagnostic laboratories in the United States and Europe, but not all of them focus solely on poultry. Since the United States is the largest producer of chicken meat in the world, there are many laboratories focused completely on diagnostic services for poultry companies. These laboratories are usually part of a state or national system (e.g. Georgia Poultry Laboratory Network (GPLN); Maryland Department of Agriculture Animal Health Laboratory, etc.), but there are also many research universities, especially in poultry producing regions, that have a diagnostic teaching laboratory (Poultry Diagnostic and Research Center (PDRC) at The University of Georgia; Poultry Research and Diagnostic Laboratory at Mississippi State University, etc.).

Other poultry producing countries also have reference diagnostic laboratories, but they are typically not dedicated to poultry. In Europe, the European Union (EU) has an international laboratory network with reference laboratories for certain poultry diseases, while multiple national laboratories operate within individual countries (http://ec.europa.eu/food/animals/animal-diseases/reference-lab_en). Both the EU and the United States also have private diagnostic laboratories to process poultry submissions. In Central/South America, most poultry diagnostics are processed by national laboratories that accept submissions from all animal species, or sent to laboratories outside of the country.

Understanding the cost and capability of a diagnostic laboratory is important when deciding on where a sample should be submitted. Different laboratories may offer different services and the laboratory closest to a disease break may not be the most appropriate place to send samples. For instance, during the avian influenza (AI) outbreak of 2014–15 in the United States, samples had to be confirmed AI positive by a National Animal Health Laboratory Network (NAHLN, https://www.nahln.org/) accredited laboratory before an intervention could take place. Most diagnostic laboratories are equipped to perform routine diagnostic assays (serology, bacteriology), but may not possess the equipment to perform newer diagnostic techniques (quantitative polymerase chain reaction (qPCR)). In other cases, a laboratory may not be designed to perform standard assays, but maintains all equipment necessary for molecular diagnostics. Location of a diagnostic laboratory must also be considered when submitting samples, because it can affect what type of submission can be made. It is simple to ship live samples within the United States, but much more complicated to bring live samples into the United States from other regions of the world in a timely manner. These factors can affect the usefulness of the information gained from sample submission, and may influence the decision to submit diagnostic samples in the future.

3 Diagnostic assays

Commercial poultry companies submit thousands of samples to diagnostic laboratories per year to assess general bird health. Diagnostic assays used to analyse submitted samples can be grouped into general categories: serology, virology, bacteriology, histopathology and molecular diagnostics. Serological evaluations routinely account for most of the sample testing, with histology typically accounting for the least. Within each of these categories are numerous assays that can be performed to assess the health status of the bird from which the sample originated. Further explanation of the most common assays are discussed in the following sections.

3.1 Serology

Serology refers to the study of blood serum, specifically measuring the immune response as evidenced by circulating antibodies after stimulation by a pathogen. Multiple assays utilize blood serum and are thus categorized as serological tests, but the enzyme-linked immunosorbent assay (ELISA) is the most frequently used one. ELISAs are designed to detect antibodies in blood serum that have been generated after infection by a pathogen (Engvall and Perlmann, 1971). For poultry, this assay has been adapted to detect antibodies against many viral and bacterial pathogens (Marquardt et al., 1980; Slaght et al., 1978; Snyder et al., 1983; Soula and Moreau, 1981; Ansari et al., 1983). In most cases, ELISA tests yield a straightforward positive or negative determination with high levels of sensitivity and specificity. The ELISA test is also relatively non-invasive, since blood serum is the substrate needed for the assay. The main drawback to the ELISA assay is timing. Since ELISA tests measure circulating antibody concentrations, the test can only be performed after infection and seroconversion by the bird, otherwise the rate of false negatives would be extremely high. Also, this assay does not discriminate between antibodies produced by the bird after vaccination or after a challenge, so determining the true health status of a flock is difficult. In total, however, ELISA a very good assay for surveillance and can provide a very clear picture of to what the flock has been exposed to, which is ideal when compiling disease or vaccination histories, as well as determining vaccination status.

3.2 Virology

Whereas serology is mainly used in passive surveillance to assess exposure over the life of a bird or flock, viral diagnostic assays are part of a more active surveillance where a specific virus is identified and/or isolated. There are many viruses that infect commercial poultry and they are widely distributed in type and clinical presentation. Routine viral diagnostics are typically performed in specific pathogen-free (SPF) embryonated eggs or cell culture (Dufour-Zavala et al., 2008), where specific lesions associated with a certain virus type can be seen. Viral diagnostic assays require a sample that contains live virus, which can be a more invasive sample to collect from a live bird (tissue sample or swab). Often, samples for virus isolation or detection are collected from birds at necropsy where a portion of the flock is being sampled as part of a general health check, or in situations where obvious clinical disease presentation warrants the sacrifice of several chickens for diagnostic testing. One advantage to viral diagnostics where a virus is isolated is that the virus can then be further characterized in terms of genotype, serotype, virulence and so on. These samples can then be used as starting material for the development of both autogenous and commercially produced vaccines if the need is sufficient (Jackwood et al., 2010; Perozo et al., 2008), or archived for further characterization at a later time.

3.3 Bacteriology

Bacteria are ever present in poultry production and are a normal part of the animal microbiota. Bacteria are typically thought of as secondary pathogens, though some do possess the virulence factors to cause clinical disease without any predisposition (Oakley et al., 2014; Jackwood et al., 2015). Bacterial diagnostics, like viral, require that samples be submitted that contain live specimens. For bacteria, however, sampling must be performed in a sterile manner so that the sample is not contaminated with bacteria that may be present in other parts of the chicken, but not causing the disease of note. Depending on the type of bacteria in question and the sample, it may be preferred to submit a whole bird or specific tissues to the diagnostic laboratory so that a diagnostic clinician can perform the necropsy and sampling to ensure purity and aseptic technique. Bacterial diagnostics also rely on the ability to grow bacteria in the laboratory, which molecular characterization of bacterial populations has shown us is a very small portion of the total bacteria present (Stewart, 2012). This makes bacteriology a targeted diagnostic assay, where the client submitting the sample often specifies which bacterial species should be targeted for growth and characterization.

While many bacterial species are present in the natural biota of the chicken and are not a concern for disease, some species are routinely monitored as they pose a threat to human health. Certain species or strains of *Salmonella, Staphylococcus, Campylobacter, Listeria* and some *E. coli* are considered possible pathogens contributing to foodborne illness in humans. Therefore, bacteriology is constantly performed to monitor relative levels of these bacteria in live poultry, as well as on the carcasses of slaughtered poultry (https://www.fsis.usda.gov/wps/wcm/connect/ad74bb8d-1dab-49c1-b05e-390a74ba7471/Chicken_from_Farm_to_Table.pdf?MOD=AJPERES). This monitoring is typically performed by governmental regulatory agencies and may or may not be associated with diagnostic laboratories that perform other routine poultry bacteriology.

3.4 Histopathology

Histopathology refers to the study of changes in tissues caused by a pathogen or disease progression. Histopathology as a diagnostic tool for identifying the aetiological agent causing disease is a classical method and has been used for hundreds of years. Depending on the type of infection, the actual pathogen of importance can be directly visualized in histopathological sections of tissues (Johnson and Reid, 1970). In most cases, however, definitive signs of disease progression are the most prominent finding. These specific lesions can then be used to further narrow the diagnostic differential list of possible pathogens. Histopathologic examination provides very qualitative diagnostic information, as lesion scores and disease progression for most poultry pathogens have been well described. Histopathology also provides insight into the severity of the disease state or the virulence of the pathogen causing the disease based on the severity of the tissue changes seen. Performing histopathological examination requires tissues, which requires the euthanization of the chicken to collect samples. Because routine necropsy is performed on both clinically diseased and apparently healthy birds, acquiring tissues for this purpose is not a challenge. This technique also requires specific reagents to fix the tissues (e.g. 10% neutral buffered formalin) to maintain their structure in the disease state, as well as specific sample preparation techniques and stains for inspection. Furthermore, a specialized avian pathologist must read the tissue sections and make the determination of what cellular changes have occurred and the extent to which they have progressed, making this a technique that not all diagnostic laboratories can perform specifically for poultry samples.

3.5 Molecular diagnostics

While the term 'molecular' generally refers to anything relating to molecules, molecular diagnostics most often refers to techniques involving molecular biology, such as polymerase chain reaction (PCR) and its derivatives. Indeed, PCR, reverse transcriptase PCR (RT-PCR) and quantitative PCR or RT-PCR (qPCR or qRT-PCR, also generically referred to as 'real-time' PCR) have become a predominant diagnostic assay tools used in all areas (Hoffmann et al., 2009; Mansour et al., 2015; Josefsen et al., 2015). Specific PCR-type assays have been developed for many poultry pathogens, including viruses and bacteria. This technology continues to progress in efficiency and simplicity and has replaced traditional diagnostic methods for some diagnostic laboratories for certain pathogens.

There are many advantages to performing a molecular diagnostic assay such as PCR. Results from these reactions are usually obtained in 1–3 days from the time of processing, making this turn around much faster than traditional assays. PCR tests can be more sensitive than traditional assays as well, detecting the presence of a pathogen below the detection level of a traditional assay (Roh et al., 2014) or detecting multiple pathogens in a sample. Using a PCR test also provides flexibility in the type of pathogen that can be detected; the equipment and reagents are the same for every assay. The only difference is the primer and/or probe used, which can be synthesized in the laboratory for any pathogen with known sequence. Live pathogen is not needed to perform this assay, making it possible to send samples from around the world to a diagnostic laboratory on FTA® cards for analysis. If a diagnostic laboratory with the capability to perform this test is not available within a country, being able to ship samples to an outside laboratory could dramatically expand a producer's ability to monitor disease in their flocks or production system.

There are also disadvantages to using molecular diagnostic assays such as PCR, and many of them relate directly to the advantages. While costs per reaction have decreased significantly since its inception, there is a significant upfront cost in equipment and reagent purchasing. This is easily amortized over the life of the machine, but requires initial capital. As mentioned above, the PCR assay is often more sensitive than a traditional diagnostic assay for detecting pathogens, but this must be taken into consideration when analysing results. Very weak positive reactions must be weighed against the probability that that amount of pathogen could cause disease. Furthermore, a positive reaction for a specific pathogen does not mean that it was the pathogen that caused the clinical presentation. PCR only detects what it is asked to detect, not everything possible in a sample. PCR also does not discriminate between infective and non-infective (live vs. dead) pathogens since it only measures the level of genetic material in a sample. While not needing a live pathogen to perform this assay can be a major benefit, it can also be a drawback in that there is no pathogen to further characterize after positive identification.

As mentioned previously, the PCR relies on primers and/or a probe for specificity, but these primers and probe are synthesized based on previously identified genomic sequence. If no or little genomic sequence of a pathogen is known, PCR processing is not useful (Morris and Gasser, 2006). Conversely, the PCR test may not be valid if known genomic sequences between subtypes of a pathogen are so closely related that sequence differences that could specify one subtype cannot be found (Roh et al., 2014). Primers and probes must also be thoroughly tested for sensitivity and specificity prior to use in a diagnostic setting. Region of the genome detected, secondary structure of the genomic region targeted, temperature and so on can all influence the effectiveness of the PCR. PCRs with primers and probes that are not optimized can lead to false results, which would greatly hinder the diagnostic process.

4 Diagnostic sampling and submission

Diagnostic assays are highly sensitive and specific and can provide valuable information on the health status of a flock. The assays are influenced, however, by the quality of the sample submitted, and the information obtained can be diminished if proper information was not provided at submission. Each type of diagnostic assay relies on a quality sample for analysis, so many factors such as sample collected, collection method, storage time, temperature, medium and transport method can negatively impact the quality of the starting material. Most pathogens are biologically sensitive in that extreme temperatures, oxygen levels or time outside the host can decrease growth potential or kill the agent. Training on proper sample collection and handling techniques is critical to obtaining the highest quality and most relevant results. This training is often provided by diagnostic laboratory personnel or it can be learned through courses specific to diagnostic sampling, but there are also many references publicly available (e.g. http://www.merckvetmanual.com/clinical-pathology-and-procedures/collection-and-submission-of-laboratory-samples/overview-of-collection-and-submission-of-laboratory-samples, https://www.youtube.com/playlist?list=PLzf8tGKj10zzE5jXlhAxw5Z7UaPJV6alB).

Submission of samples to a diagnostic laboratory can be just as critical as the sampling method used when collecting the sample to be analysed. Without a proper history, the full clinical picture cannot be obtained, which may prevent a clinician from making a proper diagnosis. Most diagnostic submission forms require a minimum of basic information, such as company name, flock identifier, physical location, specimen type and test to be performed, to process the sample. More information is always preferred, and qualifiers such as breed and type of bird, age, number affected, total number of birds on farm and flock history help to create the full clinical picture. When diagnostic laboratories or governmental agencies perform active surveillance, even more information regarding environmental conditions (temperature, humidity, wind speed and direction), location of samples in relation to other sampling sites, routes of animal transport in relation to positive pathogen identifications and so on are collected. Together, these data can help clinicians and diagnosticians develop a disease map, where all factors that may influence the onset, prevalence and spread of disease can be factored into the equation. But this equation can only be completed if accurate and ample information is provided in the beginning.

5 Epidemiology

Epidemiology is the study of patterns, causes, effects and nuances of disease conditions in specific populations. The data points used in this study are created from information obtained through disease surveillance performed by companies and individuals through sample submissions to diagnostic laboratories, university and governmental research programmes that survey wild and commercial avian species, and food safety and inspection services that inspect processing facilities and test for pathogen presence on food products. Individually, this data provides valuable information on the disease state of the flock or the individual bird from which it was collected. Together, these data create a landscape of disease in a population of flocks or birds. The sample population may be as small as the poultry producing regions of northeast Georgia or the DelMarVa peninsula, or they may be as large as all commercial poultry in the United States, the EU or Asia. The populations may spread over multiple continents and migrate seasonally, as is the case with wild waterfowl. They may also be a cross-over between multiple bird populations typically seen with HPAI outbreaks. To further complicate the picture, the population being analysed may change based on the epidemiological questions being asked, making this a complicated but extremely beneficial undertaking.

One of the most recent and high profile examples of extensive epidemiological examination is the HPAI outbreak that occurred in the United States in 2014–15 (https://www.aphis.usda.gov/aphis/ourfocus/animalhealth/animal-disease-information/avian-influenza-disease/sa_detections_by_states/hpai-2014-2015-confirmed-detections). This virus initially entered the United States in wild birds, but then spread to backyard poultry along the west coast (Ip et al., 2015). The virus then began to move east, infecting flocks of commercial poultry, mostly turkeys at first, before being found in most of the northern Midwest. The route of transmission was puzzling, since there are not any flyways that run east-west for migratory waterfowl to travel through. Epidemiological studies after the infection cycle had subsided revealing several factors that may have influenced the disease outbreak and spread across the United States. Many farms affected by the HPAI outbreak shared multiple types of farm equipment, had water (ponds, lakes) on the property in proximity to the poultry barns or even reported seeing wild birds inside the barns with the commercial poultry (Dargatz et al., 2016). Other epidemiology studies revealed that wind transmission may have played a part in the lateral spread of the virus (Torremorell et al., 2016). These determinations could not have been made without analysis of accurate and complete data sets, highlighting the importance of thorough data reporting.

Of course, HPAI is not the only disease that can be studied through epidemiology. Indeed, researchers around the world are turning to epidemiological investigations to model other disease outbreaks, to try and predict when an outbreak may occur or understand what factors are involved (Chengat Prakashbabu et al., 2017; Gonçalves and de Moraes, 2017; Basler et al., 2016). Viral diseases are the most commonly studied, as they tend to have more defined routes of transmission and influential factors can be more clearly elucidated, but more bacterial epidemiology work is underway to model infections with relevance to human health (Kirsty et al., 2013; Kaakoush et al., 2015). There are many factors that can influence the relevance or accuracy of a predictive disease model for livestock, including but not limited to pairing the data we currently have to the correct type of modelling system, applying models to data-poor situations, linking livestock populations to other species and understanding the impact of animal disease models on political and socio-economic factors (Brooks-Pollock et al., 2015). Some disease outbreaks have been accurately modelled post-outbreak (foot-and-mouth disease in the United Kingdom in 2001), but predictive models in livestock are still rare. Without greater financial support, developing predictive modelling systems that can accurately predict when factors are permissive for a disease outbreak will remain a challenge.

6 Discussion and conclusions

Understanding poultry diseases is an ongoing task, and one that may never be fully grasped. Poultry meat and eggs are one of, if not the most, the important sources of protein around the world, making poultry production a valuable commodity. Poultry are infected by numerous pathogens that cause a wide variety of clinical disease presentations. Therefore, it is critical that disease surveillance and study is continually ongoing to combat disease progression. Poultry producers are at the forefront of the effort, and the onus of diagnostic sampling and submission lies with them. The data that these submissions and clinical cases provide is invaluable to developing a disease model, and will ultimately yield dividends for them and others. The importance of quality sample submission and overabundance of case history cannot be over exaggerated. Naturally, this information is further supported by the active surveillance work performed by university, national and federal laboratories to monitor and detect diseases of pandemic potential or human health concern. When these individual points of data are combined, true epidemiological patterns of disease emerge, providing the poultry industry with historical and predictive information that can be used to improve processes and protocols to better maintain the health of flocks around the world.

7 Where to look for further information

Further information regarding appropriate sampling and diagnostic testing methods can be found in several published texts, including *Diseases of Poultry*, the *Avian Disease Manual* and *Isolation, Identification, and Characterization of Avian Pathogens*. Further information on diagnostic laboratories can be found on state and national websites such as https://www.gapoultrylab.org, https://www.aphis.usda.gov/aphis/ourfocus/animalhealth/lab-info-services/nahln/sa_labs/ct_labs and http://ec.europa.eu/food/safety/official_controls/legislation/ref-labs_en.

8 References

Ansari, A. A., Taylor, R. F. and Chang, T. S. 1983. Application of enzyme-linked immunosorbent assay for detecting antibody to mycoplasma gallisepticum infections in poultry. *Avian Dis.*, 27(1), 21–35.

Banat, G. R., Tkalcic, S., Dzielawa, J. A., Jackwood, M. W., Saggese, M. D., Yates, L., Kopulos, R., Briles, W. E. and Collisson, E. W. 2013. Association of the chicken MHC B haplotypes with resistance to avian coronavirus. *Dev. Comp. Immunol.*, 39, 430–7.

Basler, C., Nguyen, T.-A., Anderson, T. C., Hancock, T. and Behravesh, C. B. 2016. Outbreaks of human salmonella infections associated with live poultry, United States, 1990–2014. *Emerg. Infect. Diseases*, 22, 1705–11.

Brooks-Pollock, E., De Jong, M. C. M., Keeling, M. J., Klinkenberg, D. and Wood, J. L. N. 2015. Eight challenges in modelling infectious livestock diseases. *Epidemics*, 10, 1–5.

Chengat Prakashbabu, B., Thenmozhi, V., Limon, G., Kundu, K., Kumar, S., Garg, R., Clark, E. L., Srinivasa Rao, A. S. R., Raj, D. G., Raman, M., Banerjee, P. S., Tomley, F. M., Guitian, J. and Blake, D. P. 2017. Eimeria species occurrence varies between geographic regions and poultry production systems and may influence parasite genetic diversity. *Vet. Parasitol.*, 233, 62–72.

Dargatz, D., Beam, A., Wainwright, S. and Mccluskey, B. 2016. Case Series of Turkey Farms from the H5N2 Highly Pathogenic Avian Influenza Outbreak in the United States During 2015. *Avian Dis.*, 60, 467.

Dufour-Zavala, L., Swayne, D. E., Glisson, J. R., Jackwood, M. W., Pearson, J. E., Reed, W. M. and Woolcock, P. R. 2008. *A Laboratory Manual for the Isolation and Identification of Avian Pathogens*, Jacksonville, FL: American Association of Avian Pathologists.

Engvall, E. and Perlmann, P. 1971. Enzyme-linked immunosorbent assay (ELISA). Quantitative assay of immunoglobulin G. *Immunochemistry*, 8, 871–4.

Gonçalves, V. S. P. and De Moraes, G. M. 2017. The application of epidemiology in national veterinary services: Challenges and threats in Brazil. *Prev. Vet. Med.*, 137, 140–6. https://doi.org/10.1016/j.prevetmed.2016.11.018

Hoffmann, B., Beer, M., Reid, S. M., Mertens, P., Oura, C. A. L., Van Rijn, P. A., Slomka, M. J., Banks, J., Brown, I. H., Alexander, D. J. and King, D. P. 2009. Review: A review of RT-PCR technologies used in veterinary virology and disease control: Sensitive and specific diagnosis of five livestock diseases notifiable to the World Organisation for Animal Health. *Vet. Microbiol.*, 139, 1–23.

Ip, H. S., Torchetti, M. K., Crespo, R., Kohrs, P., Debruyn, P., Mansfield, K. G., Baszler, T., Badcoe, L., Bodenstein, B., Shearn-Bochsler, V., Killian, M. L., Pedersen, J. C., Hines, N., Gidlewski, T., Deliberto, T. and Sleeman, J. M. 2015. Novel Eurasian highly pathogenic avian influenza A H5 viruses in wild birds, Washington, USA, 2014. *Emerg. Infect. Dis.*, 21, 886–90.

Jackwood, M. W., Hilt, D. A., Sellers, H. S., Williams, S. M. and Lasher, H. N. 2010. Rapid heat-treatment attenuation of infectious bronchitis virus. *Avian Pathol.*, 39, 227–33.

Jackwood, M. W., Jordan, B. J., Roh, H.-J., Hilt, D. A. and Williams, S. M. 2015. Evaluating protection against infectious bronchitis virus by clinical signs, ciliostasis, challenge virus detection, and histopathology. *Avian Dis.*, 59, 368–74.

Johnson, J. and Reid, W. M. 1970. Anticoccidial drugs: Lesion scoring techniques in battery and floor-pen experiments with chickens. *Exp. Parasitol.*, 28, 30–6.

Josefsen, M. H., Bhunia, A. K., Engvall, E. O., Fachmann, M. S. R. and Hoorfar, J. 2015. Review: Monitoring Campylobacter in the poultry production chain – From culture to genes and beyond. *J. Microbiol. Methods*, 112, 118–25.

Kaakoush, N. O., Castano-Rodriguez, N., Mitchell, H. M. and Man, S. M. 2015. Global epidemiology of Campylobacter infection. *Clin. Microbiol. Rev.*, 28(3), 687–720.

Kirsty, K., Tom, H., Steven, R., Andrew, C., Paul, W. and Nicola, J. W. 2013. A longitudinal study simultaneously exploring the carriage of APEC virulence associated genes and the molecular epidemiology of faecal and systemic E. coli in commercial broiler chickens. *PLoS One*, 8(6), e67749.

Mansour, S. M. G., Ali, H., Chase, C. C. L. and Cepica, A. 2015. Loop-mediated isothermal amplification for diagnosis of 18 World Organization for Animal Health (OIE) notifiable viral diseases of ruminants, swine and poultry. *Anim. Health Res. Rev.*, 16, 89–106.

Marquardt, W. W., Johnson, R. B., Odenwald, W. F. and Schlotthober, B. A. 1980. An Indirect Enzyme-Linked Immunosorbent Assay (ELISA) for Measuring Antibodies in Chickens Infected with Infectious Bursal Disease Virus. *Avian Dis.*, 24(2), 375–85.

Morris, G. M. and Gasser, R. B. 2006. Biotechnological advances in the diagnosis of avian coccidiosis and the analysis of genetic variation in Eimeria. *Biotechnol. Adv.*, 24, 590–603.

Oakley, B. B., Lillehoj, H. S., Kogut, M. H., Kim, W. K., Maurer, J. J., Pedroso, A., Lee, M. D., Collett, S. R., Johnson, T. J. and Cox, N. A. 2014. The chicken gastrointestinal microbiome. *FEMS Microbiol. Lett.*, 360, 100–12.

Perozo, F., Villegas, P., Dolz, R., Afonso, C. L. and Purvis, L. B. 2008. The VG/GA strain of Newcastle disease virus: Mucosal immunity, protection against lethal challenge and molecular analysis. *Avian Pathol.*, 37, 237–45.

Roh, H. J., Jordan, B. J., Hilt, D. A. and Jackwood, M. W. 2014. Detection of infectious bronchitis virus with the use of real-time quantitative reverse transcriptase-PCR and correlation with virus detection in embryonated eggs. *Avian Dis.*, 58, 398–403.

Slaght, S. S., Yang, T. J., Van Der Heide, L. and Fredrickson, T. N. 1978. An Enzyme-Linked Immunosorbent Assay (ELISA) for detecting chicken anti-reovirus antibody at high sensitivity. *Avian Dis.*, 22(4), 802–05.

Snyder, D. B., Marquardt, W. W., Mallinson, E. T. and Russek, E. 1983. Rapid serological profiling by Enzyme-Linked Immunosorbent Assay. I. measurement of antibody activity titer against newcastle disease virus in a single serum dilution. *Avian Dis.*, 27(1), 161–70.

Soula, A. and Moreau, Y. 1981. Antigen requirements and specificity of a microplate Enzyme-Linked Immunosorbent Assay (ELISA) for detecting infectious bronchitis viral antibodies in chicken serum. *Arch. Virol.*, 67, 283.

Stewart, E. J. 2012. Growing Unculturable Bacteria. *J. Bacteriol.*, 194, 4151–60. https://doi.org/10.1128/jb.00345-12

Torremorell, M., Alonso, C., Davies, P. R., Raynor, P. C., Patnayak, D., Torchetti, M. & Mccluskey, B. 2016. Investigation into the Airborne Dissemination of H5N2 Highly Pathogenic Avian Influenza Virus During the 2015 Spring Outbreaks in the Midwestern United States. *Avian Dis.*, 60, 637-643.

Chapter 2

Gut health and susceptibility to enteric bacterial diseases in poultry

B. M. Hargis and G. Tellez, University of Arkansas, USA; and L. R. Bielke, Ohio State University, USA

1 Introduction

Most pathogens of poultry and other vertebrates enter the body via a mucosal portal of entry, and for the vast majority, this means either the respiratory or gastrointestinal tract (GIT). It is instructive to remember that the intestinal tract of vertebrate animals is essentially a 'tube within a tube', and that which is within the lumen of the intestinal tract is completely external to the host. Given the vast numbers of bacteria, including some opportunistic pathogens, viruses and protozoa that are ubiquitous in the digesta of commercial poultry, it is obviously advantageous that this is so. On the other hand, it is remarkable that a fragile, single cell layer of enteric epithelium is all that separates this enormously complex microbiota within the intestinal tract from the submucosa and systemic entry into the body. While the intestinal mucosa is the target for a plethora of viral and protozoal pathogens, it is also remarkable how few obligatory bacterial pathogens directly target the enteric mucosa, with most bacterial causes of enteropathies being opportunistic in nature.

http://dx.doi.org/10.19103/AS.2016.0011.16

2 Specific bacterial gut pathogens in poultry

Specific obligatory bacterial pathogens may be limited to a relatively small number of bacteria, and even these frequently exploit damage to the intestinal mucosa in terms of clinically relevant infections and disease. For example, clinically asymptomatic salmonellae in poultry, which cause small reductions in growth rate and feed efficiency that are impossible to measure under commercial conditions, are largely relegated to a category of 'opportunistic' pathogens by most poultry veterinarians. Nevertheless, these salmonellae are obligatory pathogens, deriving nutrients and energy sources primarily from the host. In fact, the relatively unusual characteristic of *Salmonella* spp. to utilize tetrathionate as an energy source allows salmonellae to prosper in the context of enteric inflammatory secretions, regardless of the cause of the inflammatory insult. This ability of salmonellae to exploit the intestinal mucosa of a host with any type of inflammatory bowel disease has been described, showing that tetrathionate leakage led to markedly increased proliferation and disease within the enteric mucosa and submucosa (Thiennimitr et al., 2012; Santos et al., 2003; Winter and Bäumler, 2011; Winter et al., 2010). Nevertheless, the principal problem for poultry producers in regard to infections with salmonellae are those of food-borne illness in humans, linked to origins of undercooked or mishandled poultry products. The large numbers of serotypes have limited the efficacy of currently available vaccines to protect from all *Salmonella* serotypes (Kuttappan et al., 2013; Boyle et al., 2007; Ohl and Miller, 2001). These topics have been well reviewed elsewhere and the reader is referred to several excellent reviews on this subject (Holt et al., 1999; Cox and Pavic, 2010; Thiennimitr et al., 2012). Moreover, the highly pathogenic serotypes/biotypes of salmonellae for poultry, *Salmonella* Pullorum and *Salmonella* Gallinarum, causing pullorum disease and fowl typhoid, respectively, are largely controlled by constant surveillance and eradication in developed countries (Shivaprasad and Barrow, 2008; Gast, 2008). While these particular salmonellae are capable of causing non-zoonotic but severely pathogenic salmonellosis in poultry without predisposing conditions, they are well reviewed in a variety of sources (Shivaprasad and Barrow, 2008; Gast, 2008; Hernández-Reyes and Schikora, 2013).

Colibacillosis is described as a pathologic condition caused by pathogenic strains of *Escherichia coli*. While pathogenic strains in mammals most commonly cause urinary tract infections or enteric disease, *E. coli* are infrequently associated as a primary causative agent for multiple diseases of poultry. Avian pathogenic *E. coli* (APEC) are associated with a long list of infectious sites and disease descriptions, including colisepticaemia, coligranuloma, air sacculitis, chronic respiratory disease, coliform cellulitis (also called inflammatory process), peritonitis, turkey osteomyelitis complex, omphalitis, salpingitis, orchitis, panophthalmitis, swollen head syndrome and even venereal colibacillosis. Expression of these diseases involves abnormal exposure routes, very high loads of APEC from faecal-derived airborne transmission, immunosuppression or combinations of the above. Within the GIT, APEC are rarely identified as primary aetiologies of enteritis. Colibacillosis of poultry has been extensively reviewed elsewhere (Dziva and Stevens, 2008; Brown, 1996; Huff et al., 2006; Barnes et al., 2008).

Clostridium perfringens is another potentially pathogenic but nearly ubiquitous contributor to the enteric microbiota of poultry. At low levels, most isolates of *C. perfringens* are not recognized as clinically contributing to poultry enteritis or other diseases. Nevertheless, overgrowth of *C. perfringens* leads to one of the most important emerging enteritis conditions of commercial poultry, necrotic enteritis. Low levels of

C. perfringens can commonly be isolated from newly hatched chicks. This is not surprising, since these Gram-positive bacteria are well-known spore formers, it is likely impossible to sufficiently sanitize eggs to eliminate this specific or opportunistic pathogen. Thus, vertical transmission from breeder flocks to commercial hatcheries without transmission to at least a few chicks in each flock is expected. Clostridial spores are exceptionally resistant to desiccation, heat, pH extremes and are nearly ubiquitous in soil, poultry litter, feed and have been shown to survive for years. For the last few decades, this disease has been largely controlled by low level feeding of antibiotic compounds, widely known as antibiotic growth promoters. However, with recent regulatory prohibitions in Europe, and voluntary consumer-driven demand in the United States and other countries, many flocks are being raised without the use of these chemicals, resulting in a resurgence of necrotic enteritis as perhaps the single most important enteric disease of commercial poultry (Timbermont et al., 2011; Williams, 2005; Dahiya et al., 2006; Opengart, 2008).

As the primary aetiology of necrotic enteritis, *C. perfringens* is not limited to only this devastating disease syndrome. This organism is also commonly associated with multiple clinical manifestations, including necrotic dermatitis, cholangiohepatitis and gizzard erosions. Some isolates of *C. perfringens* have been shown to reproduce necrotic enteritis without prior enteric integrity disruption (Chalmers et al., 2008; Lepp et al., 2013). Nevertheless, most of the clinical necrotic enteritis seen in commercial poultry operations are obligatorily preceded by predisposing enteric inflammatory conditions. Most notably, predisposing factors include coccidiosis, especially those caused by *Eimeria maxima*, but enteric inflammation due to high non-starch polysaccharide (NSP) carbohydrate sources, heat-damaged animal protein sources causing dysbiosis or viral enteritis are also predisposing factors for this disease. Necrotic enteritis has been thoroughly reviewed (Prescott et al., 1978; Chalmers et al., 2008; Lepp et al., 2013). Interestingly, the mechanism of predisposing factors causing enteritis is at least partially explained by increased goblet cell production of mucin, as discussed below. Common pathogenic *C. perfringens* isolates are mucinolytic, and capable of utilizing sialic acid and other carbon sources within mucin for energy. As a result, increased levels of mucin are actually known to support phenomenal blooms of *C. perfringens*, allowing a plethora of exotoxins to damage and induce acute cell death in the enteric epithelium (Yegani and Korver, 2008a; Hofacre, 2001; Opengart 2008). An interesting and complicating factor with clinical experience with necrotic enteritis may be associated with early infections of salmonellae. Shivaramaiah and co-workers (2011) showed that infection of chicks within the first week of life with low-virulence isolates of *Salmonella* Typhimurium (ST) resulted in approximately twice the mortality of *Salmonella*-negative chicks. Interestingly, infection at a later age did not predispose chicks to increased necrotic enteritis susceptibility, suggesting that persistent phenotypic changes within the intestinal tract during the critical formative period of the gut can be predisposing for necrotic enteritis (Shivaramaiah et al., 2011). This observation may increase the production-relevant reasons for improving control of salmonellae, in addition to the more common interest in antemortem food safety interventions.

Ulcerative enteritis, a condition of quail and sometimes other poultry, is similar to necrotic enteritis in that it is attributed to a closely related clostridial species, *Clostridium colinum*, but predisposing factors have yet to be elucidated in order to reproduce the disease (Berkhoff, 1985; Wages, 2008). Other clostridial species, such as *Clostridium septicum*, are frequently found within the GIT but not associated with enteric disease. In turkeys, enteric origin *C. septicum* has been shown to be the etiology of turkey clostridial dermatitis (cellulitis), probably related to enteric leakage as discussed below (Opengart, 2008; Tellez et al., 2009).

Clinical botulism in poultry species is almost always a result of ingestion of pre-formed botulinum toxin from anaerobic fermentation of *C. botulinum*. These cases often involve failure to quickly remove carcasses from the environment for any cause of mortality, and the ubiquitous *C. botulinum* ferments within the carcass, providing the source for direct ingestion of the pre-formed toxin, which can also be concentrated by maggots, which quickly form and are attractive to poultry (Jean et al., 1995; Dohms, 2008). However, some cases of botulism have been related to enteric blooms of *C. botulinum* in turkeys, although such cases are unusual and are often clinically related to very wet or humid environments (Maas et al., 1989; Jeffrey et al., 1992; Popp et al., 2012).

A number of opportunistic bacterial species utilize the GIT as a portal of entry for translocation across the enteric epithelial barrier. As described below, once bacteria reach the submucosa, they are distributed via the hepatic portal vein directly to the liver, but can also translocate directly via the circulation to other organs and tissues systemically. Examples include listeriosis, bacterial chondronecrosis with osteomyelitis and spondylolisthesis (Wideman et al., 2011). In these cases, the various aetiologies of these pathologic conditions are less important than the causes of intestinal barrier failure, also known as 'leaky gut', which allows for markedly enhanced transcellular and paracellular transport of these agents across the epithelium. Causes of leaky gut can include enteric pathogenic infections such as coccidiosis and viral enteritis, but also less well-known inductions including relatively short (24 hr) feed restriction (Wideman et al., 2012; Vicuña et al., 2015), high levels of NSPs from cereal grains such as wheat, bakery by-product meal, rye, heat-damaged animal protein supplements or physiologic stress induction (Wideman et al., 2015; Tellez et al., 2015; Latorre et al., 2014). These predisposing causes for a variety of systemic bacterial infections are described below.

3 Symbiosis between prokaryotes and the host in the gut

While the terms 'microbiota' and 'microbiome' are often used interchangeably, the term 'microbiome' relates more specifically to the presence of specific nucleic acid sequences indicative of phyla or genus-level organisms, regardless of viability. While the term microbiota more accurately refers to the viable organisms actually present within this organ, accurate analysis of viable organisms must reasonably include a very large number of currently non-culturable organisms. In the context of this chapter, we will predominately describe the microbiota as related to the bacterial microflora actually undergoing full life cycle proliferation within the intestinal lumen or associated with the enteric mucosa.

A common human propensity is to regard all microorganisms as 'harmful', essentially equating bacteria to pathogenic 'germs'. Nothing could be further from the truth. The number of beneficial bacterial species far exceeds the number of pathogenic species, and many of the known bacteria are in fact useful or even indispensable for the continued existence of life on Earth. Prokaryotic microorganisms are widespread in all environments on Earth, establishing diverse interactions with many eukaryotic taxa (Bronstein et al., 2006; Gnad et al., 2010). The cooperative mutualism between species have had a central role in the generation and maintenance of life on Earth (Kikuchi et al., 2009; Jones et al., 2010). Prokaryotes and eukaryotes are involved in diverse forms of mutualism (Saridaki and Bourtzis, 2010). Throughout millions of years, animals have developed various means for supporting complex and dynamic consortia of microorganisms during their life cycle

(Dale and Moran, 2006; Moran, 2007; Fraune and Bosch, 2010; Dethlefsen et al., 2007). One example of such a beneficial group of microorganisms is the incredibly complex and abundant ensemble of microbes that harbours in the GIT of metazoans (Neish, 2009). The GIT is more densely populated with microorganisms than any other organ and is an interface where microflora may have a pronounced impact on animal biology, and over 100 bacterial phyla have been identified in the microbiome of mammals (Maslowski and Mackay, 2010; Musso et al., 2010; Arumugam et al., 2011; Morgan and Huttenhower, 2012). Nevertheless, as in other vertebrates, the intestinal microflora of chickens is also dominated by the phyla Firmicutes and Bacteroidetes, representing over 90% of the microbiome (Bjerrum et al., 2006; Wise and Siragusa, 2007). Interestingly, most of these are obligate anaerobes, with Proteobacteria and Actinobacteria also reported in poultry as representing the remainder (Oakley et al., 2014). At the Class level, Clostridia are significant contributors of short-chain fatty acids (Lund et al., 2010; Luo et al., 2013). Less than 10% of the microbiome can be cultured with conventional bacteriology techniques, but molecular and sequencing tools have been used to have a better understanding of these multifaceted groups of bacteria, revealing that the variability of microbiome between metazoan individuals can be compared to a thumbprint (Mariat et al., 2009). More than 50 genera and at least 500–1000 different species are distributed along the length of the GIT (Neish, 2009; Sharma et al., 2010; Schiffrin and Blum, 2002). The bacterial population of the human caecum and colon is numerically ~10^{13} cfu/g, a number that is bigger than the total number of somatic cells in a proportion 10:1, with over 300 000 genes in the microbiome versus an average of 30 000 genes in a mammal (Sharma et al., 2010; Neish, 2009; Xu and Gordon, 2003). This mass of bacteria represents about 40–55% of solid stool matter and weights ~1 kg in humans (Neish, 2009; Blaser, 2006). While similar estimates for poultry have not been published, one can reasonably expect that similarly impressive numbers might be found in poultry, although the reduced length and abbreviated fermentation might make the relative bacterial contribution smaller.

Colonization of the microbiome in metazoans begins at birth, and is followed by progressive assembly of a complex and dynamic microbial society, which in a state of eubiosis maintains a perfect harmony or homeostasis (Martin et al., 2010; Di Mauro et al., 2013). In commercial broiler operations repeated, use of the same litter across multiple flocks, becomes an important source for intestinal colonization of chicken gastrointestinal microbiome (Torok et al., 2009). This is especially true when we consider that chicks coming from hatcheries are primarily only exposed to potentially pathogenic bacteria associated with the bacterial bloom that accompanies increased humidity beginning at pipping within the hatcher cabinet. As discussed below, the respiratory portal of entry plays a crucial role in dissemination of many enteropathogens.

4 Impact of bacterial pathogens on host biology

Presumably, the assembly of gut microflora is regulated by elaborate and combinatorial host–microbial and microbial–microbial interactions predicated on principles refined over the course of evolution (Xu and Gordon, 2003; Fraune and Bosch, 2010; Xu et al., 2007). Comparison of rodents raised without exposure to any microorganisms to those colonized with an assembly of microbiota revealed a wide range of host functions affected by indigenous microbial communities. For example, the microbiota directs the assembly of the gut-associated lymphoid tissue (GALT) (Martin et al., 2010; Sekirov et al., 2010) helps educate the immune system (McFall-Ngai, 2007), affects the integrity of the intestinal mucosal barrier (Duerkop et al., 2009), modulates proliferation and differentiation of its epithelial lineages (Moran, 2007), regulates angiogenesis (Sekirov et al., 2010), modifies the activity of the enteric nervous system (Tlaskalová-Hogenová et al., 2011; Sherman et al., 2009) and plays a key role in extracting and processing nutrients consumed in the diet (Walter et al., 2011; Bergman, 1990). The microflora can metabolize proteins and protein degradation products, sulphur-containing compounds, and endogenous and exogenous glycoproteins (Qiu et al., 2012; Fuller and Brooker, 1974). Some organisms grow on intermediate products of fermentation such as H_2, lactate, succinate, formate and ethanol. Converting these to end products, including short-chain fatty acids, is a process which has direct benefits on digestive physiology (Dass et al., 2007; Tellez et al., 2006, 2012). For this reason, dietary ingredients have a profound effect on the composition of gut microflora, which in turn, regulates the physiology of all animals (Fraune and Bosch, 2010). As such, nutritional components of the diet are of critical importance not only for meeting nutrient requirements of the host, but also shaping the profile of the microbiome, which in turn, will determine the balance between health and disease. As an example, several studies have shown the effect of the Western diet in promoting an obesogenic gut microbiota, insulin sensitivity, inflammatory markers, diabetes, cancer, metabolic disorders and glucose tolerance in humans (Cani and Delzenne, 2009; Tagliabue and Elli, 2012; Shen et al., 2012; Cani et al., 2008; Elson and Cong, 2012; Saleh and Elson, 2011; Cani and Delzenne, 2009). Some researchers believe that the alarming increase in human autoimmune diseases in the West may be due to a disruption in the ancient relationship between our bodies and a healthy microbiome (Salzman, 2011). Undoubtedly, both diet and microbiota contribute to both eubiosis and dysbiosis in poultry, although there is much yet to be understood specifically in poultry species (Tellez et al., 2006; Teirlynck et al., 2011; Tellez et al., 2014; Latorre et al., 2015).

A delayed microbial colonization of the gut mucosa, the largest immune organ of the body, could cause significant changes in the immune system possibly having long-term impacts on systemic immunity (Thavagnanam et al., 2008; Guibas et al., 2013; Martin et al., 2010; Mshvildadze et al., 2008; Collado et al., 2012; Neu et al., 2011; Salminen et al., 2004; Stevceva et al., 2000). For instance, some effects of the microbiome are mediated through immune regulation, particularly through balanced control of pro-inflammatory and anti-inflammatory cytokines (Borchers et al., 2009; Braat et al., 2004). Moreover, several animal and human studies have provided unequivocal evidence that specific bacterial strains are capable of stimulating multiple aspects of innate immunity (Maslowski and Mackay, 2010; Feng and Elson, 2010; Salzman, 2011; Alvarez-Olmos and Oberhelman, 2001; Reveneau et al., 2002; Jounai et al., 2012; Farnell et al., 2006) as well as to increase humoral immunity (Parvez et al., 2006; Joint, 2001; Kalliomaki et al., 2001; Ouwehand et al., 2002). Very interestingly, through a process of 'crosstalk' with the mucosal immune system, the microbiota negotiates mutual growth, survival and inflammatory control of the intestinal ecosystem (Salzman, 2011). While we are only beginning to understand the role of diet and microbiota in poultry, it seems clearly likely that similar relationships will ultimately be elucidated.

5 Dysbiosis and disease in poultry

In 400 BC, Hippocrates proclaimed that 'death sits in the bowels' and 'bad digestion is the root of all evil'. Today, the term dysbiosis refers to a shift in the microbiome alignment resulting in a disproportion between beneficial and harmful bacteria that have a negative impact in the GIT, including reduced nutrient digestibility, increased gut permeability and upregulated inflammatory responses (Myers, 2004; Chan et al., 2013). The eubiosis/dysbiosis status of the microbiome defines health or disease, and the fragile balance can be altered by any kind of stress that induces inflammation such as environment, diet, therapeutic antibiotics and pathogens (Choct, 2009; Bäckhed, 2011; Stecher et al., 2013).

It is interesting to note that dysbiosis is a condition reported in all metazoans. For several decades now, important economic losses worldwide have been reported from apiculture, particularly to a condition known as colony collapse disorder, where colonies suddenly disappear with no clear explanations. Honeybee colonies can be affected by abiotic factors such as environmental pollution or insecticide applications. Additionally, biotic stresses can cause colony losses, including bacterial, fungal, parasites and several viruses with alteration of the gut microbial community and have been shown to be linked to health and fitness disequilibrium (Hamdi et al., 2011). In domestic animals, dysbiosis was practically unknown until the ban on antimicrobial growth promoters in animal feed in Europe in 2006 (Teirlynck et al., 2011). Today, the condition is recognized as one of the most challenging problems, particularly in broilers and in newly weaned pigs, where dysbiosis is associated with intestinal inflammation, shortening of small intestinal villi and bacterial translocation (BT) (Teirlynck et al., 2009; Latorre et al., 2015; Vicuña et al., 2015). Under experimental conditions in broilers, dysbiosis, characterized by typical inflammation and villus shortening, can be induced by feeding a diet containing high amounts of NSPs in the absence of NSP-degrading enzymes (Tellez et al., 2014, 2015).

6 Inflammation and effects on anaerobiosis in poultry

A healthy GIT has very low oxygen availability and is colonized mostly by obligate anaerobes. Any kind of stress, regardless of its origin, such as (1) biological – viruses, bacteria, parasites; (2) environmental – temperature, ammonia, ventilation; or (3) dietary ingredients, will result in inflammation, which will modify the anaerobic degree of the GIT (Kallapura et al., 2014e). During dysbiosis caused by inflammation or stress, an increase in reactive oxygen species (ROS) has been associated with a significant reduction in obligate anaerobes of Firmicutes (mostly Class Clostridia). Interestingly, the result of these changes is also linked with significant increases in facultative anaerobes from the Phylum Proteobacteria, particularly Class Gammaproteobacteria, where Enterobacteriaceae bacteria predominate (Rigottier-Gois, 2013). During infection, host organisms detect the presence of infection by recognizing specific elements on pathogens known as pathogen-associated molecular patterns (PAMPs) such as lipopeptides, peptidoglycans and teichoic acids. The endotoxin lipopolysaccharide (LPS) is a major component of the outer membranes of Gram-negative bacteria and a chief member of PAMPs, which is largely responsible for signal transduction of the nuclear factor-kB (NF-kB) and subsequent release of cytokines and other pro-inflammatory mediators leading to beneficial or detrimental consequences. The presence of by-products generated during the inflammatory response, promotes the growth of anaerobic facultative bacteria able to use these products as terminal electron acceptors for anaerobic respiration, a phenomenon that benefits them immensely, and which will be discussed below. On the other hand, inflammation during viral infections such as Marek's disease virus has been reported to induce dysbiosis (Perumbakkam et al., 2014). Hence, the fragile microbiome can be altered by pathogenic eukaryotes, prokaryotes and viruses (Wylie et al., 2012; Minot et al., 2013).

7 Positive and negative effects of mucins

As a remarkable chemical and physical defence barrier, mucins are large glycoproteins that cover epithelial surfaces of the intestine and form a mucus layer to protect epithelial cells from gut health challenge. There are two major types of mucins: membrane-bound and secreted (Williams et al., 2001; Toribara et al., 1993). In chickens, three transmembrane mucins (MUC4, MUC13 and MUC16) and four gel-forming mucins (MUC6, MUC2, MUC5ac and MUC5b) have been identified (Lang et al., 2006). In mammals, MUC2, the mucin secreted by goblet cells, is the most abundant mucin in the intestine, and its deficiency has been reported to increase BT and inflammation (Wei et al., 2012; Kim and Ho, 2010). Evolutionary studies suggest that mucins share a common ancestor, since their domain structures are well conserved in metazoans (Kimbrell and Beutler, 2001; Lang et al., 2007). All mucins (MUC) contain at least one PTS domain, a region rich in proline, threonine and serine (Jiang et al., 2013; Kim and Ho, 2010). Chicken MUC2 has been reported to be remarkably similar to human and mouse outside of the central PTS domain, but is highly divergent within this central repetitive structure (Jiang et al., 2013; Zhang et al., 2015). The highly glycosylated PTS domain, once hydrated post-secretion, allows for different electrostatic bonding and effective 'pore' sizes within the mucin layers that create different micro-environments for bacteria. Jiang et al. (2013) noted that alternate splicing within the PTS domain of MUC2 may lead to differences observed in mucin composition. Although the physiological implications and disease associations of MUC on various mucosal surfaces are well understood, there are still many questions as to how and why the gene architecture of this family contributes to diverse protein modifications that show diverse biological effects between metazoans in health and disease (Clark et al., 2006; Forder et al., 2012; Liu et al., 2012; Hollingsworth and Swanson, 2004). MUC2 gene expression has been used as a marker for gut health in poultry and other species (Forder et al., 2012; Chen et al., 2015).

Perhaps, the bad and ugly parts of mucins are that some bacteria take advantage of this barrier. Mucus colonization is an indispensable step for mucosal pathogens such as *Campylobacter jejuni* or *C. perfringens*. It has been reported that mucins curb the pathogenicity of *C. jejuni* in a species-specific manner and likely help determine whether these bacteria become pathogenic, as in humans, or adopt a commensal mode of existence, as in chickens and other animals. Therefore, understanding the interaction of these pathogens with mucus and mucins has the potential to lead to development of intervention strategies that take advantage of the propensity of these bacteria to colonize the mucus layer of the GIT.

8 Gut infections and dietary disruption

The intestinal epithelium constitutes the largest and most important barrier against external environmental agents and has two critical functions: prevent the entry of harmful intraluminal microorganisms, antigens and toxins, and enable selective translocation of dietary nutrients and electrolytes into circulation (Salzman, 2011; Elson and Cong, 2012; Salminen and Isolauri, 2006). One of the basic properties of GALT is oral tolerance to harmless components of microbiota and diet (Kau et al., 2011). Inappropriate immunological reactions against food compounds can lead to the breakdown of oral tolerance and the development of intestinal immune disorders (Stepniak and Koning, 2006). Likewise, composition of the diet also has a tremendous impact in digestibility and gut health of chickens (Dunsford et al., 1989; Hrncir et al., 2008; Maslowski and Mackay, 2010). A specific example is rye-based diets versus traditional corn-based diets, where different cereals are used as the principal source of energy. The inclusion of rye in poultry diets has been fraught with problems, principally related to the production of sticky droppings, malabsorption syndrome, elevated feed conversion and intestinal bacterial overgrowth (Campbell et al., 1983; Bedford and Schulze, 1998; Shirzadi et al., 2010). Very recently, high concentrations of NSP in these alternative grains have been convincingly shown to be the principle cause of dysbiosis, leaky gut and poor feed conversion when used in poultry diets (Tellez et al., 2014; Latorre et al., 2015; Vicuña et al., 2015).

In recent years, nutrition research has moved from classical epidemiology and physiology to molecular biology and genetics. Modern nutritional research is aimed at health promotion, disease prevention and performance improvement (Kussmann et al., 2008). Hence, nutritional sciences are discovering the application of the so-called '-omics' sciences (Dimitrov, 2011). Nutritional genomics is a recent offshoot of this genetic revolution. As a consequence of these ambitious objectives, the disciplines 'nutrigenetics' and 'nutrigenomics' have evolved (Van Ommen and Stierum, 2002). In particular, nutrigenomics is the junction between health, diet and genomics, where it can be seen as the combination of molecular nutrition and genomics, addressing the inverse relationship, which is how diet influences gene transcription, protein expression and metabolism (Trujillo et al., 2006; Kussmann et al., 2006; Garcia-Cañas et al., 2010). With emerging '-omics' technologies, scientists now have tools to investigate how dietary food components affect physiological functions and the underlying cellular and molecular mechanisms implicated in the digestive process (Ghosh and Poisson, 2009; Dimitrov, 2011). Nutrition-related genomics technology has revolutionized the field of nutrition providing unprecedented opportunities for increasing our understanding of how nutrients modulate gene and protein expression to influence cellular metabolism (Subbiah, 2007).

When integrated with other '-omics' technologies in a biological system approach, novel nutrition-based intervention strategies are expected to provide an effective alternative for disease control (Trulillo et al., 2006; Higgins et al., 2011). Several studies in poultry have been conducted to investigate the effects of probiotics and phytonutrients on the translational regulation of genes associated with immunology, pathogen control, physiology and metabolism using high-throughput microarray analysis and *in vivo* disease challenge models (Tellez et al., 1993; Kiarie et al., 2013; Sherryll Lynn et al., 2013). Chicken has long been an important animal system for discoveries in developmental biology, immunology and microbiology (Stern, 2004; Burt, 2007). With sequencing of the genome and development of high-throughput tools for exploration of functional elements of the genome, chicken has attained a model organism status (Stern, 2005). An increase in genomic resources, easy access to embryos and the application of RNA interference showed that it will be easy and quick to use chicken embryos to screen the function of many genes during embryonic development (Stern, 2004). Thus, it seems likely that chicken will increasingly be a system of choice for many vertebrate biologists, especially in the field of human sciences with interests in gene function (Aggarwal and Lee, 2003; Cogburn et al., 2007).

Thousands of years of evolution shaped the digestive system of jungle fowl and wild pigs to handle the dietary ingredients encountered in an efficient manner. More recently, through intensive genetic selection, nutrition and health programmes, we have altered the biology and growth potential of chickens and pigs among other productive animals (Fuller et al., 1995). In the wild, the diets of these animals would typically consist of many different ingredients, few of which would ever reach >30% of total intake on a lifetime basis. The range in types and relative quantities of ingredients that can be presented to modern commercial monogastric animals, while complex, tend to result in a diet in which 2 or 3 ingredients may constitute >75% of intake. Such change is driven by least-cost formulation processes, and endeavours to provide maximum nutrient density at a minimal cost (Bedford and Schulze, 1998). Corn is usually the main source of energy in poultry diets, but at times it is difficult to formulate least-cost diets using only corn, and unconventional grains or bakery by-products are substituted on a least-cost basis. When chickens are fed alternative grains, such as wheat or rye, which are high in NSP, poor performance and unmanageable litter conditions caused by sticky droppings are reported (Campbell et al., 1983; Choct et al., 1995). Wheat and rye can contain high concentrations of NSP, leading to reduced digestibility. In addition, high NSP diets have also been associated with necrotic enteritis (Hofacre, 2001; Timbermont et al., 2011; Annett et al., 2002). NSPs are comprised mainly of highly branched arabinoxylans, which are associated with an increase in digesta viscosity, and are responsible for poor digestibility through interference with the movement of particles and solutes across the intestinal lumen. This prevents access of digestive enzymes to the endosperm contents, and reducing intestinal absorption of sodium and calcium (Fengler and Marquardt, 1988). In addition, increased digesta viscosity reduces conjugated bile acid, affecting fat emulsification and fat digestibility. Recent poultry studies have suggested that the significant reduction in bone strength and mineralization in high NSP diets in poultry (Table 1) are also associated with malabsorption of minerals and fat-soluble vitamins (Tellez et al., 2014, 2015). This may be similar to conditions such as gluten intolerance in humans (Campbell et al., 1983; Bianchi and Bardella, 2008; Capriles et al., 2009; Kotake et al., 2009; Wideman et al., 2011). In

Table 1 Evaluation of bone breaking strength and bone composition parameters in chickens fed with corn- or rye-based diets in Experiments 1 and 2 (Adapted from Tellez et al., 2014)

Diet	Tibia strength Load at yield (kg/mm^2)[1]	Tibia diameter (mm)	Total ash from tibia (%)	Calcium (% of ash)	Phosphorus (% of ash)
	Experiment 1				
Corn	5.04 ± 0.011^a	3.34 ± 0.17^a	55.01 ± 0.41^a	29.48 ± 0.27^a	18.15 ± 0.12^a
Rye	1.58 ± 0.009^b	1.61 ± 0.28^b	34.87 ± 0.35^b	18.48 ± 0.27^b	13.15 ± 0.12^b
	Experiment 2				
Corn	6.14 ± 0.01^a	4.55 ± 0.32^a	65.61 ± 0.81^a	37.65 ± 0.07^a	21.35 ± 0.52^a
Rye	2.58 ± 0.03^b	1.82 ± 0.78^b	30.87 ± 0.75^b	21.32 ± 0.46^b	15.67 ± 0.29^b

[1]Tibias from twelve chickens were collected to evaluate bone quality. Data is expressed as mean ± standard error.
[a–b]Superscripts within columns indicate significant difference at $p < 0.05$

one of the presented studies, viscosity of the gut in chickens fed rye was so extreme, the supernatant being more semi-solid than fluid, that it alone could be directly responsible for the reported poor performance (Table 2).

Alterations in gut permeability are connected with BT in the portal and/or systemic circulation in several types of leaky gut syndromes (Seki and Schnabl, 2012; Ilan, 2012). Similarly, fluorescein isothiocyanate dextran (FITC-d, 3–5 kDa) is a large molecule which does not usually leak through the intact GIT barrier. However, when conditions disrupt the tight junctions between epithelial cells, FITC-d molecules can enter circulation, as demonstrated by an increase in trans-mucosal permeability associated with chemically induced disruption of tight junctions by elevated serum levels of FITC-d after oral administration (Yan et al., 2009). Prolonged feed passage rate increases the time available for digesta-associated bacteria to multiply prior to evacuation in the faeces, and provides more substrate availability in the distal parts of the small intestine for microbial fermentation (Choct et al., 1995; Murphy et al., 2009; Kiarie et al., 2013). Increased BT and increased serum FITC-d leakage from the gut (2.5 h after gavage) were associated with the increased intestinal viscosity in chickens fed with rye when compared with chickens fed with corn (Table 1). Variations in the composition of the microbiome within different segments of the alimentary tract are influenced by the environment, by the diet and by the host (Ouwehand et al., 2002; Xu and Gordon, 2003; Mazmanian et al., 2008). Since poultry have little or no intrinsic enzymes capable of hydrolysing NSP, exogenous xylanases as additives are used in an attempt to reduce this anti-nutritive factors (Bedford and Schulze, 1998; Bedford and Classen, 1993; Bedford et al., 1991). Previously, we have evaluated the inclusion of selected direct-fed microbial (DFM) candidates that produce exogenous enzymes (protease, phytase, lipase, xylanase and cellulose) in high NSP diets, including rye, wheat, barley and oat. In those studies, a significant reduction in both viscosity and *C. perfringens* proliferation was observed between high NSP control non-treated diets or the same diets supplemented with DFM (Latorre et al., 2014, 2015). The incorporation of one or more '-omics' techniques, in particular, assessment of the microbiome, will provide a better understanding of how dietary food components can affect physiological functions and the fundamental cellular and molecular mechanisms implicated in the digestive process of high NSP diets in chickens.

Table 2 Evaluation of body weight, intestinal viscosity, serum FITC-d and liver bacterial translocation in chickens fed with corn or rye in Experiments 1 and 2 (Adapted from Tellez et al., 2014)

Diet	Body weight (g)	Intestinal viscosity (cP Log_{10})	Serum FITC-d (μg/mL)	Bacterial translocation (CFU Log_{10})
	Experiment 1			
Corn	283.21 ± 10.57^a	0.30 ± 0.04^b	0.20 ± 0.01^b	0.00 ± 0.00^b
Rye	110.69 ± 5.21^b	2.84 ± 0.57^a	0.42 ± 0.05^a	1.35 ± 0.45^a
	Experiment 2			
Corn	301.46 ± 10.57^a	0.23 ± 0.35^b	0.31 ± 0.03^b	0.00 ± 0.00^b
Rye	140.89 ± 5.21^b	2.90 ± 0.83^a	0.52 ± 0.07^a	2.40 ± 0.73^a

Data is expressed as mean ± SE. Intestinal viscosity is (in centipoise, cP = 1/100 dyne sec/cm^2) from 5 chickens. Serum FITC-d (2.5 h post-gavage) and liver bacterial translocation from 12 chickens.
[a–b]Superscripts within columns indicate significant difference at $p < 0.05$

9 Airborne infection by enteric bacteria among poultry

Poultry are a major reservoir for food-borne *Salmonella enterica* serovars, most notably ST, *S.* Enteritidis (SE), *S.* Heidelberg, *S.* Kentucky and *S.* Senftenberg as the most prevalent serovars in the U.S. poultry (Centers for Disease Control and Prevention 2013; Basler et al., 2015). In chickens, infections with host-specific serovars *Salmonella* Gallinarum and *Salmonella* Pullorum cause septicaemia and expression of fowl typhoid and pullorum disease, respectively (Shivaprasad and Barrow, 2008), whereas infections with non-host-specific serovars generally display no clinical signs. In the last decade, significant progress has been made in the knowledge of *Salmonella* invasion and pathogenesis in mammalian hosts (Santos et al., 2003; Ohl and Miller, 2001; Zhang et al., 2003). However, specific information regarding *Salmonella* invasion and colonization mechanisms and interactions with host cells of poultry is limited and poorly defined (Cox and Pavic, 2010).

The natural route of transmission of zoonotic pathogens, such as *Salmonella*, is long considered faecal–oral (White et al., 1997; Galanis et al., 2006). However, published studies have also suggested that airborne transmission of *Salmonella* in poultry is possible (Fallschissel et al., 2009; Baskerville et al., 1992; Leach et al., 1999; Wathes et al., 1988). The frequent recovery of *Salmonella* from dust and bioaerosols from infected poultry is well-documented, and bioaerosolized *Salmonella* of a very small droplet size is capable of reaching the lower respiratory tract of chicks (Chinivasagam et al., 2009; Cox and Pavic, 2010). *Salmonella* is viable in laboratory-generated aerosols for more than 2–4 h depending on the overall prevailing relative humidity, temperature of the air and strain of *Salmonella* (Wathes et al., 1988). A detailed study by Hayter and Besch (1974) showed that the site of deposition of particles critically depends on their size. Large respired particles with a size of 4–7 μm are trapped and removed by the mucociliary ladder, but particles smaller than this are deposited at deeper areas (Hayter and Besch, 1974). Aerosols of size <4 μm should be able to carry *Salmonella*, considering that most salmonellae measure between 0.7 and 1.5 μm diameters and lengths from 2 to 4 μm, and hence would not face the mucociliary challenge to potentially infect poultry. Several studies have examined particle analysis to determine the relationship between the particles in the air, both size and number, and the dose of *Salmonella* with estimates ranging from 10 to 1.2×10^4 CFU of *Salmonella* per m^3 of air, depending on study conditions (Baskerville et al., 1992).

Our laboratory has recently showed that transmission by the faecal–respiratory route is a viable portal of entry for *Salmonella* (Kallapura et al., 2014a,b,c,d). This mode of infection could explain some clinical impressions of relatively low-dose infectivity under field conditions in relation to the high oral challenge dose that is typically required for infection through the oral route in laboratory studies. Recently published results compared intra-tracheal (IT) *versus* oral (OR) administration of two no-typhoid serovars of *Salmonella enterica*, SE and ST with relatively different invasiveness (Kallapura et al., 2014c). Neonatal chicks were infected via the respiratory route at a very low dose, resulting in caecal colonization that was equivalent to levels of *Salmonella* recovered from a higher OR challenge. These results may be important, considering previous studies demonstrating fan-driven spread of *Salmonella* within hatching cabinets and hatchery incubators (Berrang et al., 1995). The ability of *Salmonella* to reach the liver and spleen, and colonize caeca when administered IT, suggested a possible route of systemic infection (Kallapura et al., 2014a,b,c).

Salmonella Senftenberg, a serovar that is more resistant to the environmental stresses, is frequently isolated from hatcheries and raw feed materials, and is adapted to colonize and persist in poultry houses (Galanis et al., 2006). The prevailing theory of *Salmonella* enteropathogenesis is that bacterial invasion of the intestinal epithelium is essential for virulence and that this requires the virulence-associated genomic region, *Salmonella* pathogenicity island 1 (SPI-1; Stecher et al., 2013). However, *S.* Senftenberg strains isolated from food-borne outbreaks have been reported to lack the invasion-associated locus SPI-1, indicating it may not be essential for intestinal inflammatory disease (Hu et al., 2008). Yet, despite its lack of invasiveness, *S.* Senftenberg has emerged as a human pathogen with the ability to acquire important pathogenic loci from other bacteria, such as *Yersinia* high pathogenicity island (HPI), which encodes a *Yersinia* bacterin-mediated iron acquisition system present in highly pathogenic strains of *Yersinia* and several members of the Enterobacteriaceae family (Petermann et al., 2008). Several studies have illustrated the presence of the HPI in *S.* Senftenberg and the importance of poultry and pigs as a reservoir and vehicle for the dissemination of zoonotic pathogens (Carniel, 2001; Hacker et al., 1997). In a recent study, it was remarkable to observe that IT delivery of *S.* Senftenberg was the only route able to colonize the caeca of chickens, when compared with oral or intra-venous (IV) challenge in a dose response fashion in both trials (Kallapura et al., 2014b). This is in alignment with our previous results, suggesting that *Salmonella*, depending on the isolate, was far more likely to be recovered from caecal tonsils when administered IT, presumably involving a systemic dissemination and infection in chickens (Kallapura et al., 2014a,c). One possible way of systemic spread and caecal colonization of *S.* Senftenberg, Enteritidis or Typhimurium evaluated in our previous studies is that, when given IT, *Salmonella* conceivably reach non-ciliated regions of the respiratory tract and may be able to either translocate directly into the blood or be directly phagocytized through several documented mechanisms for removing particulate matter from the non-ciliated avian respiratory tract (West and Mathieu, 1993). The mode in which *Salmonella* infect and systemically spread by directly entering the blood and causing bacteraemia is quite conceivable considering previous studies proposing disrupted epithelial barrier of the developing bronchus-associated lymphoid tissue structure, endocytosis by epithelial cells and further transport to the interstitial matrix and high prevalence of blood–gas barrier breaks and epithelial–epithelial cell connection breaks (West and Mathieu-Costello, 1995). Alternatively, various reports have described the presence of an efficient phagocytic system in chickens, involving the phagocytic epithelial lining, interstitial macrophages in the blood–gas barrier and the infiltrating free avian respiratory macrophages (Menash and Brain, 1982). Continued survival of bacteria entering blood is considered unlikely, as reports suggest that blood-borne bacteria are rapidly cleared from the circulation and the clearance rate is enhanced significantly in the presence of opsonizing factors in the blood (Qureshi, 2003). Passively immunized turkey poults, when challenged with *E. coli*, rapidly reduced the bacterial load by approximately 5 Log_{10} within 10 min post-IV challenge, in contrast to non-immunized poults (Arp, 1982). In addition, a study by Kiama and colleagues (2008) demonstrated that chicken lung lavaged erythrocytes readily phagocytized 1.5 μm diameter polystyrene particles compared to rat erythrocytes that showed absolutely no signs of phagocytosis. Even though little is known about such interactions, phagocytic activity by chicken erythrocytes is attributed to the nucleated properties (Kiama et al., 2008). The role(s) that inherently phagocytic erythrocytes play, particularly in the defence of the pulmonary system and the avian body in general, needs to be confirmed with further studies.

Salmonellae have an established intracellular lifestyle and are known to systemically disseminate via macrophages when given orally, which could potentially infect chickens via a similar pathway following respiratory exposure (Vazquez-Torres and Fang, 2001). In our published studies, *S.* Senftenberg was recovered, in a dose-related fashion, from the liver and spleen of chickens after IV administration. However, *S.* Senftenberg was not recovered after enrichment of caeca, even with a high dose by either oral or IT administration route in any experiment (Kallapura et al., 2014b). Based on these results, we conducted a series of field trials in North and South Americas (the United States, Mexico, Colombia and Argentina) to evaluate the association between caecal and tracheal recovery of *Salmonella* in chickens and turkeys from commercial flocks (Kallapura et al., 2014a). Environmental humidity and temperature were measured to evaluate their effects on frequency of isolation from organs. In all trials, *Salmonella* was recovered from tracheal samples. Three of four trials, in which both trachea and caeca were sampled, showed that incidence of *Salmonella* recovery was higher in tracheal samples, confirming that tracheal contamination can be an indicator of *Salmonella* infection under commercial poultry conditions.

10 Future trends and conclusion

Due to its complex nature, an ongoing relationship between host and microbiome, the GIT is a critical organ for health maintenance in poultry. Changes in nutrition can have profound effects on microbiota, and at the same time, changes in microbiota can have devastating effects on nutrient availability and absorption in the host. Developing research technologies, especially -omics, have provided tools by which scientists can begin to understand this complex relationship, and this emerging field is continually revealing the importance of the microbiome in the health of mammals. The markedly expanding -omics fields have allowed a 'shotgun' approach to discovery of molecular differences in various states of disease or health, dietary influences, methods of prophylaxis and treatment, and the discovery of biomarkers for subtle but important physiological or pathophysiologic differences. Such profiling is supported by the rapidly emerging technologies within genomics, transcriptomics and proteomics. However, there are a number of limitations of -omics approaches. For example, transcriptome analysis provides patterns of changes in gene expression that often poorly correlate with patterns of gene expression (Gutstein et al., 2008). Moreover, it has been estimated that about a third of expressed transcripts do not get translated into proteins (Lundberg, et al., 2010) and the half-life of proteins can differ by orders of magnitude (Gutstein et al., 2008). Preliminary investigations within our laboratories with proteomic analysis of chicken enteric mucosa with selected enteric stressors have resulted in a virtual tsunami of analysis possibilities using canonical analysis, which are entirely dependent upon database annotations, many of which are derived from work done in disparate species. For all of the shotgun approaches, markedly different results are dependent upon sample preparation, errors in measurement and sample degradation, each of which can lead to different possible outcomes.

Recent reports on microbiome analysis of GIT populations in chickens vary greatly, with bacteroides present in ranges of 2–60% of total content in caeca across multiple tests (Kohl, 2012; Wei et al., 2013; Xiao et al., 2016). This variation was also wide in Firmicutes, ranging from 20 to 90% of total content within a single data set (Xiao et al., 2016). Stanley and co-workers (2013) tested microbiome composition of three different sets of broilers from the same source, reared in the same facility under the same conditions, and found that, through multiple metrics analyses, all three animal sets were heavily, if not completely, separated with very little overlap. These reports of widely varying answers suggest that, rather than searching for an ideal set of microbiota within the GIT or attempting to define a characteristic pattern within an animal species, perhaps -omics analysis would be better suited for asking specific questions of how changes and shifts are affected or do affect aspects of health and nutrient absorption.

Nevertheless, this rapidly evolving area of '-omics' is likely to rapidly lead to huge breakthroughs in our understanding of health and pathogenesis as related to gut health, although it seems likely that many false trails will be started as improved methodologies and understanding evolve.

11 References

Aggarwal, K., and H. K. Lee. 2003. Functional genomics and proteomics as a foundation for systems biology. *Briefings In Functional Genomics Proteomics* 2:175–84.

Alvarez-Olmos, M. I., and R. A. Oberhelman. 2001. Probiotic agents and infectious diseases: a modern perspective on a traditional therapy. *Clinical Infectious Diseases* 32:1567–76.

Annett, C., J. Viste, M. Chirino-Trejo, H. Classen, D. Middleton and E. Simko. 2002. Necrotic enteritis: effect of barley, wheat and corn diets on proliferation of *Clostridium perfringens* type A. *Avian Pathology* 31:598–601.

Arp, L.H. 1982. Effect of passive immunization on phagocytosis of blood-borne Escherichia coli in spleen and liver of turkeys. *American journal of veterinary research* 43(6): 1034–40.

Arumugam, M., J. Raes, E. Pelletier, D. Le Paslier, T. Yamada, D. R. Mende, G. R. Fernandes, J. Tap, T. Bruls, J.-M. Batto, et al. 2011. Enterotypes of the human gut microbiome. *Nature* 473:174–80.

Arvola, T., K. Laiho, S. Torkkeli, H. Mykkänen, S. Salminen, L. Maunula and E. Isolauri. 1999. Prophylactic *Lactobacillus* GG reduces antibiotic-associated diarrhea in children with respiratory infections: a randomized study. *Pediatrics* 104:e64–e64.

Bäckhed, F. 2011. Programming of host metabolism by the gut microbiota. *Annals of Nutrition and Metabolism* 58:44–52.

Barnes, H. J., L. K. Nolan and J. Vaillancourt. 2008. Colibacillosis. *Diseases of Poultry*, 12th edn., Blackwell Publishing Company, Ames, pp. 691–737.

Baskerville, A., T. Humphrey, R. Fitzgeorge, R. Cook, H. Chart, B. Rowe and A. Whitehead. 1992. Airborne infection of laying hens with *Salmonella enteritidis* phage type 4. *The Veterinary Record* 130:395.

Basler, C., Forshey, T.M., Machesky, K., Erdman, C.M., Gomez, T.M., Brinson, D.L., Nguyen, T.A., Behravesh, C.B., Bosch, S. and Centers for Disease Control and Prevention (CDC). 2015. Notes from the field: multistate outbreak of human *Salmonella* infections linked to live poultry from a mail-order hatchery in Ohio--February-October 2014. *MMWR Morb. Mortal. Wkly. Rep.* 64:258.

Bedford, M., and H. Classen. 1993. An *in vitro* assay for prediction of broiler intestinal viscosity and growth when fed rye-based diets in the presence of exogenous enzymes. *Poultry Science* 72:137–43.

Bedford, M., H. Classen and G. Campbell. 1991. The effect of pelleting, salt, and pentosanase on the viscosity of intestinal contents and the performance of broilers fed rye. *Poultry Science* 70:1571–7.

Bedford, M., and H. Schulze. 1998. Exogenous enzymes for pigs and poultry. Nutrition Research Reviews 11:91–114.

Bergman, E. 1990. Energy contributions of volatile fatty acids from the gastrointestinal tract in various species. *Physiological reviews* 70:567–90.

Berkhoff, H. A. 1985. *Clostridium colinum* sp. nov., nom. rev., the causative agent of ulcerative enteritis (quail disease) in quail, chickens, and pheasants. *International Journal of Systematic and Evolutionary Microbiology* 35:155–9.

Berrang, M., N. Cox and J. Bailey. 1995. Measuring air-borne microbial contamination of broiler hatching cabinets. The Journal of Applied Poultry Research 4:83–7.

Bianchi, M.-L., and M. Bardella. 2008. Bone in celiac disease. Osteoporosis international 19:1705–16.

Bjerrum, L., R. Engberg, T. D. Leser, B. B. Jensen, K. Finster and K. Pedersen. 2006. Microbial community composition of the ileum and cecum of broiler chickens as revealed by molecular and culture-based techniques. *Poultry Science* 85:1151–64.

Blaser, M. J. 2006. Who are we? Indigenous microbes and the ecology of human diseases. EMBO reports 7:956.

Borchers, A. T., C. Selmi, F. J. Meyers, C. L. Keen and M. E. Gershwin. 2009. Probiotics and immunity. Journal of gastroenterology 44:26–46.

Boyle, E. C., J. L. Bishop, G. A. Grassl and B. B. Finlay. 2007. *Salmonella*: from pathogenesis to therapeutics. Journal of Bacteriology 189:1489–95.

Braat, H., J. van den Brande, E. van Tol, D. Hommes, M. Peppelenbosch and S. van Deventer. 2004. *Lactobacillus rhamnosus* induces peripheral hyporesponsiveness in stimulated CD4+ T cells via modulation of dendritic cell function. *The American Journal of Clinical Nutrition* 80:1618–25.

Bronstein, J. L., R. Alarcón and M. Geber. 2006. The evolution of plant-insect mutualisms. New Phytologist 172:412–28.

Brown, S. A. 1996. Fluoroquinolones in animal health. J. Vet. Pharmacol. Ther. 19:1–14.

Burt, D. 2007. Emergence of the chicken as a model organism: implications for agriculture and biology. *Poultry Science* 86:1460–71.

Campbell, G., L. Campbell and H. Classen. 1983. Utilisation of rye by chickens: Effect of microbial status, diet gamma irradiation and sodium taurocholate supplementation. *British Poultry Science* 24:191–203.

Cani, P. D., R. Bibiloni, C. Knauf, A. Waget, A. M. Neyrinck, N. M. Delzenne and R. Burcelin. 2008. Changes in gut microbiota control metabolic endotoxemia-induced inflammation in high-fat diet-induced obesity and diabetes in mice. *Diabetes* 57:1470–81.

Cani, P. D., and N. M. Delzenne. 2009. The role of the gut microbiota in energy metabolism and metabolic disease. *Current Pharmaceutical Design* 15:1546–58.

Capriles, V. D., L. A. Martini and J. A. G. Areas. 2009. Metabolic osteopathy in celiac disease: importance of a gluten-free diet. *Nutrition Reviews* 67:599–606.

Carniel, E. 2001. The *Yersinia* high-pathogenicity island: an iron-uptake island. *Microbes and Infection* 3:561–9.

Centers for Disease Control and Prevention (CDC). 2013. Outbreak of *Salmonella* Heidelberg infections linked to a single poultry producer – 13 states, 2012–13. MMWR Morb. Mortal. Wkly. Rep. 62:553–6.

Chalmers G., Bruce, H. L., Hunter, D. B., Parreira, V. R., Kulkarni, R. R., Jiang, Y. F., Prescott, J. F., Boerlin, P. Multilocus sequence typing analysis of *Clostridium perfringens* isolates from necrotic enteritis outbreaks in broiler chicken populations. *Journal of Clinical Microbiology* 46(12):3957–64.

Chan, Y. K., M. Estaki and D. L. Gibson. 2013. Clinical consequences of diet-induced dysbiosis. *Annals of Nutrition and Metabolism* 63:28–40.

Chen, J., G. Tellez, J. D. Richards and J. Escobar. 2015. Identification of potential biomarkers for gut barrier failure in broiler chickens. *Frontiers in Veterinary Science* 2:14.

Chinivasagam, H., T. Tran, L. Maddock, A. Gale and P. Blackall. 2009. Mechanically ventilated broiler sheds: a possible source of aerosolized *Salmonella, Campylobacter*, and *Escherichia* coli. *Applied and Environmental Microbiology* 75:7417–25.

Choct, M. 2009. Managing gut health through nutrition. *British Poultry Science* 50:9–15.

Choct, M., R. J. Hughes, R. P. Trimble, K. Angkanaporn and G. Annison. 1995. Non-starch polysaccharide-degrading enzymes increase the performance of broiler chickens fed wheat of low apparent metabolizable energy. *The Journal of Nutrition* 125:485–92.

Clark, J. A., S. M. Doelle, M. D. Halpern, T. A. Saunders, H. Holubec, K. Dvorak, S. A. Boitano and B. Dvorak. 2006. Intestinal barrier failure during experimental necrotizing enterocolitis: protective effect of EGF treatment. *American Journal of Physiology-Gastrointestinal and Liver Physiology* 291:G938–G949.

Cogburn, L., T. Porter, M. Duclos, J. Simon, S. Burgess, J. Zhu, H. Cheng, J. Dodgson and J. Burnside. 2007. Functional genomics of the chicken – a model organism. *Poultry Science* 86:2059–94.

Collado, M. C., M. Cernada, C. Baüerl, M. Vento and G. Pérez-Mart'inez. 2012. Microbial ecology and host-microbiota interactions during early life stages. *Gut Microbes* 3:352–65.

Cox, J., and A. Pavic. 2010. Advances in enteropathogen control in poultry production. *Journal of Applied Microbiology* 108:745–55.

Dahiya, J., D. Wilkie, A. Van Kessel and M. Drew. 2006. Potential strategies for controlling necrotic enteritis in broiler chickens in post-antibiotic era. *Animal Feed Science and Technology* 129:60–88.

Dale, C., and N. A. Moran. 2006. Molecular interactions between bacterial symbionts and their hosts. *Cell* 126:453–65.

Dass, N., A. John, A. Bassil, C. Crumbley, W. Shehee, F. Maurio, G. Moore, C. Taylor and G. Sanger. 2007. The relationship between the effects of short-chain fatty acids on intestinal motility *in vitro* and GPR43 receptor activation. *Neurogastroenterology and Motility* 19:66–74.

Dethlefsen, L., M. McFall-Ngai and D. A. Relman. 2007. An ecological and evolutionary perspective on human-microbe mutualism and disease. *Nature* 449:811–18.

Di Mauro, A., J. Neu, G. Riezzo, F. Raimondi, D. Martinelli, R. Francavilla, et al. 2013. Gastrointestinal function development and microbiota. *Italian Journal of Pediatrics* 39:1–7.

Dimitrov, D. V. 2011. The human gutome: nutrigenomics of the host-microbiome interactions. *Omics: A Journal of Integrative Biology* 15:419–30.

Dohms, J. E. 2008. Botulism. *Diseases of Poultry*, 12th edn. Blackwell Publishing Company, Ames, pp. 879–85.

Duerkop, B. A., S. Vaishnava and L. V. Hooper. 2009. Immune responses to the microbiota at the intestinal mucosal surface. *Immunity* 31:368–76.

Dunsford, B. R., D. Knabe and W. Haensly. 1989. Effect of dietary soybean meal on the microscopic anatomy of the small intestine in the early-weaned pig. *Journal of Animal Science* 67:1855.

Dziva, F., and M. P. Stevens. 2008. Colibacillosis in poultry: unravelling the molecular basis of virulence of avian pathogenic *Escherichia coli* in their natural hosts. *Avian Pathology* 37:355–66.

Elson, C. O., and Y. Cong. 2012. Host-microbiota interactions in inflammatory bowel disease. *Gut Microbes* 3:332–44.

Fagerland, J.A. and Arp, L.H., 1993. Structure and development of bronchus-associated lymphoid tissue in conventionally reared broiler chickens. *Avian diseases*, pp.10–18.

Fallschissel, K., P. Kämpfer and U. Jäckel. 2009. Direct detection of *Salmonella* cells in the air of livestock stables by real-time PCR. *Ann Occup Hyg* 53:859–68.

Farnell, M., A. Donoghue, F. S. De Los Santos, P. Blore, B. Hargis, G. Tellez and D. Donoghue. 2006. Upregulation of oxidative burst and degranulation in chicken heterophils stimulated with probiotic bacteria. *Poultry Science* 85:1900–6.

Feng, T., and C. Elson. 2010. Adaptive immunity in the host-microbiota dialog. *Mucosal Immunology* 4:15–21.

Forder, R. E. A., G. S. Nattrass, M. S. Geier, R. J. Hughes and P. I. Hynd. 2012. Quantitative analyses of genes associated with mucin synthesis of broiler chickens with induced necrotic enteritis. *Poultry Science* 91:1335–41.

Fraune, S., and T. C. Bosch. 2010. Why bacteria matter in animal development and evolution. *Bioessays* 32:571–80.

Fuller, R., and B. Brooker. 1974. Lactobacilli which attach to the crop epithelium of the fowl. *The American Journal of Clinical Nutrition* 27:1305–12.

Fuller, R.J., Gregory, R.D., Gibbons, D.W., Marchant, J.H., Wilson, J.D., Baillie, S.R. and Carter, N., 1995. Population declines and range contractions among lowland farmland birds in Britain. *Conservation Biology*, 9(6), pp.1425–41.

Galanis, E., M. E. Patrick, N. Binsztein, A. Cieslik, T. Chalermchikit, A. Aidara-Kane, A. Ellis, F. J. Angulo and H. C. Wegener. 2006. Web-based surveillance and global *Salmonella* distribution, 2000-2002. *Emerging Infectious Disease* 12:381–8.

Garcia-Cañas, V., C. Simó, C. León and A. Cifuentes. 2010. Advances in Nutrigenomics research: novel and future analytical approaches to investigate the biological activity of natural compounds and food functions. *Journal of Pharmaceutical and Biomedical Analysis* 51:290–304.

Gast R. K. 'Parathyphoid Infections'. Diseases of Poultry, 12th edn. Blackwell Publishing Company, Ames (2008): 636–65.

Ghosh, D., and L. M. Poisson. 2009. 'Omics' data and levels of evidence for biomarker discovery. *Genomics* 93:13–16.

Gnad, F., F. Forner, D. F. Zielinska, E. Birney, J. Gunawardena and M. Mann. 2010. Evolutionary constraints of phosphorylation in eukaryotes, prokaryotes, and mitochondria. *Molecular & Cellular Proteomics* 9:2642–53.

Guibas, G. V., G. Moschonis, P. Xepapadaki, E. Roumpedaki, O. Androutsos, Y. Manios and N. G. Papadopoulos. 2013. Conception via in vitro fertilization and delivery by Caesarean section are associated with paediatric asthma incidence. *Clinical & Experimental Allergy* 43:1058–66.

Gutstein, H.B., J.S. Morris, S.P. Annangudi and J.V. Sweedler. 2008. Microproteomics: analysis of protein diversity in small samples. *Mass Spectrometry Reviews* 27(4): 316–30.

Hacker, J., G. Blum-Oehler, I. Mühldorfer and H. Tschäpe. 1997. Pathogenicity islands of virulent bacteria: structure, function and impact on microbial evolution. *Molecular Microbiology* 23:1089–97.

Hamdi, C., A. Balloi, J. Essanaa, E. Crotti, E. Gonella, N. Raddadi, et al. 2011. Gut microbiome dysbiosis and honeybee health. *Journal of Applied Entomology* 135:524–33.

Hayter, R., and E. Besch. 1974. Airborne-particle deposition in the respiratory tract of chickens. *Poultry Science* 53:1507–11.

Hernández-Reyes, C., and A. Schikora. 2013. *Salmonella*, a cross-kingdom pathogen infecting humans and plants. *FEMS Microbiology Letters* 343:1–7.

Higgins, S., A. Wolfenden, G. Tellez, B. Hargis and T. Porter. 2011. Transcriptional profiling of cecal gene expression in probiotic-and *Salmonella*-challenged neonatal chicks. *Poultry Science* 90:901–13.

Hofacre, C. L. 2001. Necrotic enteritis, currently a billion dollar disease: is there anything new on the horizon. Pages 79–86 in Science and Technology in the Feed Industry: Proceedings of Alltech's 17th Annual Symposium. T. P. Lyons and K. A. Jacques, ed. Nottingham University Press, Nottingham, UK.

Hollingsworth, M. A., and B. J. Swanson. 2004. Mucins in cancer: protection and control of the cell surface. *Nature Reviews Cancer* 4:45–60.

Holt, P. S., R. K. Gast, R. Porter and H. D. Stone. 1999. Hyporesponsiveness of the systemic and mucosal humoral immune systems in chickens infected with *Salmonella enterica* serovar enteritidis at one day of age. *Poultry Science* 78:1510–17.

Hrncir, T., R. Stepankova, H. Kozakova, T. Hudcovic and H. Tlaskalova-Hogenova. 2008. Gut microbiota and lipopolysaccharide content of the diet influence development of regulatory T cells: studies in germ-free mice. *BMC Immunology* 9:65.

Hu, Q., B. Coburn, W. Deng, Y. Li, X. Shi, Q. Lan, B. Wang, B. K. Coombes and B. B. Finlay. 2008. *Salmonella enterica* serovar Senftenberg human clinical isolates lacking SPI-1. *Journal of Clinical Microbiology* 46:1330–6.

Huff, G., W. Huff, N. Rath and G. Tellez. 2006. Limited treatment with beta-1, 3/1, 6-glucan improves production values of broiler chickens challenged with *Escherichia coli*. *Poultry Science* 85:613–18.

Ilan, Y. 2012. Leaky gut and the liver: a role for bacterial translocation in nonalcoholic steatohepatitis. *World Journal of Gastroenterology* 18:2609.

Jean, D., G. Fecteau, D. Scott, R. Higgins and S. Quessy. 1995. *Clostridium botulinum* type C intoxication in feedlot steers being fed ensiled poultry litter. *The Canadian Veterinary Journal* 36:626–8.

Jeffrey, J. S., Droual, R., Meteyer, C. U., Galey, F. D., Kinde, H., Medina, H. 1992. Recurrent transient paresis in a turkey flock. *Avian Diseases* 1:760–5.

Jiang, Z., T. J. Applegate and A. C. Lossie. 2013. Cloning, annotation and developmental expression of the chicken intestinal MUC2 gene. *PloS ONE* 8:e53781.

Joint, F. 2001. WHO Expert consultation on evaluation of health and nutritional properties of probiotics in food including powder milk with live lactic acid bacteria. *Córdoba, Argentina* 1–4.

Jones, E. O., A. White and M. Boots. 2010. The evolution of host protection by vertically transmitted parasites. *Proceedings of the Royal Society: Biological Sciences*: rspb20101397.

Jounai, K., K. Ikado, T. Sugimura, Y. Ano, J. Braun and D. Fujiwara. 2012. Spherical lactic acid bacteria activate plasmacytoid dendritic cells immunomodulatory function via TLR9-dependent crosstalk with myeloid dendritic cells. *PloS ONE* 7:e32588.

Kallapura G., M. J. Morgan, N. R. Pumford, L. R. Bielke, A. D. Wolfenden, O. B. Faulkner, J. D. Latorre, A. Menconi, X. Hernandez-Velasco, V. A. Kuttappan, B. M. Hargis and G. Tellez. 2014a. Evaluation of Respiratory Route as a Viable Portal of Entry for Salmonella in Poultry via Intratracheal Challenge of Salmonella Enteritidis and Salmonella Typhimurium. *Poultry Science.* 93 :340–6.

Kallapura, G., A. Botero, S. Layton, L. Bielke, J. Latorre, A. Menconi, X. Hernández-Velasco, D. Bueno, B. Hargis and G. Téllez. 2014b. Evaluation of recovery of Salmonella from trachea and ceca in commercial poultry. *The Journal of Applied Poultry Research* 23:132–6.

Kallapura, G., M. Kogut, M. Morgan, N. Pumford, L. Bielke, A. Wolfenden, et al. 2014c. Fate of Salmonella Senftenberg in broiler chickens evaluated by challenge experiments. *Avian Pathology*:1–16.

Kallapura G., N. R. Pumford, L. R. Bielke, B. M. Hargis and G. Tellez. 2014.d Evaluation of Respiratory Route as a Viable Portal of Entry for Salmonella in Poultry. *Veterinary Medicine: Research and Reports* 5: 59–73.

Kallapura G., N. R. Pumford, X. Hernandez-Velasco, B. M. Hargis and G. Tellez. 2014e. Mechanisms Involved in Lipopolysaccharide Derived ROS and RNS Oxidative Stress and Septic Shock. *Journal of Microbiology Research and Reviews.* 2 (1): 6–11.

Kalliomaki, M., S. Salminen, H. Arvilommi, P. Kero, P. Koskinen, E. Isolauri, et al. 2001. Probiotics in primary prevention of atopic disease: a randomised placebo-controlled trial. *Lancet* 357:1076–9.

Kau, A. L., P. P. Ahern, N. W. Griffin, A. L. Goodman and J. I. Gordon. 2011. Human nutrition, the gut microbiome and the immune system. *Nature* 474:327–36.

Kiama, S.G., Adekunle, J.S. and Maina, J.N., 2008. Comparative in vitro study of interactions between particles and respiratory surface macrophages, erythrocytes, and epithelial cells of the chicken and the rat. *Journal of Anatomy*, 213(4), pp.452–63.

Kiarie, E., L. F. Romero and C. M. Nyachoti. 2013. The role of added feed enzymes in promoting gut health in swine and poultry. *Nutrition Research Reviews* 26:71–88.

Kikuchi, Y., T. Hosokawa, N. Nikoh, X.-Y. Meng, Y. Kamagata and T. Fukatsu. 2009. Host-symbiont co-speciation and reductive genome evolution in gut symbiotic bacteria of acanthosomatid stinkbugs. *BMC Biology* 7:2.

Kim, Y. S., and S. B. Ho. 2010. Intestinal goblet cells and mucins in health and disease: recent insights and progress. *Current Gastroenterology Reports* 12:319–30.

Kimbrell, D. A., and B. Beutler. 2001. The evolution and genetics of innate immunity. *Nature Reviews Genetics* 2:256–67.

Kohl, K. D. 2012. Diversity and function of the avian gut microbiota. *Journal of Comparative Physiology B* 182:591–602.

Kotake, S., Y. Nanke, T. Yago, M. Kawamoto and H. Yamanaka. 2009. Human osteoclastogenic T cells and human osteoclastology. *Arthritis & Rheumatism* 60:3158–63.

Kussmann, M., F. Raymond and M. Affolter. 2006. OMICS-driven biomarker discovery in nutrition and health. *Journal of Biotechnology* 124:758–87.

Kussmann, M., S. Rezzi and H. Daniel. 2008. Profiling techniques in nutrition and health research. *Current Opinion in Biotechnology* 19:83–99.

Kuttappan, V., O. Faulkner, A. Wolfenden, G. Tellez, B. Hargis and L. Bielke. 2013. Novel Recombinant Food Safety Vaccines against *Salmonella* and *Campylobacter* in Broiler Chickens. *International Journal of Biotechnology and Bioengineering Research*. 6:589–96.

Lang, T., Hansson, G.C. and Samuelsson, T., 2006. An inventory of mucin genes in the chicken genome shows that the mucin domain of Muc13 is encoded by multiple exons and that ovomucin is part of a locus of related gel-forming mucins. *BMC genomics*, 7(1), p.1.

Lang, T., G. C. Hansson and T. Samuelsson. 2007. Gel-forming mucins appeared early in metazoan evolution. *Proceedings of the National Academy of Sciences* 104:16209–14.

Latorre, J. D., X. Hernandez-Velasco, M. H. Kogut, J. L. Vicente, R. Wolfenden, A. Wolfenden, B. M. Hargis, V. A. Kuttappan and G. Tellez. 2014. Role of a *Bacillus subtilis* Direct-Fed Microbial on Digesta Viscosity, Bacterial Translocation, and Bone Mineralization in Turkey Poults Fed with a Rye-Based Diet. *Frontiers in Veterinary Science* 1:26.

Latorre, J. D., X. Hernandez-Velasco, V. A. Kuttappan, R. E. Wolfenden, J. L. Vicente, A. D. Wolfenden, et al. 2015. Selection of *Bacillus* spp. for cellulase and xylanase production as direct-fed microbials to reduce digesta viscosity and *Clostridium perfringens* proliferation using an *in vitro* digestive model in different poultry diets. *Frontiers in Veterinary Science* 2

Leach, S. A., A. Williams, A. C. Davies, J. Wilson, P. D. Marsh and T. J. Humphrey. 1999. Aerosol route enhances the contamination of intact eggs and muscle of experimentally infected laying hens by *Salmonella typhimurium* DT104. *FEMS Microbiology Letters* 171:203–7.

Lepp, D., Gong, J., Songer, J. G., Boerlin, P., Parreira, V. R., Prescott, J. F. 2013. Identification of accessory genome regions in poultry *Clostridium perfringens* isolates carrying the netB plasmid. *Journal of Bacteriology* 195(6):1152–66.

Li, C., Guo, S., Gao, J., Guo, Y., Du, E., Lv, Z. and Zhang, B., 2015. Maternal high-zinc diet attenuates intestinal inflammation by reducing DNA methylation and elevating H3K9 acetylation in the A20 promoter of offspring chicks. *The Journal of Nutritional biochemistry*, 26(2), pp.173–83.

Liu, D., S. Guo and Y. Guo. 2012. Xylanase supplementation to a wheat-based diet alleviated the intestinal mucosal barrier impairment of broiler chickens challenged by *Clostridium perfringens. Avian Pathology* 41:291–8.

Lund, M., L. Bjerrum and K. Pedersen. 2010. Quantification of *Faecalibacterium prausnitzii*-and Subdoligranulum variabile-like bacteria in the cecum of chickens by real-time PCR. *Poultry Science* 89:1217–24.

Lundberg, E., L. Fagerberg, D. Klevebring, I. Matic, T. Geiger, J. Cox, C. Algenas, J. Lundberg, M. Mann and M. Uhlem. 2010 Defining the transcriptome and proteome in three functionally different cell lines. *Molecular and Systemic Biology* 6, 450–9.

Luo, Y., H. Peng, A.-D. G. Wright, S. Bai, X. Ding, Q. Zeng, et al. 2013. Broilers fed dietary vitamins harbor higher diversity of cecal bacteria and higher ratio of *Clostridium, Faecalibacterium*, and *Lactobacillus* than broilers with no dietary vitamins revealed by 16S rRNA gene clone libraries. *Poultry Science* 92:2358–66.

Maas, M. R., K. A. Glass and M. P. Doyle. 1989. Sodium lactate delays toxin production by *Clostridium botulinum* in cook-in-bag turkey products. *Applied and Environmental Microbiology* 55:2226–9.

Mariat, D., O. Firmesse, F. Levenez, V. Guimaruaes, H. Sokol, J. Dore, G. Corthier and J. Furet. 2009. The *Firmicutes/Bacteroidetes* ratio of the human microbiota changes with age. *BMC Microbiology* 9:123.

Martin, R., A. Nauta, K. Ben Amor, L. M. Knippels, J. Knol and J. Garssen. 2010. Early life: gut microbiota and immune development in infancy. *Beneficial Microbes* 1:367–82.

Maslowski, K. M., and C. R. Mackay. 2010. Diet, gut microbiota and immune responses. *Nature Immunology* 12:5–9.
Mazmanian, S.K., Round, J.L. and Kasper, D.L., 2008. A microbial symbiosis factor prevents intestinal inflammatory disease. *Nature*, 453(7195), pp.620–5.
McFall-Ngai, M. 2007. Adaptive immunity: care for the community. *Nature* 445:153.
Mensah, G.A. and Brain, J. D., 1982. Deposition and clearance of inhaled aerosol in the respiratory tract of chickens. *Journal of Applied Physiology*,53(6): 1423–8.
Minot, S., A. Bryson, C. Chehoud, G. D. Wu, J. D. Lewis and F. D. Bushman. 2013. Rapid evolution of the human gut virome. *Proceedings of the National Academy of Sciences* 110:12450–5.
Moran, N. A. 2007. Symbiosis as an adaptive process and source of phenotypic complexity. *Proceedings of the National Academy of Sciences* 104:8627–33.
Morgan, X. C., and C. Huttenhower. 2012. Chapter 12: Human Microbiome Analysis. *PLOS Computational Biology* 8:e1002808.
Mshvildadze, M., J. Neu and V. Mai. 2008. Intestinal microbiota development in the premature neonate: establishment of a lasting commensal relationship? *Nutrition Reviews* 66:658–63.
Murphy, T., J. McCracken, M. McCann, J. George and M. Bedford. 2009. Broiler performance and in vivo viscosity as influenced by a range of xylanases, varying in ability to affect wheat *in vitro* viscosity. *British Poultry Science* 50:716–24.
Musso, G., R. Gambino and M. Cassader. 2010. Obesity, Diabetes, and Gut Microbiota: The hygiene hypothesis expanded? *Diabetes Care* 33:2277–84.
Myers, S. P. 2004. The causes of intestinal dysbiosis: a review. *Alternative Medicine Review* 9:180–97.
Neish, A. S. 2009. Microbes in gastrointestinal health and disease. *Gastroenterology* 136:65–80.
Neu, J., et al. 2011. Cesarean versus vaginal delivery: long term infant outcomes and the hygiene hypothesis. *Clinics in Perinatology* 38:321–31.
Oakley, B. B., H. S. Lillehoj, M. H. Kogut, W. K. Kim, J. J. Maurer, A. Pedroso, M. D. Lee, S. R. Collett, T. J. Johnson and N. A. Cox. 2014. The chicken gastrointestinal microbiome. *FEMS Microbiology Letters* 360:100–12.
Ohl, M. E., and S. I. Miller. 2001. *Salmonella*: a model for bacterial pathogenesis. *Annual Review of Medicine* 52:259–74.
Opengart, K. 'Necrotic Enteritis'. Diseases of Poultry, 12th edn. Blackwell Publishing Company, Ames (2008): 872–9.
Opengart, K. 'Gangrenous Dermatitis'. Diseases of Poultry, 12th edn. Blackwell Publishing Company, Ames (2008): 885–9.
Ouwehand, A., E. Isolauri and S. Salminen. 2002. The role of the intestinal microflora for the development of the immune system in early childhood. *European Journal of Nutrition* 41.
Parvez, S., K. Malik, S. Ah Kang and H.-Y. Kim. 2006. Probiotics and their fermented food products are beneficial for health. *Journal of Applied Microbiology* 100:1171–85.
Perumbakkam, S., H. D. Hunt and H. H. Cheng. 2014. Marek's disease virus influences the core gut microbiome of the chicken during the early and late phases of viral replication. *FEMS Microbiology Ecology* 90:300–12.
Petermann, S. R., J. S. Sherwood and C. M. Logue. 2008. The *Yersinia* high pathogenicity island is present in *Salmonella enterica* Subspecies I isolated from turkeys. *Microbial Pathogenesis* 45:110–14.
Popp, C., R. Hauck, W. Gad and H. M. Hafez. 2012. Type C botulism in a commercial turkey farm: a case report. *Avian Diseases* 56:760–3.
Prescott, J. F., R. Sivendra and D. A. Barnum. 1978. The use of bacitracin in the prevention and treatment of experimentally-induced necrotic enteritis in the chicken. *The Canadian Veterinary Journal* 19:181–3.
Qiu, R., J. Croom, R. Ali, A. Ballou, C. Smith, C. Ashwell, H. Hassan, C.-C. Chiang and M. Koci. 2012. Direct fed microbial supplementation repartitions host energy to the immune system. *Journal of Animal Science* 90:2639–51.
Qureshi, M. 2003. Avian macrophage and immune response: An overview. *Poultry Science* 82:691–8.

Reveneau, N., M.-C. Geoffroy, C. Locht, P. Chagnaud and A. Mercenier. 2002. Comparison of the immune responses induced by local immunizations with recombinant *Lactobacillus plantarum* producing tetanus toxin fragment C in different cellular locations. *Vaccine* 20:1769–77.
Rigottier-Gois, L. 2013. Dysbiosis in inflammatory bowel diseases: the oxygen hypothesis. *The ISME Journal* 7:1256–61.
Saleh, M., and C. O. Elson. 2011. Experimental inflammatory bowel disease: insights into the host-microbiota dialog. *Immunity* 34:293–302.
Salminen, S., G. Gibson, A. McCartney and E. Isolauri. 2004. Influence of mode of delivery on gut microbiota composition in seven year old children. *Gut* 53:1388–9.
Salminen, S., and E. Isolauri. 2006. Intestinal colonization, microbiota, and probiotics. *The Journal of Pediatrics* 149:S115–S120.
Salzman, N. H. 2011. Microbiota-immune system interaction: an uneasy alliance. *Current Opinion in Microbiology* 14:99–105.
Santos, R., R. Tsolis, A. Bäumler and L. Adams. 2003. Pathogenesis of *Salmonella*-induced enteritis. *Brazilian Journal of Medical and Biological Research* 36:03–12.
Saridaki, A., and K. Bourtzis. 2010. *Wolbachia*: more than just a bug in insect's genitals. *Current Opinion in Microbiology* 13:67–72.
Schiffrin, E., and S. Blum. 2002. Interactions between the microbiota and the intestinal mucosa. *European Journal of Clinical Nutrition* 56:S60–4.
Seki, E., and B. Schnabl. 2012. Role of innate immunity and the microbiota in liver fibrosis: crosstalk between the liver and gut. *The Journal of Physiology* 590:447–58.
Sekirov, I., S. L. Russell, L. C. M. Antunes and B. B. Finlay. 2010. Gut microbiota in health and disease. *Physiological Reviews* 90:859–904.
Sharma, R., C. Young and J. Neu. 2010. Molecular modulation of intestinal epithelial barrier: contribution of microbiota. *Journal of Biomedicine and Biotechnology* 2010.
Shen, J., M. S. Obin and L. Zhao. 2012. The gut microbiota, obesity and insulin resistance. *Molecular Aspects of Medicine*.
Sherman, P. M., J. C. Ossa and K. Johnson-Henry. 2009. Unraveling mechanisms of action of probiotics. *Nutrition in Clinical Practice* 24:10–14.
Sherryll Lynn, L., H.-V. Xochitl, C. Shivaramaiah, X. Jorge, M. Anita, L. Juan David, et al. 2013. The effect of a *lactobacillus*-based probiotic for the control of necrotic enteritis in broilers. *Food and Nutrition Sciences*
Shirzadi, H., H. Moravej and M. Shivazad. 2010. Influence of non starch polysaccharide-degrading enzymes on the meat yield and viscosity of jejunal digesta in broilers fed wheat/barley-based diet. *African Journal of Biotechnology* 9:1517–22.
Shivaprasad H. L. and P. A. Barrow. 2008. 'Pullorum Disease and Fowl Typhoid' Diseases of Poultry, 12th edn. Blackwell Publishing Company, Ames: 620–36.
Shivaramaiah, S., R. Wolfenden, J. Barta, M. Morgan, A. Wolfenden, B. Hargis and G. Téllez. 2011. The role of an early *Salmonella* typhimurium infection as a predisposing factor for necrotic enteritis in a laboratory challenge model. *Avian Diseases* 55:319–23.
Stanley, D., R. J. Hughes and R. J. Moore. 2014. Microbiota of the chicken gastrointestinal tract: influence on health, productivity and disease. *Applied Microbiology and Biotechnology* 98:4301–10.
Stecher, B., L. Maier and W.-D. Hardt. 2013. 'Blooming' in the gut: how dysbiosis might contribute to pathogen evolution. *Nature Reviews Microbiology* 11:277–84.
Stepniak, D., and F. Koning. 2006. Celiac disease—sandwiched between innate and adaptive immunity. *Human Immunology* 67:460–8.
Stern, C. D. 2004. The chick embryo-past, present and future as a model system in developmental biology. *Mechanisms of Development* 121: 1011–13.
Stern, C. D. 2005. The chick: a great model system becomes even greater. *Developmental Cell* 8:9–17.
Stevceva, L., A. Abimiku and G. Franchini. 2000. Targeting the mucosa: genetically engineered vaccines and mucosal immune responses. *Genes & Immunity* 1.

Subbiah, M. 2007. Nutrigenetics and nutraceuticals: the next wave riding on personalized medicine. *Translational Research* 149:55–61.

Tagliabue, A., and M. Elli. 2012. The role of gut microbiota in human obesity: Recent findings and future perspectives. *Nutrition, Metabolism and Cardiovascular Diseases.*

Teirlynck, E., L. Bjerrum, V. Eeckhaut, G. Huygebaert, F. Pasmans, F. Haesebrouck, J. Dewulf, R. Ducatelle and F. Van Immerseel. 2009. The cereal type in feed influences gut wall morphology and intestinal immune cell infiltration in broiler chickens. *British Journal of Nutrition* 102: 1453–61.

Teirlynck, E., M. Gussem, J. Dewulf, F. Haesebrouck, R. Ducatelle and F. Van Immerseel. 2011. Morphometric evaluation of 'dysbacteriosis' in broilers. *Avian Pathology* 40:139–44.

Tellez, G., S. Higgins, A. Donoghue and B. Hargis. 2006. Digestive physiology and the role of microorganisms. *Journal of Applied Poultry Research* 15:136–44.

Tellez, G., L. Jaeger, C. Dean, D. Corrier, J. DeLoach, J. Williams and B. Hargis. 1993. Effect of prolonged administration of dietary capsaicin on *Salmonella enteritidis* infection in leghorn chicks. *Avian Diseases* 143–8.

Tellez, G., J. D. Latorre, V. A. Kuttappan, B. M. Hargis and X. Hernandez-Velasco. 2015. Rye Affects Bacterial Translocation, Intestinal Viscosity, Microbiota Composition and Bone Mineralization in Turkey Poults.

Tellez, G., J. D. Latorre, V. A. Kuttappan, M. H. Kogut, A. Wolfenden, X. Hernandez-Velasco, B. M. Hargis, W. G. Bottje, L. R. Bielke and O. B. Faulkner. 2014. Utilization of rye as energy source affects bacterial translocation, intestinal viscosity, microbiota composition, and bone mineralization in broiler chickens. *Nutrigenomics* 5:339.

Tellez, G., C. Pixley, R. Wolfenden, S. Layton and B. Hargis. 2012. Probiotics/direct fed microbials for *Salmonella* control in poultry. *Food Research International* 45:628–33.

Tellez, G., N. R. Pumford, M. J. Morgan, A. D. Wolfenden and B. M. Hargis. 2009. Evidence for Clostridium septicum as a primary cause of cellulitis in commercial turkeys. *Journal of Veterinary Diagnostic Investigation* 21:374–7.

Thavagnanam, S., J. Fleming, A. Bromley, M. Shields and C. Cardwell. 2008. A meta-analysis of the association between Caesarean section and childhood asthma. *Clinical & Experimental Allergy* 38:629–33.

Thiennimitr, P., S. E. Winter and A. J. Bäumler. 2012. *Salmonella*, the host and its microbiota. *Current Opinion in Microbiology* 15:108–14.

Timbermont, L., F. Haesebrouck, R. Ducatelle and F. Van Immerseel. 2011. Necrotic enteritis in broilers: an updated review on the pathogenesis. *Avian Pathology* 40:341–7.

Tlaskalová-Hogenová, H., R. \vSt\vepánková, H. Kozáková, T. Hudcovic, L. Vannucci, L. Tu\vcková, et al. 2011. The role of gut microbiota (commensal bacteria) and the mucosal barrier in the pathogenesis of inflammatory and autoimmune diseases and cancer: contribution of germ-free and gnotobiotic animal models of human diseases. *Cellular & Molecular Immunology* 8:110–20.

Toribara, N., A. Roberton, S. Ho, W. Kuo, E. Gum, J. Hicks, J. Gum, J. Byrd, B. Siddiki and Y. Kim. 1993. Human gastric mucin. Identification of a unique species by expression cloning. *Journal of Biological Chemistry* 268:5879–85.

Torok, V. A., R. J. Hughes, K. Ophel-Keller, M. Ali and R. Macalpine. 2009. Influence of different litter materials on cecal microbiota colonization in broiler chickens. *Poultry Science* 88:2474–81.

Trujillo, E., C. Davis and J. Milner. 2006. Nutrigenomics, proteomics, metabolomics, and the practice of dietetics. *Journal of the American Dietetic Association* 106:403–13.

Van Ommen, B., and R. Stierum. 2002. Nutrigenomics: exploiting systems biology in the nutrition and health arena. *Current Opinion in Biotechnology* 13:517–21.

Vazquez-Torres, A., and F. C. Fang. 2001. Oxygen-dependent anti-*Salmonella* activity of macrophages. *Trends in Microbiology* 9:29–33.

Vicuña, E., V. Kuttappan, R. Galarza-Seeber, J. Latorre, O. Faulkner, B. Hargis, G. Tellez and L. Bielke. 2015. Effect of dexamethasone in feed on intestinal permeability, differential white blood cell counts, and immune organs in broiler chicks. *Poultry Science*:pev211.

Wages, D. P. 'Ulcerative Enteritis (Quail Disease)'. 2008. Diseases of Poultry, 12th edn. Blackwell Publishing Company, Ames : 867–72.

Walter, J., R. A. Britton and S. Roos. 2011. Host-microbial symbiosis in the vertebrate gastrointestinal tract and the *Lactobacillus reuteri* paradigm. *Proceedings of the National Academy of Sciences* 108:4645–52.

Wathes, C., W. Zaidan, G. Pearson, M. Hinton and N. Todd. 1988. Aerosol infection of calves and mice with *Salmonella typhimurium*. *The Veterinary Record* 123:590–4.

Wei, X., Z. Yang, F. E. Rey, V. K. Ridaura, N. O. Davidson, J. I. Gordon and C. F. Semenkovich. 2012. Fatty acid synthase modulates intestinal barrier function through palmitoylation of mucin 2. *Cell Host & Microbe* 11:140–52.

Wei, S., M. Morrison and Z. Yu. 2013. Bacterial census of poultry intestinal microbiome. *Poultry Science* 92:671–83.

West, J.B. and Mathieu-Costello, O., 1995. Vulnerability of pulmonary capillaries in heart disease. *Circulation*, 92(3), pp.622–31.

White, P. L., A. R. Baker and W. O. James. 1997. Strategies to control *Salmonella* and *Campylobacter* in raw poultry products. *Rev. - Off. Int. Epizoot.* 16:525–41.

Wideman, R., K. Hamal, J. Stark, J. Blankenship, H. Lester, K. Mitchell, G. Lorenzoni and I. Pevzner. 2012. A wire-flooring model for inducing lameness in broilers: Evaluation of probiotics as a prophylactic treatment. *Poultry Science* 91:870–83.

Wideman, R. F., et al. 2011. Bone circulatory disturbances in the development of spontaneous bacterial chondronecrosis with osteomyelitis: a translational model for the pathogenesis of femoral head necrosis. *Frontiers in Endocrinology* 3:183.

Wideman, R., A. Al-Rubaye, Y. Kwon, J. Blankenship, H. Lester, K. Mitchell, I. Pevzner, T. Lohrmann and J. Schleifer. 2015. Prophylactic administration of a combined prebiotic and probiotic, or therapeutic administration of enrofloxacin, to reduce the incidence of bacterial chondronecrosis with osteomyelitis in broilers. *Poultry Science* 94:25–36.

Williams, R. 2005. Intercurrent coccidiosis and necrotic enteritis of chickens: rational, integrated disease management by maintenance of gut integrity. *Avian Pathology* 34:159–80.

Williams, S. J., D. H. Wreschner, M. Tran, H. J. Eyre, G. R. Sutherland and M. A. McGuckin. 2001. Muc13, a novel human cell surface mucin expressed by epithelial and hemopoietic cells. *Journal of Biological Chemistry* 276:18327–36.

Winter, S. E., and A. J. Bäumler. 2011. A breathtaking feat: to compete with the gut microbiota, *Salmonella* drives its host to provide a respiratory electron acceptor. *Gut Microbes* 2:58–60.

Winter, S. E., A. M. Keestra, R. M. Tsolis and A. J. Bäumler. 2010. The blessings and curses of intestinal inflammation. *Cell Host Microbe* 8:36–43.

Wise, M., and G. Siragusa. 2007. Quantitative analysis of the intestinal bacterial community in one-to three-week-old commercially reared broiler chickens fed conventional or antibiotic-free vegetable-based diets. *Journal of Applied Microbiology* 102:1138–49.

Wylie, K. M., G. M. Weinstock and G. A. Storch. 2012. Emerging view of the human virome. Transl Res 160:283–90.

Xiao, Y., Y. Xiang, W. Zhou, J. Chen, K. Li and H. Yang. 2016. Microbial community mapping in intestinal tract of broiler chicken. *Poultry Science* 372.

Xu, J., and J. I. Gordon. 2003. Honor thy symbionts. *Proceedings of the National Academy of Sciences* 100:10452–9.

Xu, J., M. A. Mahowald, R. E. Ley, C. A. Lozupone, M. Hamady, E. C. Martens, et al. 2007. Evolution of symbiotic bacteria in the distal human intestine. *PLoS Biology* 5:e156.

Yan, Y., V. Kolachala, G. Dalmasso, H. Nguyen, H. Laroui, S. V. Sitaraman and D. Merlin. 2009. Temporal and spatial analysis of clinical and molecular parameters in dextran sodium sulfate induced colitis. *PloS ONE* 4:e6073.

Yegani, M., and D. R. Korver. 2008a. Factors affecting intestinal health in poultry. *Poultry Science* 87:2052–63.

Yegani, M., and D. Korver. 2008b. Factors affecting intestinal health in poultry. *Poultry Science* 87:2052–63.

Zhang, Q., S. D. Eicher and T. J. Applegate. 2015. Development of intestinal mucin 2, IgA, and polymeric Ig receptor expressions in broiler chickens and Pekin ducks. *Poultry Science*: 94:172–80.

Zhang, S., R. A. Kingsley, R. L. Santos, H. Andrews-Polymenis, M. Raffatellu, J. Figueiredo, J. Nunes, R. M. Tsolis, L. G. Adams and A. J. Bäumler. 2003. Molecular pathogenesis of *Salmonella enterica* serotype Typhimurium-induced diarrhea. *Infection and Immunity* 71:1–12.

Chapter 3

Viruses affecting poultry

Venugopal Nair, Pirbright Institute, UK

1 Introduction

During the past 50 years, the poultry industry has shown significant all-round progress in increasing efficiency in production, feed conversion and management practices for intensive production. However, despite this progress, there are many challenges for sustaining the growth of this important sector. One of the greatest threats to the sustainability and growth of the poultry industry is the impact of infectious diseases, particularly those caused by multiple viral pathogens (Fig. 1), many of which are endemic in many countries. The impact of these infectious diseases comes from serious economic losses from increased mortality, loss of productivity, multiple organ pathology, immunosuppression and cancer. Precise estimates of the economic impact of many of these diseases on the poultry industry are difficult to obtain, but losses due to infectious diseases are thought to account for on average around 20% of productivity.

As in the case of other infectious diseases, prevention and control of avian viral diseases are based on the interconnected processes of (a) *surveillance* through continuous monitoring for disease prevalence and health status, (b) *bioexclusion* by reducing the exposure through increased hygiene and biosecurity measures, and (c) *biocontainment* through quarantine measures or emergency slaughter for preventing the spread of, and increasing resistance to the diseases mainly through vaccination. Unlike bacterial, protozoan or fungal diseases, there are no available treatment options and hence the fight against avian viral diseases is achieved largely

http://dx.doi.org/10.19103/AS.2016.0011.17

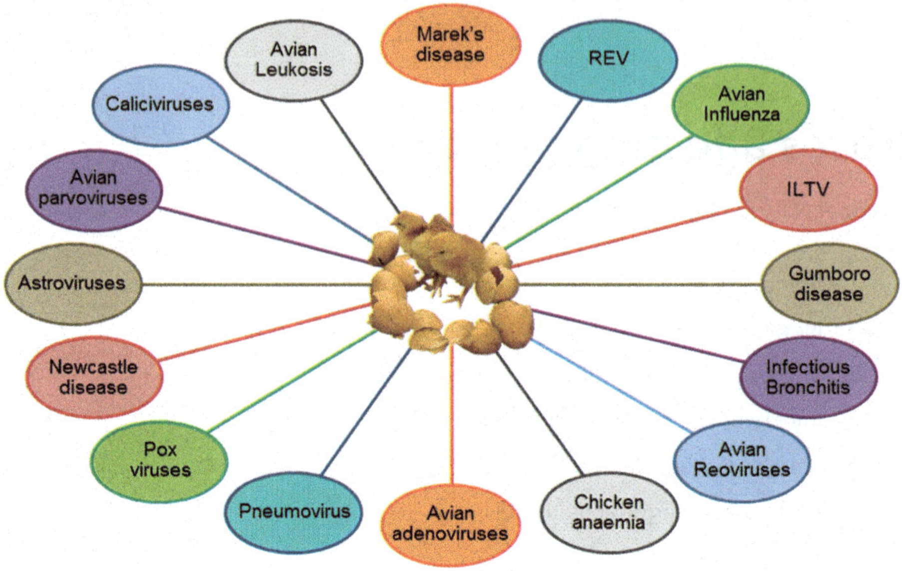

Figure 1 Major viral pathogens that threaten poultry.

through vaccination. Modern poultry production is heavily reliant on the use of multiple vaccines to prevent losses from a number of diseases. Although vaccination-based control strategies have been largely successful in controlling disease losses and sustaining growth of the poultry industry over the past several decades, a number of continuing challenges to the vaccination-based control strategy still remain. Major threats for the success of the vaccination strategy come from the immunosuppressive effects of some of the viral pathogens that interfere with the vaccine-induced immune responses. In addition, the early exposure of young chicks to high levels of viral pathogens from the heavily contaminated poultry houses well before the neonatal immune system had the time to mount an immune response also contributes to the reduced efficiency of vaccines. Moreover, some of the currently used vaccines having standard classical strains fail to induce significant protective responses against the increasing diversity of many viruses, some of which may be driven by the vaccines themselves (Ganar et al., 2014; Read et al., 2015). Research efforts for better understanding of the virus–host molecular interactions associated with viral pathogenesis, and development of vaccines that induce strong broad protection against a diverse number of viral pathogens are critical in sustaining effective control of many of the avian viral diseases. In terms of the widespread distribution and the devastating effects on avian health and economic losses from the disease, major viral diseases that continue to cause significant threats to avian health around the world are Newcastle disease (ND), infectious bronchitis (IB), infectious bursal disease (IBD), Marek's disease (MD), infectious laryngotracheitis (ILT) and avian influenza (AI). This chapter will provide an overall review on the recent advances on the studies of these major diseases.

2 Newcastle disease (ND)

ND is one of the most devastating diseases affecting the poultry industry worldwide (Ganar et al., 2014). It is caused by the highly contagious ND virus (NDV), which is a member of the genus *Avulavirus* (also known as APMV-1, avian paramyxovirus serotype-1) in the family Paramyxoviridae. They are an extremely diverse group of viruses, and, based on disease severity, NDV strains are classified into velogenic (highly virulent), mesogenic (moderately virulent), lentogenic (mildly virulent) and asymptomatic pathotypes (Alexander et al., 2012). The clinical signs and pathology of ND can range from inapparent infections to severe disease with high mortality, depending on the virus strains and the susceptibility of the host species. Around 250 bird species have been reported to be susceptible to NDV, and virus infections in feral and pet birds can influence virus spread and initiation of major outbreaks and panzootics in poultry. ND is a major global threat to the poultry industry as it has been reported in most countries. While it is endemic in majority of the countries, including some of largest poultry-producing nations, several countries maintain disease-free status despite the accidental incursions of the virus. In NDV-free countries, statutory surveillance measures are in place and outbreaks of NDV are notifiable to the OIE (*Office International des Epizooties*) and are associated with considerable economic impact including international trade restrictions.

NDV is a pleomorphic enveloped virus of about 200–300 nm in diameter with a non-segmented, negative sense, single-stranded RNA genome that encodes six genes for the essential proteins N (nucleocapsid), M (matrix protein), P (phosphoprotein), F (fusion protein), HN (haemagglutinin-neuraminidase protein) and L (large polymerase protein). Among these, F and HN glycoproteins, involved in virus attachment and membrane fusion, are the most important in the protection through the induction of neutralising antibodies. While the non-glycosylated M protein is involved in virus assembly and budding, the V protein generated by RNA editing of the P gene functions as a virulence factor by interfering with the interferon pathway (Park et al., 2003).

NDV infection is initiated at the respiratory epithelial cells where the virus glycoproteins bind to the sialic acid-containing receptors. NDV infectivity is dependent on the cleavage of F protein precursor F0 to F1–F2 subunits by non-viral proteases at specific sequence motifs around the cleavage site. The monobasic amino acid sequence motif G-R/K-Q-G-R-L in low-virulence NDV strains are only cleaved by extracellular trypsin-like proteases present in the respiratory and intestinal tract. However, virulent strains have multibasic amino acid sequence motifs R/G/K-R-Q/K-K/R-R-F that are cleaved by intracellular furin-like proteases ubiquitous in most host tissues. This difference in protease activation is the major determinant of systemic replication associated with the severe disease by virulent NDV (reviewed by (Ganar et al., 2014)).

The manifestations of the disease vary immensely and are dependent on a number of factors including the immune status and the susceptibility of the host species, as well as on the virulence, dose and the route of exposure to the virus. With an average incubation period of 2–15 days, birds infected with the highly virulent viscerotropic NDV strains show up to 100% mortality in 3 or 4 days, after being listless and depressed for 2 days. Infected birds may also show symptoms of conjunctivitis, facial swelling, nasal secretions and respiratory distress. Infection with neurotropic strains results in neurological symptoms including tremors, torticollis and paralysis of wings and legs, with up to 50% mortality.

Lesions seen in the infected birds are not typical but include enlargement of spleen and necrosis and haemorrhage in multiple organs.

Rapid diagnosis of NDV outbreaks is very important to control the spread and effective management of the disease. It is also essential to distinguish the low virulent and vaccine strains from the highly virulent pathotypes. Virological and serological tests are valuable in the diagnosis but are inadequate for the differentiation of vaccine strains. However, a number of molecular techniques based on the nucleic acid sequence differences have been developed for such differentiation (Desingu et al., 2015; Miller et al., 2010; Saba Shrivan and Mardani, 2014).

Control measures of ND are instituted with the objective of preventing susceptible birds from getting infected, or reducing the number of susceptible birds by vaccination. In NDV-free countries, legislation aimed at preventing the introduction of the virus through infected birds or products is in place (Miller and Koch, 2013). With the demonstration of psittacine birds as potential sources of panzootics in the 1970s, many countries have instituted increased regulatory measures on the importation of live birds, eggs and poultry products. Implementation of strict biosecurity measures and prevention of contact of domestic poultry with other birds cannot be overemphasized. In many countries, ND is controlled by culling of infected and exposed birds and restricting movement of birds and bird products within a defined area of the outbreak. Detailed recommendations on ND have been published (OIE, 2012).

Vaccination plays a major role in the control of ND, particularly when it is used along with good management and biosecurity practices. Although vaccines are helpful in reducing morbidity and mortality, they do not completely prevent virus shedding. Vaccination policy varies from country to country. Although prophylactic vaccination is allowed in most countries, a number of Scandinavian countries do not allow this practice. Formulations of both live and inactivated vaccines derived from low virulent strains are the most commonly used vaccines, although some countries with high endemic ND continue to use live mesogenic strains. Mass vaccinations of live ND vaccines applied through drinking water, sprays or aerosols are cheap to administer; however, they always do not induce protective antibodies in high proportion of birds. Ocular delivery provides the best route for the induction of antibodies. Recombinant vaccines based on viral vectors such as the herpesvirus of turkeys (HVT) and fowl pox virus (FPV) expressing NDV glycoproteins have also been used. Some of these recombinant vaccines are administered *in ovo*, have the advantage of preventing respiratory reactions and are not affected by maternal antibodies. In addition, NDV itself is used as a recombinant viral vector for the expression of protective antigens from other avian pathogens such as avian influenza (AI) (Cornelissen et al., 2010) and IBD (Huang et al., 2004).

3 Infectious bronchitis (IB)

Infectious bronchitis (IB) is another highly contagious and economically important acute upper respiratory tract disease in chickens and is caused by IB virus (IBV), an avian coronavirus. IBV has a worldwide occurrence and is transmitted by inhalation or direct contact with infected birds or materials. In addition to causing respiratory illness, the disease also results in loss of weight, reduced feed efficiency and predisposing birds to secondary infections. In layer and breeder chickens, IBV infection often results in reduced egg production and a huge decline in the eggshell quality. All these manifestations, together with morbidity often close to 100%, reveal that IB remains a major threat to achieve sustainable production (Cook et al., 2012; Jackwood, 2012).

IBV is an enveloped virus with a particle size of approximately 120nm in diameter and club-shaped surface spikes giving a crown-like appearance, and hence the name coronaviruses. IBV has a large RNA genome with an organization consisting of 5′UTR-leader-1a/1ab-S-3a-3b-E-M-5a-5b-N-3′UTR (Jackwood and de Wit, 2013) and encodes a number of structural and non-structural proteins. Among these, the most immunogenic viral protein is the S glycoprotein, which mediates host cell attachment and membrane fusion (Promkuntod et al., 2014, 2015; Wickramasinghe et al., 2014). However, the S protein shows extensive sequence variation accounting for the diversity among IBV serotypes and poor cross-protection (Dhama et al., 2014; Promkuntod et al., 2015; Toro et al., 2012).

The pathogenicity of IBV varies significantly between strains, and the clinical outcome depends on variables such as the virus strains, the sex and age of the chickens, immune status and presence of maternal antibodies, environmental factors and co-infections with other bacterial pathogens. Infection is initiated through the respiratory tract and virus replicates in many types of epithelial cells, including those of the respiratory tract, the reproductive organs and the kidneys. Affected birds show signs of respiratory infection including gasping, coughing, sneezing, tracheal rales and nasal discharge. Nephropathogenic strains showed depression, ruffled feathers, wet droppings and mortality. Pathological lesions include the presence of serous, catarrhal or caseous exudate in the trachea, nasal passages and sinuses. Infection with nephropathogenic strains produces swollen and pale kidneys with deposition of urates in the tubules and ureters.

Diagnosis of IB can be made by serological and virological tests to detect viruses or viral antigens (Jackwood and de Wit, 2013). More recently, a number of molecular techniques that can detect IBV-specific nucleic acids have been developed (Roh et al., 2014; Saba Shrivan and Mardani, 2014). Clinical presentation of IB may resemble other respiratory diseases such as infectious laryngotracheitis (ILT), ND or infectious coryza, and differential diagnosis may require the use of laboratory tests.

Control of IB would require improvement in the ventilation and heating, cleaning and disinfection of the poultry house and equipment to reduce the effects of secondary complications. Vaccination is an important aspect of control, but the use of appropriate vaccines in relation to the circulating strains of the virus is critical, as there is extensive diversity among IBV strains due to sequence variations in the spike protein (Toro et al., 2012; Wickramasinghe et al., 2014). Live vaccines attenuated by repeated passages in embryos are widely used in several countries. However, continued use of single classical vaccine (e.g. Massachusetts) strains in countries where a number of heterologous strains are circulating can result in less than optimum protection. In such circumstances, broadening the protection using multiple vaccine strains can be helpful (Jackwood and de Wit, 2013). Live vaccines are applied by mass vaccination methods using coarse sprays or in drinking water. High levels of maternal antibodies may interfere with the vaccine-induced immunity, particularly if the vaccines are administered in the hatchery, although this was not observed in all cases. Recent advances in the use of reverse genetics technology have enabled the development of novel vaccines with heterologous spike proteins, offering hope for more effective 'designer' vaccines for different situations (Armesto et al., 2011; Bentley et al., 2013; Britton et al., 2012; Keep et al., 2015).

4 Infectious bursal disease (IBD)

Infectious bursal disease (IBD) is another acute, highly contagious infection of young chickens characterised by the destruction of the bursa of Fabricius, the primary lymphoid organ involved in B-cell maturation and differentiation. The major impact of the disease is the immunosuppressive effect, which significantly increases the susceptibility to other infections and interferes with the responses to vaccines (Eterradossi and Saif, 2013). Being caused by the IBD virus (IBDV), a member of the Birnaviridae family, the disease has a worldwide distribution, occurring in almost all poultry-producing areas.

IBDV has a non-enveloped virion containing a bisegmented double-stranded RNA genome that encodes five viral proteins designated VP1–VP5. Of these, the capsid protein VP2 is the essential protective antigen. IBDV shows variations and a number of phenotypic and molecular genetic approaches have been developed for characterising these viruses. IBDV strains also vary in virulence and are described as vIBDV or vvIBDV (Li et al., 2015). IBDV strains can be grown in the laboratory using embryonated eggs and tissue culture, although some of the virulent strains need adaptation through several passages.

Pathogenesis of IBDV is mostly due to direct effect on the B cells and the bursa of Fabricius. However, the effects of the virus on the avian immune system, particularly the T-lymphocytes and IFN-gamma play indirect role (Ingrao et al., 2013; Wang et al., 2014). A number of changes in the global transcriptome in infected B cells have also been determined (Lee et al., 2015). Early clinical signs include watery diarrhoea, anorexia, depression, severe prostration and death, with mortality rates that could go up to 100%. In severe cases, infected bursa of Fabricius show necrotic foci and haemorrhages, with extensive microscopic atrophic lesions of the follicles.

Diagnosis of acute IBD is often not difficult from the clinical signs and lesions. Confirmatory diagnosis can be made by laboratory tests such as virus isolation from the bursal tissues in chicken embryos or cell culture, as well as RT-PCR-based molecular tests to detect viral nucleic acids (Tomas et al., 2012). Serological tests are also used to monitor the infection and vaccination levels of infected flocks.

Control of IBD is very important for sustaining a healthy flock because of the extensive immunosuppressive effects of the virus on avian health. With the relative resistance of the virus in the environment, usual biosecurity and hygienic measures alone are not enough to control IBD. However, thorough cleaning and disinfection of the poultry houses between introductions of flocks can significantly reduce the levels of exposure to the virus. In practice, the control of IBD in poultry flocks is essentially achieved by vaccination, and recent advances in the IBD vaccination strategy have been reviewed (Muller et al., 2012). Live vaccines used against IBD consist of strains of varying virulence and are designated as 'mild', 'intermediate' or 'intermediate plus' and could induce different levels of bursal damage. Vaccination with 'immune complex' vaccines comprising a mixture of live vaccine strains and IBDV-specific antibodies given by *in ovo* vaccination or into chicks at the hatchery has also shown to be effective even in the presence of maternal antibodies. With the advent of the recombinant vaccine technology, subunit vaccines using the protective VP2 protein produced in different expression systems were shown to be protective. However, major success with the new generation of vaccines has come from the live replicating HVT-based vaccine. Such vaccines give very strong protection and also help in differentiating vaccinated and infected birds. Vaccination of the breeder flocks to produce high levels of maternal antibodies that protect chicks from early immunosuppressive infections is also important.

5 Marek's disease (MD)

Marek's disease (MD), named after the Hungarian pathologist Jozsef Marek, is a common neoplastic disease of chickens characterised by the development of lymphomas and paralysis associated with infiltration of mononuclear cells into the peripheral nerves. It is caused by the highly contagious MD virus (MDV). The infection of susceptible unvaccinated birds commonly results in tumour-associated mortality ranging from 70 to 100% (Osterrieder et al., 2006). Infection is initiated by the inhalation of dust from the contaminated poultry house environment. After several rounds of early infection in the cells of the haematopoietic system, MDV establishes a latent infection in the T lymphocytes, some of which are eventually transformed resulting in lymphomas in multiple organs.

MDV is a member of the genus *Mardivirus* in the family Herpesviridae. There are three species: *Gallid herpesvirus* 2 (serotype 1), *Gallid herpesvirus* 3 (serotype 2) and *Meleagrid herpesvirus* 1 (herpesvirus to turkeys, HVT). While the serotype 1 strains include all the pathogenic strains (and vaccine strains such as the CVI988), the serotype 2 and the HVT are not pathogenic and are used exclusively as vaccines (Schat and Nair, 2013). Based on virulence, pathogenic strains of MDV are also grouped as mild (mMDV), virulent (vMDV), very virulent (vvMDV) and very virulent plus (vv+MDV) pathotypes (Witter, 1997). Complete genome sequences of several strains of MDV have been determined (Osterrieder et al., 2006). With a large double-stranded DNA genome organised as unique long (U_L) and unique short (U_S) regions flanked by internal and terminal repeat (IR_L, TR_L, IR_S and TR_S) regions, MDV encodes more than 100 proteins and several microRNAs (Schat and Nair, 2013). Some of the virus-encoded genes such as Meq, vIL-8, viral lipase, pp38, vTR (virus-encoded telomerase RNA) and miR-M4 have been shown to have significant effect on viral pathogenesis (Nair, 2013).

MD has a worldwide distribution and continues to cause problems in many countries with huge economic losses (Dunn and Gimeno, 2013). MDV is a relatively stable virus that can survive for long periods in the dust and dander from the infected birds. Latently infected birds can shed the virus through the feather follicle debris for extensive periods. Due to this ability of MDV to persist both within the host and outside the host for long periods, poultry house environment can remain a permanent source of infection, and young birds get infected almost immediately by the airborne route.

Incubation period can vary with the levels of exposure and virulence of the virus. Susceptible birds as young as 3–4 weeks may show signs of the disease. The pathogenesis of the disease is complex and includes stages of acute cytolytic stage and latency involving multiple cell types (Schat and Nair, 2013). While the virus is cell-associated in the lymphocytes, infected feather follicle epithelial cells continue to produce cell-free virus that is shed with the dander and dust that form the source of infection. Birds with lymphomas may have immunosuppression, loss of weight, paleness, anorexia and diarrhoea, and depending on the involvement of the nerves may show blindness or paralysis of the wings or legs. Some vvMDV and vv+MDV strains cause acute disease in very young chicks, sometimes referred to 'early mortality syndrome'. Lesions of MD include visceral lymphomas in multiple organs, peripheral nerve enlargement from lymphoid cell infiltration, gross lesions in the eye from mononuclear infiltration of the iris and atherosclerotic lesions and splenomegaly.

Diagnosis of MD can be made mostly from the symptoms and lesions. There are several diagnostic approaches for detecting viruses and viral nucleic acids, and for differentiating virulent and vaccine strains (Schat and Nair, 2013). It is important to note that the virus is ubiquitous, and the detection of the viral proteins and nucleic acids does not always mean the disease. Since the recognition of different pathotypes of MDV, tests for pathotyping of virus isolates have been developed using birds vaccinated with different regimes (Witter, 1997).

Control of MD is essentially achieved through vaccination, although increased biosecurity and hygienic practices and improved genetic resistance to the disease can contribute significantly in reducing losses from the disease. All the MD vaccines are derived from live attenuated strains belonging to the three serotypes. While HVT vaccines are widely used in broilers, CVI988 (Rispens) strain is more effective against more virulent pathotypes (Schat and Nair, 2013). With the widespread use of recombinant vaccines in poultry (Mahony et al., 2015), poultry population globally will be exposed to more doses of HVT vaccines. While this would be useful to improve the immunity against MD and to prevent losses, recent report of the potential role of vaccines in driving virulence of MDV strains (Read et al., 2015) can have consequences in the sustainability of a vaccination-based control strategy.

6 Infectious laryngotracheitis (ILT)

Infectious laryngotracheitis (ILT) is a viral respiratory disease of chickens that causes serious production losses due to a combination of depressed egg production and mortality. This disease can manifest itself in a variety of forms ranging from mild to severe with symptoms of sinusitis, mucoid tracheitis, conjunctivitis, low mortality and unthriftiness. ILT was formally described first in the mid-1920s but was probably around earlier than this date. It was also known as avian diphtheria, with the name ILT being adopted in 1931. The first recorded occurrence of ILT in the United Kingdom was in 1935.

ILT is caused by *Gallid herpesvirus* 1, a member of the genus *Iltovirus* in the subfamily Alphaherpesvirinae. The virus initiates infection by attachment to cell receptors and then enters the cell by fusion of the envelope of the virus with the host cell's membrane before releasing its nucleocapsid into the cell. The transcription and replication of viral DNA occur in the nucleus, leading to the formation of new virus particles. ILT virus (ILTV) produces pocks on the chorioallantoic membrane of the chicken embryo and syncytia on chicken embryo-derived liver, lung and kidney cell cultures.

Spread of ILTV occurs mainly through the oropharyngeal route by infectious droplets shed by infected birds, especially in the early stages of the infection. Recovered birds could remain latently infected, serving as carriers. ILTV can survive outside the host for several weeks when the environment is cold, especially in the presence of biological materials such as blood, mucus and faeces. All these factors contribute to the efficient transmission of ILTV. The incubation period of the disease is around 6–12 days, and the clinical signs can vary from asymptomatic to acute forms, depending on the virulence of the virus and the susceptibility of the birds. In the acute forms of ILT, signs of obstruction of the trachea and dyspnoea are prominent. Signs of conjunctivitis and exudation from the eyes and nostrils are common. Generally, the course of the natural outbreaks lasts for about 10–14 days. Severe forms of the disease can cause 90–100% morbidity, with the reported mortality as high as 70%. Affected birds show gross pathology involving the respiratory tract, particularly in the larynx and the trachea, which show haemorrhages and diphtheritic lesions.

Diagnosis of severe forms of the disease is often not difficult, although it is important to differentiate from other respiratory infectious diseases, such as ND. Confirmatory diagnosis may require histopathological studies to demonstrate specific lesions and the characteristic intranuclear inclusion bodies in respiratory and conjunctival epithelial cells. Other methods of diagnosis include virus isolation from clinical materials such as the nasal swabs or tissue homogenates from the respiratory tract. Both embryonated eggs using the chorioallantoic membrane inoculation and tissue culture cells from the embryonic liver or chicken kidney are useful for virus isolation. More recently, rapid diagnosis using PCR-based methods of diagnosis has also been developed.

Control of ILT would require a co-ordinated approach including a rapid diagnosis, prevention of further spread using quarantine measures and vaccination. In the case of outbreaks, strict implementation of quarantine and hygienic measures are very important to prevent the spread of the virus infection. Vaccination with live attenuated vaccines has been shown to be effective in developing resistance in susceptible populations. However, as vaccine strains can lead to latent infections, vaccination is generally recommended only in geographical areas where the virus is endemic. This is also because of the incidence of a number of vaccine-related outbreaks of ILT. More recently, challenges from the use of vaccines due to recombination of ILTV strains have also been reported in the field (Coppo et al., 2013). However, the advent of recombinant vaccines based on herpesvirus of turkeys and fowl pox vectors expressing ILTV genes has given the opportunity for safer vaccines.

7 Avian influenza (AI)

Avian influenza (AI), also known as the 'fowl plague', is one of the devastating poultry diseases with rapid onset and acute mortality that could wipe out flocks, sometimes within hours after onset of clinical disease. Apart from the major impact on avian health, AI also impacts human health due of its zoonotic potential. These influenza viruses belong to the family Orthomyxoviridae and have segmented negative-stranded RNA genome that encodes multiple proteins including the matrix (M), nucleocapsid (NC), haemagglutinin (H) and neuraminidase (N). All Influenza viruses are grouped broadly into types A, B and C based on the characteristics of the M and NC antigens. Only type A viruses have been isolated from birds, and AI viruses belong to 16 H and 9 N subtypes. Designations of AI virus isolates are derived from various criteria placed in the order of (1) antigenic type, (2) host of origin, (3) geographical location, (4) reference number, (5) year of isolation and (6) H and N subtypes. Example of one such designation is *A/turkey/England/199/79 (H7N7)*.

Control of AI is very difficult because of a number of factors including the large viral diversity, multiple host range including wild migratory birds and the evolutionary potential of viruses from antigenic shifts and drifts (Machalaba et al., 2015). Viruses also vary in their virulence, and are broadly grouped in low and high pathogenicity AI (LPAI and HPAI) viruses based on the absence or presence of multiple basic amino acids at the cleavage site within the haemagglutinin protein. HPAI strains can originate *de novo* from LPAI viruses of H5 and H7 subtypes and can spread rapidly among poultry farms. Spread of AI occurs mainly through contaminated faeces, and most outbreaks are initiated through migratory birds and waterfowl species. LPAI strains such as H9N2 are widespread in poultry populations in several countries, often causing widespread and persistent outbreaks in commercial poultry. However, the degree of the spread among poultry is highly variable and depends on the density of bird populations and the biosecurity measures. However, in the last decade or so, the spread of LPAI virus H9N2 or the HPAI virus H5N1 across many countries and the 2015 outbreak of the H5 subtypes in North America have become very rapid and complex, making the control methods very challenging. All the available evidence from different outbreaks suggests the role of wild birds in the primary introduction of AI viruses. Subsequent spread of the viruses to secondary farms is largely attributable to failure in the biosecurity measures.

Clinical signs of the disease can vary depending on the virus strain and the general health status and age of birds. The incubation period can also vary and can be from hours to days, depending on age, sex, species affected, concurrent infections and pathogenicity of virus. Respiratory distress, coughing, sneezing, rales, depression, sinusitis, nervous disorder and diarrhoea are some of the signs observed. Rapid mortality even up to 100% can occur in infections with some of the HPAI subtypes. Infections with LPAI viruses may only cause mild respiratory disease, but may also be associated with watery eyes, excessive lacrimation, oedema of head and face, cyanosis and drop in egg production and shell quality. Lesions include mucous in trachea, air saculitis, swollen head and wattles, sinusitis and enteritis. Confirmatory diagnosis can be made using laboratory tests to detect viruses, viral nucleic acids or specific antibodies. Isolation and identification of virus using tracheal or cloacal swabs in embryonated eggs or tissue culture, haemagglutination inhibition tests as well as PCR-based diagnostic tests are used for accurate diagnosis and subtyping of AI viruses.

The critical factor for prevention of AI is the strict implementation of biosecurity to prevent the introduction of the virus to the farms either through wild birds or through farm workers. National-level surveillance and rapid reporting of suspected outbreaks to institute quarantine measures to prevent spread is very important to control epidemics. Most of the developed countries institute a control strategy comprising of depopulation of susceptible poultry species in a significant area surrounding the outbreak to restrict the spread of the virus and thorough disinfection of the premises before allowing restocking after a period of time. In countries with high degree of endemicity of AI, similar depopulation strategy may not be an option, and vaccination is used for the disease control. Vaccines widely in use in countries such as in Southeast Asia include both conventional inactivated vaccines and the vectored recombinant vaccines.

8 Summary

The last five decades have witnessed remarkable growth of the global poultry industry with much improved efficiency in the production of valuable protein critical for achieving global food security. Major challenge for the continued sustainability of growth in this sector comes from infectious diseases, particularly from a plethora of avian viruses. The global nature of the poultry industry also contributes to the widening landscape of distribution of pathogens. Introduction of diseases such as AI to new areas by migratory birds demonstrate the difficulty in disease control. International trade restrictions, although aimed at preventing global spread of pathogens, inflict huge economic impact to the industry. Many of the avian pathogens continue to emerge and re-emerge and show huge diversity often posing immense challenges to the traditional methods of control, particularly on the vaccination-based control strategies. These are evident in diseases such as MD, where biological arms race between pathogens and vaccines appears to drive virulence. Pathogen diversity also poses problems in diagnosis, although the advent of new generation sequencing and PCR-based diagnostic tests are trying to tackle the issues. In this context, development of DIVA (differentiating infected from vaccinated animals) tests and rapid diagnostic tools that could be deployed in the field are vital for curbing outbreaks. Finally, it is important to examine the heavy reliability on vaccination strategy for sustainable control. Certainly in the case of diseases such as AI, eradication of the virus by depopulation of infected birds is preferred over vaccination in several countries, as it is generally believed by experts that vaccination can lead to endemic distribution. Recent success of recombinant vectored vaccines against a number of avian diseases is a welcome change as some of the more traditional vaccines were becoming less effective in inducing protection.

9 Where to look for further information

For further details on the diseases outlined here please refer to the appropriate chapters in the textbook *Diseases of Poultry*, 13th ed., Wiley-Blackwell, Iowa.

10 References

Alexander, D. J., Aldous, E. W. and Fuller, C. M. 2012. The long view: a selective review of 40 years of Newcastle disease research. *Avian Pathol.* 41, 329–35.

Armesto, M., Evans, S., Cavanagh, D., Abu-Median, A. B., Keep, S. and Britton, P. 2011. A recombinant avian infectious bronchitis virus expressing a heterologous spike gene belonging to the 4/91 serotype. *PLoS ONE* 6, e24352.

Bentley, K., Armesto, M. and Britton, P. 2013. Infectious bronchitis virus as a vector for the expression of heterologous genes. *PLoS ONE* 8, e67875.

Britton, P., Armesto, M., Cavanagh, D. and Keep, S. 2012. Modification of the avian coronavirus infectious bronchitis virus for vaccine development. *Bioeng Bugs* 3, 114–19.

Cook, J. K., Jackwood, M. and Jones, R. C. 2012. The long view: 40 years of infectious bronchitis research. *Avian Pathol.* 41, 239–50.

Coppo, M. J., Noormohammadi, A. H., Browning, G. F. and Devlin, J. M. 2013. Challenges and recent advancements in infectious laryngotracheitis virus vaccines. *Avian Pathol.* 42, 195–205.

Cornelissen, L. A., de Vries, R. P., de Boer-Luijtze, E. A., Rigter, A., Rottier, P. J. and de Haan, C. A. 2010. A single immunization with soluble recombinant trimeric hemagglutinin protects chickens against highly pathogenic avian influenza virus H5N1. *PLoS ONE* 5, e10645.

Desingu, P. A., Singh, S. D., Dhama, K., Kumar, O. R., Singh, R. and Singh, R. K. 2015. A rapid method of accurate detection and differentiation of Newcastle disease virus pathotypes by demonstrating multiple bands in degenerate primer based nested RT-PCR. *J. Virol. Methods* 212, 47–52.

Dhama, K., Singh, S. D., Barathidasan, R., Desingu, P. A., Chakraborty, S., Tiwari, R. and Kumar, M. A. 2014. Emergence of Avian Infectious Bronchitis Virus and its variants need better diagnosis, prevention and control strategies: a global perspective. *Pak. J. Biol. Sci.* 17, 751–67.

Dunn, J. R. and Gimeno, I. M. 2013. Current status of Marek's disease in the United States and worldwide based on a questionnaire survey. *Avian Dis.* 57, 483–90.

Eterradossi, N. and Saif, Y. M. 2013. Infectious bursal disease, in D. E. Swayne, J. R. Glisson, L. R. McDougald, L. K. Nolan, D. L. Saurez and V. Nair (eds), *Diseases of Poultry*, 13th ed., Wiley-Blackwell, Iowa, pp. 219–46.

Ganar, K., Das, M., Sinha, S. and Kumar, S. 2014. Newcastle disease virus: current status and our understanding. *Virus Res.* 184, 71–81.

Huang, Z., Elankumaran, S., Yunus, A. S. and Samal, S. K. 2004. A recombinant Newcastle disease virus (NDV) expressing VP2 protein of infectious bursal disease virus (IBDV) protects against NDV and IBDV. *J. Virol* 78, 10054–63.

Ingrao, F., Rauw, F., Lambrecht, B. and van den Berg, T. 2013. Infectious Bursal Disease: a complex host-pathogen interaction. *Dev. Comp. Immunol.* 41, 429–38.

Jackwood, M. W. 2012. Review of infectious bronchitis virus around the world. *Avian Dis.* 56, 634–41.

Jackwood, M. W. and de Wit, S. 2013. Infectious bronchitis, in D. E. Swayne, J. R. Glisson, L. R. McDougald, L. K. Nolan, D. L. Saurez and V. Nair (eds), *Diseases of Poultry*, 13th ed., Wiley-Blackwell, Iowa, pp. 139–59.

Keep, S. M., Bickerton, E. and Britton, P. 2015. Transient dominant selection for the modification and generation of recombinant infectious bronchitis coronaviruses. *Methods Mol. Biol.* 1282, 115–33.

Lee, C. C., Kim, B. S., Wu, C. C. and Lin, T. L. 2015. Bursal transcriptome of chickens protected by DNA vaccination versus those challenged with infectious bursal disease virus. *Arch. Virol.* 160, 69–80.

Li, K., Courtillon, C., Guionie, O., Allee, C., Amelot, M., Qi, X., Gao, Y., Wang, X. and Eterradossi, N. 2015. Genetic, antigenic and pathogenic characterization of four infectious bursal disease virus isolates from China suggests continued evolution of very virulent viruses. *Infect. Genet. Evol.* 30, 120–7.

Machalaba, C. C., Elwood, S. E., Forcella, S., Smith, K. M., Hamilton, K., Jebara, K. B., Swayne, D. E., Webby, R. J., Mumford, E., Mazet, J. A., Gaidet, N., Daszak, P. and Karesh, W. B. 2015. Global avian influenza surveillance in wild birds: a strategy to capture viral diversity. *Emerg. Infect. Dis.* 21, e1–7.

Mahony, T. J., Hall, R. N., Walkden-Brown, S., Meers, J., Gravel, J. L., West, L., Hardy, V., Islam, A. F., Fowler, E. V. and Mitter, N. 2015. Genomic deletions and mutations resulting in the loss of eight genes reduce the in vivo replication capacity of Meleagrid herpesvirus 1. *Virus Genes.* 51, 85–95.

Miller, P. J., Decanini, E. L. and Afonso, C. L. 2010. Newcastle disease: evolution of genotypes and the related diagnostic challenges. *Infect. Genet. Evol.* 10, 26–35.

Miller, P. J. and Koch, G. 2013. Newcastle disease, in D. E. Swayne, J. R. Glisson, L. R. McDougald, L. K. Nolan, D. L. Saurez and V. Nair (eds), *Diseases of Poultry*, 13th ed., Wiley-Blackwell, Iowa, pp. 89–138.

Muller, H., Mundt, E., Eterradossi, N. and Islam, M. R. 2012. Current status of vaccines against infectious bursal disease. *Avian Pathol.* 41, 133–9.

Nair, V. 2013. Latency and tumorigenesis in Marek's disease. *Avian Dis.* 57, 360–5.

OIE. 2012. Newcastle disease, http://www.oie.int/fileadmin/Home/eng/Health_standards/tahm/2.03.14_NEWCASTLE_DIS.pdf.

Osterrieder, N., Kamil, J. P., Schumacher, D., Tischer, B. K. and Trapp, S. 2006. Marek's disease virus: from miasma to model. *Nat. Rev. Microbiol.* 4, 283–94.

Park, M. S., Garcia-Sastre, A., Cros, J. F., Basler, C. F. and Palese, P. 2003. Newcastle disease virus V protein is a determinant of host range restriction. *J. Virol.* 77, 9522–32.

Promkuntod, N., Thongmee, S. and Yoidam, S. 2015. Analysis of the S1 gene of the avian infectious bronchitis virus (IBV) reveals changes in the IBV genetic groups circulating in southern Thailand. *Res. Vet. Sci.* 100, 299–302.

Promkuntod, N., van Eijndhoven, R. E., de Vrieze, G., Grone, A. and Verheije, M. H. 2014. Mapping of the receptor-binding domain and amino acids critical for attachment in the spike protein of avian coronavirus infectious bronchitis virus. *Virology* 448, 26–32.

Read, A. F., Baigent, S. J., Powers, C., Kgosana, L. B., Blackwell, L., Smith, L. P., Kennedy, D. A., Walkden-Brown, S. W. and Nair, V. K. 2015. Imperfect vaccination can enhance the transmission of highly virulent pathogens. *PLoS Biol.* 13, e1002198.

Roh, H. J., Hilt, D. A. and Jackwood, M. W. 2014. Detection of infectious bronchitis virus with the use of real-time quantitative reverse transcriptase-PCR and correlation with virus detection in embryonated eggs. *Avian Dis.* 58, 398–403.

Saba Shrivan, A. and Mardani, K. 2014. Molecular detection of infectious bronchitis and Newcastle disease viruses in broiler chickens with respiratory signs using Duplex RT-PCR. *Vet. Res. Forum* 5, 319–23.

Schat, K. A. and Nair, V. 2013. Marek's disease, in D. E. Swayne, J. R. Glisson, L. R. McDougald, L. K. Nolan, D. L. Saurez and V. Nair (eds), *Diseases of Poultry*, 13th ed., Wiley-Blackwell, Iowa, pp. 515–52.

Tomas, G., Hernandez, M., Marandino, A., Panzera, Y., Maya, L., Hernandez, D., Pereda, A., Banda, A., Villegas, P., Aguirre, S. and Perez, R. 2012. Development and validation of a TaqMan-MGB real-time RT-PCR assay for simultaneous detection and characterization of infectious bursal disease virus. *J. Virol. Methods* 185, 101–7.

Toro, H., van Santen, V. L. and Jackwood, M. W. 2012. Genetic diversity and selection regulates evolution of infectious bronchitis virus. *Avian Dis.* 56, 449–55.

Wang, S., Teng, Q., Jia, L., Sun, X., Wu, Y. and Zhou, J. 2014. Infectious bursal disease virus influences the transcription of chicken gammac and gammac family cytokines during infection. *PLoS ONE* 9, e84503.

Wickramasinghe, I. N., van Beurden, S. J., Weerts, E. A. and Verheije, M. H. 2014. The avian coronavirus spike protein. *Virus Res.* 194, 37–48.

Witter, R. L. 1997. Increased virulence of Marek's disease virus field isolates. *Avian Dis.* 41, 149–63.

Chapter 4

Parasites affecting poultry

Larry McDougald, University of Georgia, USA

1 Introduction

Keeping poultry safe from disease is a key factor in efficient production and sustainability. Diseases of poultry take many forms, the best known being bacterial and viral. Parasites are a diverse group, including protozoa, roundworms, flatworms, insects and mites. Parasites are common. Indeed, it is uncommon for poultry *not* to have parasites at some time. The impact of parasitic diseases, in contrast to bacterial and viral diseases, depends on the *intensity of infection*, or in other words, the number of invading parasites. The aim of disease management is to keep parasites in low numbers.

The most severe parasitic diseases are caused by protozoans. These are coccidiosis and histomonosis (histomoniasis). However, as we will see, the worm parasites interact with these in important ways. Coccidiosis is probably the best known among the poultry

http://dx.doi.org/10.19103/AS.2016.0011.18

parasites, and has been kept in check for the past 60 or so years by drugs added to the feed. To a lesser extent, live vaccines have been used. A poultry producer in the 1960s (attribute unknown) said 'give me a bird with good genetics, a nutritionally complete feed, and a way to control coccidiosis'. Those close to the production of poultry will recognize the importance of this short list for today's production, regardless of the size of an operation.

To prevent and control parasitic diseases, we must know and understand the organisms and develop clever strategies. Whether one keeps a dozen backyard chickens or manages farms with 5–10 barns with up to 250 000 birds, the problems are remarkably similar. Poultry reared outdoors (free range) may be exposed to a greater diversity of parasites because of their contact with insects and other invertebrates, which often serve as intermediate hosts for certain parasites. A combination of recognition (the signs of disease), identification of the parasites (the cause of this disease), management strategies (to make flocks less prone to this disease) and treatments (prevention by medication, vaccination and treatment of infections) will help us to minimize the impact of parasites, regardless of the production template.

2 Coccidiosis

A group of protozoan parasites called Apicomplexa includes such diverse diseases of animals and humans as toxoplasmosis, malaria and coccidiosis of poultry. Coccidiosis can be considered a complex in diseases, as there are several species, all of the genus *Eimeria*. All of them live in the gut. Multiple species are often involved in outbreaks. Coccidia are somewhat unique among poultry diseases, as it is well known that each species of bird has its own coccidial species. There are several species of coccidians (*Eimeria*) infecting chickens, several others in turkeys, others in pheasants, chukars, bobwhite quail, guinea fowl and so on. See Table 1 for a list of important coccidia organized by species of host bird.

2.1 Biology of Coccidia

Coccidia have a short, direct life cycle (no intermediate host is required), although as protozoans go, it lives a fairly complicated life (Fig. 1). The life cycle is well known:

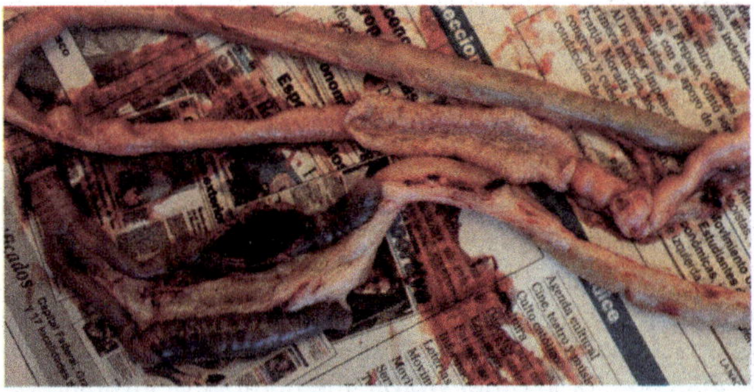

Figure 1 Gross lesions of coccidiosis in chickens: Species of *Eimeria* can generally be recognized by the location and appearance of lesions in the intestine. Note the whitish lesions in the duodenum (*E. acervulina*) and the bloody cecal material (*E. tenella*). Mixed infections are common in commercial poultry.

Table 1 Species of Coccidia (*Eimeria*) found in common types of poultry

Bird host	Coccidian species	Pathogenicity
Chicken (*Gallus domesticus*)	*Eimeria acervulina*	Moderate
	E. maxima	High
	E. brunetti	High
	E. necatrix	High (often lethal)
	E. tenella	High (often lethal)
	E. praecox	Low
	E. mitis	Moderate
Turkey (*Gallopavo meleagridis*)	*E. adenoeides*	High (often lethal)
	E. meleagrimitis	Moderate
	E. gallopavonis	High
	E. meleagridis	Low
	E. auburnensis	Moderate
	E. duodenalis	Moderate
	*E. dispersa**	Moderate
Ringneck pheasant (*Alectoris phasiani*)	*E. phasiani*	High
	E. colchici	High (often lethal)
	E. pacifica	Low–moderate
	E. duodenalis	Low–moderate
	E. tetartooimia	Low–moderate
Chukar partridge (*Alectoris chukar*)	*E. kofoidi*	High (often lethal)
	E. legionensis	Moderate
Bobwhite quail (*Colinus virginianus*)	*E. colini*	High
	E. lettyae	Low

* Although described as a parasite of turkeys, *E. dispersa* is known to infect at least six species of galliformes to some extent. This is the only species of *Eimeria* known to infect more than one host species.

1. **Oocysts** (a thick-walled cyst and the only stage existing outside the host) become sporulated and infective after exposure to suitable temperature, air and moisture (ordinary room temperature is adequate, even if broken up by periods of cooler temperatures). Infective oocysts can be easily produced for experimental use; droppings of poultry are mixed with water and preservative (2% potassium dichromate) and aerated with an aquarium pump for 2–3 days. Microscopic examination will reveal the cell divisions taking place within the oocyst, until it contains four sporocysts, with each containing two sporozoites.

2. **Sporozoites infect the gut:** When the oocyst is consumed by a suitable bird host, the hard outer wall is crushed in the gizzard. Exposure to trypsin and bile in the small intestine causes the release of sporozoites, which infect intestinal cells. The sporozoites of different species home in on specific parts of the gut. In most cases, all of the remaining life cycle will take place at that site.

3. Schizogony and gametogony: The sporozoite quickly divides in a process known as schizogony. This is a multiple fission resulting in hundreds of daughter cells called merozoites. These escape and infect other nearby cells. Two to four generations of schizogony are common. The second-generation merozoites develop into cells called gametocytes, and undergo a process of fertilization much like that of higher animals. The fertilized gametocyte develops a hard outer shell, becoming a new oocyst. These are released into the faeces to begin sporulation. It is not unusual to see schizonts, merozoites, gametocytes or oocysts, by microscopic examination of smears from the gut. Time for completion of the cycle is about 5–7 days, although the patent period may persist for days or even weeks. Many thousands or even millions of oocysts are produced for each oocyst ingested by the bird.

2.2 Identification of Coccidia

The oocysts and other developing stages of *Eimeria* can be easily seen by microscopic examination of intestinal contents or faeces from birds (Fig. 2). Coccidia of birds can be identified by a combination of biological traits, including the size and appearance of the oocyst and developmental stages, the location of lesions in the intestine and the appearance of the lesions. Some species are easily identified by these traits: *E. tenella* is the only species causing bloody lesions in the caeca, *E. maxima* causes red-orange mucus and fluid in the midgut, *E. acervulina* lesions are white patches on the mucosa of the duodenum and *E. necatrix* may cause bloody lesions from the duodenum to the large intestine. The time required for completion of the cycle and the release of fresh oocysts in the droppings is called the prepatent period and is specific to each species of coccidia. One species, *E. praecox*, can be distinguished in experimental infections because oocysts are found in the droppings before 80 hours, the shortest of any of the species. The life cycle of coccidia is said to be 'self-limiting' because of the progression through each stage until completion with the oocyst. Some of the biological characteristics of coccidia useful in diagnosis are summarized in Table 2.

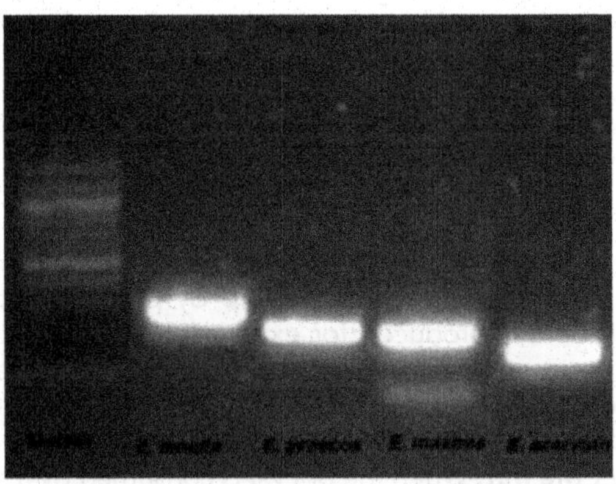

Figure 2 Diagnosis and identification of Coccidia species by polymerase chain reaction (PCR). In this example several species were detected in the same sample, using species-specific primers. (Photo provided by Dr. Robert Beckstead).

Table 2 Anticoccidials commonly used as feed additives in poultry rations

Generic name	Original trade name*	Use level in feed (ppm)
Synthetic chemical products		
Amprolium	Amprol	125–250
Clopidol	Coyden	125
Decoquinate	Deccox	30
Nicarbazin	Nicarb	125
Sulfadimethoxine + ormetoprim	Rofenaid	125+75
Zoalene	Zoamix	125
Diclazuril	Clinicox	1
Robenidine	Robenz	33
Halofuginone	Stenorol	3
Polyether ionophores		
Lasalocid	Avatec	75–125
Maduramicin	Cygro	5–6
Monensin	Coban, Elancoban	100–121
Narasin	Monteban	54–72
Narasin + nicarbazin**	Maxiban	54–90
Salinomycin	Biocox, Coccistac	44–66
Semduramycin	Aviax	25

* The actual product name will vary, as all present day anticoccidials are marketed generically in many countries.
** This product is a combination of an ionophore and a synthetic chemical.

2.3 Molecular biology

Diagnosticians often identify coccidia only on the basis of gross pathology. Understandably, the species causing indistinct lesions are missed or under-represented in reports. Examination of field samples with molecular techniques reveal a more extensive prevalence of coccidian species. Molecular primers are available for diagnosis of coccidian species by polymerase chain reaction. These tests are reliable and rapid, and highly specific (Fig. 2). Primers have been designed for species infecting chickens, turkeys, bobwhite quail, chukar partridges and ringneck pheasants.

2.4 Pathogenic effects of Coccidia

Species of coccidia vary considerably in their effect on the host. In chickens, morbidity and death may result from infections with *E. tenella* or *E. necatrix*. Both of these species cause considerable blood loss at the site of infection, but death is probably a result of some type of toxic shock. Other species cause diarrhoea and fluid loss, reduced absorption of nutrients in the gut and loss of appetite, resulting in poor weight gain and non-uniformity. The skin of chickens may be noticeably pale after recovery from coccidiosis, because of reduced absorption of xanthophyll pigments. The intestinal species in chickens are known to interact with *Clostridium perfringens*, the bacterium causing necrotic enteritis, resulting in considerable mortality and morbidity.

Species of *Eimeria* causing high mortality include *E. adenoeides* and *E. gallopavonis* in turkeys, *E. koffoidi* in the chukar, *E. lettyae* in bobwhite quail, and *E. phasiani* and *E. colchici* in ringneck pheasants. Some species are considered non-pathogenic, such as *E. meleagridis* in turkeys and *E. praecox* in chickens.

2.5 Environmental interaction with Coccidia

Oocysts are somewhat resistant to the environment, accounting for their persistence in poultry facilities after an outbreak, but can be killed by temperatures below freezing. The oocysts will also perish if exposed to temperatures above 25°C, possibly accounting for the reduced threat of coccidiosis in summer months. Surveys have shown that coccidia are found in most poultry facilities. Parasites in the litter generally increase in numbers until a flock is about 5 weeks old and decline afterwards. If parasite numbers are low, infections may pass unnoticed. There is no consensus on the conditions contributing to serious outbreaks. Severe disease may strike when the litter condition is very dry, or very wet. Outbreaks may occur in new, well-managed, facilities (the so-called 'new house syndrome'), as well as old, poorly managed farms. The practice of keeping poultry of different ages on the same farm is thought to contribute to the persistence of disease exposure.

2.6 Age and susceptibility of birds to coccidiosis

Coccidiosis is mostly a disease of young animals, although birds of any age can be infected if there was no prior exposure. The risk is the same in small and large flocks, because the birds must be confined for brooding when young. Even free-range birds are kept in groups where they can be exposed to coccidiosis during and after brooding.

2.7 Coccidiosis in caged layers

It would seem that hens kept in wire cages without contact with the floor would be protected from coccidiosis infections. However, severe coccidiosis is sometimes diagnosed in such flocks, particularly in the modern stacked-cage facilities. Flies are common, as they breed in the droppings. Oocysts are often picked up mechanically by flies, and are transferred to the hens when the flies are eaten. Layers are adept at snatching flies out of the air. Infections with small numbers of oocysts in the upper tiers of the cages are magnified quickly throughout the entire unit, resulting in clinical coccidiosis.

2.8 Diagnosis of coccidiosis

Coccidiosis may be suspected in a flock if birds are droopy, off feed and showing other signs like diarrhoea. Only with the chicken coccidia *E. tenella* and *E. necatrix* will you see blood in the faeces. When coccidiosis is suspected, necropsy of some sick birds and examination of the faeces with a microscope is recommended. Large poultry companies usually employ technical staff capable of diagnosing disease. Some useful biological characteristics of coccidia useful in identifying species of *Eimeria* in chickens are listed in Table 2. Even small flock owners can take advantage of veterinary diagnostic labs. The procedures are fairly straightforward. At necropsy:

1 Open the intestinal tract with scissors and look for lesions and other signs such as watery contents or excess mucus.
2 Note the appearance and severity of lesions.
3 What is the location of the lesions? Typical observations focus on four parts of the gut, the duodenum, the midgut, jejunum and caeca.
4 Confirm the presence of coccidia by microscopic examination of material from the lesions. What is the size of oocysts? Are other stages present, such as merozoites or schizonts?
5 Comparison with a diagnostic chart or table of characteristics.

It is generally agreed that most poultry flocks have some infection with coccidia during their growing period. However, mild infections are considered unimportant and do not require treatment or other intervention.

2.9 Control of coccidiosis

Prevention is considered the only effective approach to coccidiosis management. Treatment of outbreaks is impractical and only partly effective, owing to the lack of highly effective treatment products, and the fact that most of the damage is already done by the time diagnosis and treatment could be started. Two methods are commonly used to prevent coccidiosis in chickens and turkeys: vaccination and chemoprevention. Environmental management has traditionally been touted as important in preventing coccidiosis, although it is not clear how this works.

1. Vaccination of chickens takes advantage of the self-limiting nature of the life cycle and the rapidity of immune protection. A live vaccine was produced in the 1950s, but has only recently seen much use. The vaccine consists of live oocysts, which are administered in small doses to chicks at hatch, or soon thereafter. The coccidia develop normally in the intestine, stimulating the T-cell immune response. Low levels of infection used for vaccination do not cause measurable pathogenic effects. A vaccine has been available at times for turkeys, but not for other poultry species. Most of the vaccine usage in the United States has been with non-attenuated strains of *Eimeria,* although an attenuated product is available. Products containing attenuated lines of coccidia are used in Europe and other countries where the use of virulent organisms in vaccines is prohibited. We can expect the use of vaccines against coccidiosis to increase, as producers adjust to 'drug-free' or 'all-natural' production for a significant portion of the market. Unfortunately for the small-flock owner, vaccines against *Eimeria* are only available in large dosage sizes, and are administered by specialized equipment.

2. Chemoprevention: Drugs to prevent coccidiosis, commonly called coccidiostats or anti-coccidials, have been used throughout the world since the concept was invented in the late 1940s. A list of currently available products is found in Table 3.

In the United States, products are sold only to feed mills, under strict regulation by the Food and Drug Administration (or other government agencies in various countries). As the dosage is usually very small, only a few grams/ton of feed, the products are supplied as premixes, which comprised of a measured dose of the active ingredient, diluted by some inert ingredients, so the premixes can be conveniently mixed into multi-tonne batches of

Table 3 Biological characteristics useful in diagnosis of some *Eimeria* species in chickens*

Trait	*E. acervulina*	*E. tenella*	*E. mitis*	*E. brunetti*	*E. maxima*	*E. praecox*	*E. necatrix*
Location of gross lesions:	Duodenum	Ceca	Lower gut, yolk stalk to Ceca	Lower gut, yolk stalk to cloaca	Midgut, from duodenum to Ileum	Duodenum	Midgut, or may be seen in all sections in severe infection
Appearance of lesions:	Whitish plaques, coalescing in heavy infections	bloody mucosa, forming dark cecal core, damage visible on serosal surface	pale lower gut, thinning mucosa	pale lower gut in low inf., necrotic mucosa in heavy inf.	Thickened intestinal wall, 'ballooning', mucus and fluid, petechiae seen from serosal surface	No distinct lesions	blood in lumen 'ballooning', light and dark petechiae visible serosally
Oocyst size: (average)	18.3 x 14.6 μ	22 x 19 μ	15.6 x 14.2 μ	24.6 x 18.8 μ	30.5 x 20.7 μ	21.3 x 17.1 μ	20.4 x 17.2 μ
Microscopy:	Oocysts in duodenal lesions	Oocysts in ceca	Small round oocysts in lower gut	Large oval oocysts in lower gut	large oocysts in gametocytes in disrupted mucosa	round oocyst in duodenal epithelium	large schizonts in disrupted midgut, large oocysts in ceca
Prepatent period (hrs):**	97	115	93	120	121	83	138
Location in tissue sections (requires histopathology):							
	epithelial, lining the villi	subepithelial, in cecal glands	epithelial	subepithelial	schizonts are subepithelial	epithelial	schizogony in midgut, gametes and oocysts in cecal glands

* Not all species can be distinguished by these characteristics and would require PCR or other advanced techniques.
** Requires experimental infection with sporulated oocysts.

feed. Commonly, premixes are formulated to mix into feed at 1 lb/US ton, or internationally at ½ kg/metric ton. Anti-coccidials are added to the feed at very low levels (1–125 mg/kg, depending on the product), usually in the starter and grower, followed by a few days without medication in the finisher feed. Strategies for use of these products vary. A single product may be given through the grow-out, or products may be alternated in the starter and grower. The latter practice is commonly called a shuttle programme or dual programme.

Prevention of coccidiosis with drugs is highly successful. However, limitations include drug resistance, drug side effects and even limitations of feed mixing technology, although the latter has been lessened by accurate computer-directed mixing programmes and improvements in milling technology. Small-flock owners may purchase 'medicated feed', but usually the only anti-coccidial it contains will be amprolium, a safe but relatively weak product.

2.10 Management and disinfection for control of coccidiosis

A debate rages on whether chemicals should be used to keep disease in check in food-producing animals. We are woefully ignorant on *effective* alternatives. We have always been told that 'good management' is the key to controlling disease and maintenance of flock health. However, these are puzzling words. Often, the very conditions that are good for nurturing young animals also favour the survival and propagation of parasites. Disinfection, while important in the prevention of bacterial and viral diseases, may not reduce the risk of parasitic diseases. Management practices vary greatly in some countries: For instance, poultry industry buildings in Europe commonly have concrete floors, while those in the United States have floors of packed earth. Because of a shortage of good litter materials, it is a common practice to allow broiler litter to build up through several flocks, with only 'top dressing' with fresh litter between flocks. There is no good evidence that differences in sanitation and litter management change the prevalence or severity of coccidiosis in broilers. In cage layers, there is evidence that better sanitation and insect control could result in reduce exposure to infection.

2.11 Directions for further research on control of coccidiosis

Numerous products have been offered for litter treatment or soil treatment to kill coccidian oocysts. However, such products do not prevent the introduction of coccidian into a flock, and there is no good evidence that they are effective in practice. Possible new approaches would include breeding for improved T-cell response (primarily responsible for immunity to coccidiosis), or development of agents to improve the immune response. In the latter case, numerous plant extracts are known to stimulate immunity when used in concert with anti-coccidial vaccines and could be administered in feed or *in ovo* (McDougald et al., 2008). The science of vaccination has developed considerably in recent years, suggesting that improved coccidiosis vaccines could be likely, depending on the level of research.

3 Non-Eimerian Coccidia

Close relatives of the *Eimeria* include *Sarcocystis*, *Toxoplasma* and *Cryptosporidium*. Neither *Sarcocystis* nor *Toxoplasma* is thought to be important in poultry production, as neither is common. *Sarcocystis* is a two-host parasite, often alternating between a predator and its prey. With bird parasites, the alternate host is usually a mammal. Waterfowl and scavenging birds are often infected. A common site of infection in birds is the breast muscle; numerous white cysts can be easily seen in the muscle as the skin is stripped away. A high percentage of some wild flocks may be infected. There is little knowledge of the pathogenicity or other importance of these parasites in poultry.

3.1 Cryptosporidium

Little was known of these parasites until recently because of their diminutive size, generally 4–6 microns. These parasites live in the respiratory and/or intestinal tracts, and are associated with morbidity and mortality. Surveys show that *C. baileyi* is common in chickens and turkeys. Another species, *C. meleagridis* in turkeys, is thought to be synonymous with *C. parvum* of mammals. Like other coccidia, *C. baileyi* has a distinct life cycle. The oocyst, contains four sporozoites, hatches in the digestive tract and infects epithelial cells. After several generations of multiple division (merogony), the macro and microgametocytes are produced, and finally oocysts. Unique to the cryptosporidia, however, the oocyst can hatch inside the host and reinfect. Surveys show up to 27% of flocks positive, and up to 40% of birds within a flock. Infections have been associated with high mortality and morbidity, although this is apparently rare. Clinical outbreaks are sporadic.

There is no treatment for cryptosporidosis, although antibiotics may be used to reduce secondary infections.

4 Flagellates

Common flagellate parasites of poultry are summarized in Table 4.

5 Blackhead disease (histomonosis, histomoniasis)

The agent of this disease, *Histomonas meleagridis,* is an anaerobic, non-spore-forming protozoan. Related parasites cause disease in humans and other animals, principally *Dientamoeba fragilis, Giardia lamblia* and several species of *Trichomonas* or *Tetratrichomonas.* Blackhead disease (a misnomer, as there is no evidence of any discolouration specific to these infections) was first noticed in North America in the early 1890s, when large numbers of domestic turkeys in Rhode Island became infected and died. Within a few years, a burgeoning industry was destroyed. It is estimated that 20 million domestic turkeys were produced annually in New England in 1900, and this fell to fewer than 3 million by 1920. Unfortunately, we are facing another crisis in control of this parasite. As of 2016 the last anti-histomonal product was deregistered in the United States, leaving the industry with no drugs for prevention or treatment. The frequency of outbreaks in parts of the United States and Europe and the severity of the disease, often reaching near 100% mortality, place this disease as one of the most important diseases in the turkey industry.

Blackhead disease was brought to America when the Chinese ringneck pheasant was imported from Asia in the 1880s. The pheasant was highly successful in its new environment and quickly became popular for hunting as it spread across the country. The

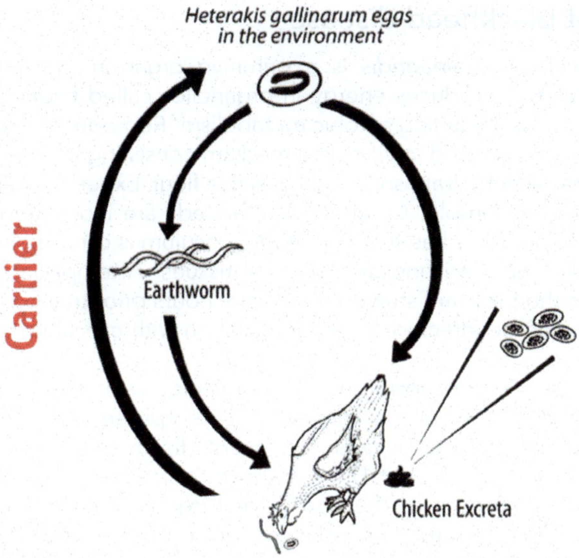

Figure 3 Blackhead disease: Epidemiology of *Histomonas meleagridis* includes the cecal worm, which maintains the protozoan in a viable state until it is consumed by a gallinaceous bird. Chickens usually get their infection in this way. Turkeys may not only become infected from consuming cecal worm eggs, but may also become infected from direct contact with their flock mates. The earthworm can ingest cecal worm eggs and keep the larvae alive until it is eaten by a bird, serving as another potential link in a complex pattern.

Table 4 Flagellated protozoan parasites of poultry

Species	Hosts	Pathogenicity
Histomonas meleagridis (blackhead disease)	All galliformes	Varies with host species. Highest in turkeys and peafowl, low in ringneck pheasant, moderate in chickens and guinea fowl.
Tetratrichomonas gallinae (trichomoniasis, canker, frounce, etc.	Turkeys, doves, chickens, pigeons, raptors, and many other birds	Varies with host species, moderately pathogenic in many species.
Spironucleus meleagridis (hexamitiasis)	Turkey, pheasant, quail, chukar and peafowl	Low to moderate
Chilomastix gallinarum (no common name)	Turkey	Low
Cochlosoma anatis (no common name)	Turkey, chukar, ducks	Moderate, may interact with intestinal bacteria

pheasant is barely affected by blackhead, although it is an excellent host for the caecal worm, *Heterakis gallinarum*. Caecal worms are the primary vector for blackhead disease, and protect the protozoan parasite inside its eggs.

5.1 Biology of blackhead disease

At the cellular level, *H. meleagridis* is a primitive organism, anaerobic and lacking mitochondria. Instead, it produces energy in organelles called hydrogenosomes, which are not sensitive to drugs targeting aerobic metabolism. Reproduction is by binary fission. Histomonads are easily cultured *in vitro,* in a medium consisting of a cell culture medium, a horse serum and a carbohydrate source such as rice flour. Experimental infection of birds with cultured material is simple: Cultured histomonads are inoculated directly into the cloaca by pipette. Experience has shown that the inoculum is carried directly to the caeca by reverse peristalsis, where it invades mucosal tissues. Attempts at oral infection are unsuccessful unless the birds are starved for several hours prior to inoculation. Classically, turkeys or chickens can be infected by oral inoculation with caecal worm eggs harvested from chickens.

Like many parasites, *H. meleagridis* has a complicated relationship with intermediate hosts. Figure 3 illustrates what we know about its epidemiology at present. Some of this differs from older texts because of recent discoveries. The caecal worm (*H. gallinarum*) is key to the survival of *H. meleagridis* outside the bird host. Most, if not all, caecal worms are carriers of the blackhead organism in its ova. Caecal worms become infected with the protozoan and incorporate the cells into its eggs. The eggs, shed in the faeces, may remain viable for months or years, providing a reservoir for infection.

In some environments, earthworms are known to ingest caecal worm eggs. The eggs hatch, allowing the larvae to invade the tissues of the worm, where they remain viable for months or years. When an earthworm is eaten by a chicken or turkey, the larvae within may begin development and release the histomonads, resulting in backhead disease results. Beyond caecal worms and earthworms, little is known about the possible vectors of histomonads. Because of the seasonality of outbreaks in turkeys, it is suspected that arthropods such as beetles, grasshoppers or flies may contribute (mechanically) to the spread of blackhead disease.

5.2 Susceptibility of birds to blackhead disease

While *Histomonas* will infect any species of gallinaceous bird, the disease is quite different in different hosts. In turkeys (not the original host), the disease is severe, often destroying entire flocks. The parasites are introduced into a flock by caecal worm eggs, which may be brought in by mechanical means (on shoes, contaminated equipment or possibly by unknown invertebrates). In some cases, where chickens had been raised on the farm before it was converted to producing turkeys, residual contamination of caecal worm eggs provides a continuing threat of infection. After a few birds have become infected, the disease spreads throughout the flock without the aid of other vectors or intermediate hosts. Birds in direct contact are at risk. It is not known precisely how this takes place, but we know it is difficult to induce infections by the oral route, and also that physical barriers prevent further spread.

5.3 Pathology of blackhead disease

Histomonads first infect the caeca, invading the mucosa. The lumen form is both amoeboid and flagellated. Next, they transform into tissue forms and lose the flagellum.

During the caecal phase, large caseous cores develop, consisting of sloughed tissues and coagulated blood proteins (Fig. 4). Even though this is severe, birds may recover from an infection if it is confined to the caeca. Tissue forms are carried in the blood stream to other tissues. While histomonads have been found in many tissues, the liver is most frequently targeted. Liver function is destroyed in a few days, resulting in death (Fig. 5). Bright yellow material in the faeces signals the loss of liver function, so that bile salts (mainly bilirubin) are excreted in the urine rather than the intestine. Death follows within a day or two after the appearance of yellow droppings.

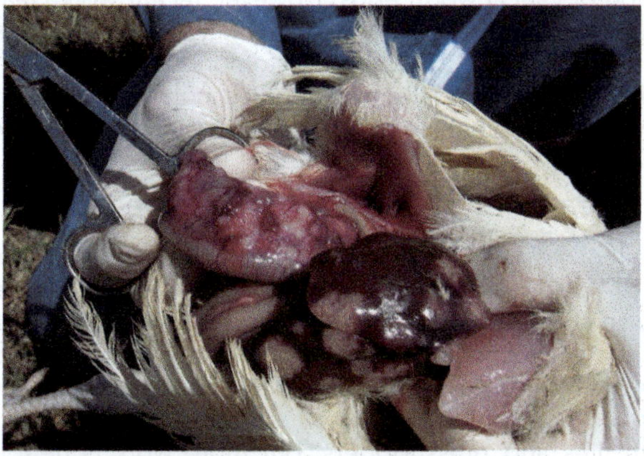

Figure 4 Blackhead (cecal lesions): The ceca are greatly enlarged and engorged with a firm cecal core consisting of sloughed cecal tissues, blood, and serous proteins.

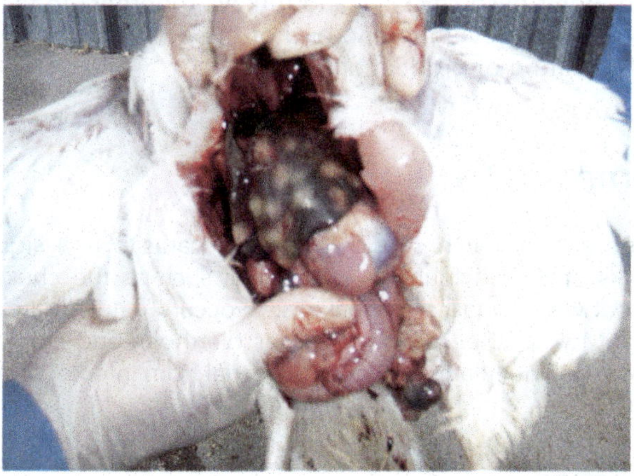

Figure 5 Blackhead disease (liver lesions): The liver becomes mottled with lesions as the disease progresses. The size and appearance of such lesions varies considerably in size and number. Most of the death losses from blackhead are a result of damage to the liver.

Infections in turkeys are essentially a dead end for the parasites. In the first place, turkey will likely die of infection. Secondly, turkey is a poor host for caecal worms, so there are usually no worm eggs produced. Histomonads cannot survive outside the host more than a few minutes, so there is no reservoir of infection to be passed on to future flocks.

5.4 Epidemiology of blackhead disease

The work of Tyzzer in the 1920s suggested that *H. meleagridis* could be transmitted only by the ingestion of caecal worm eggs. However, recent work clearly demonstrated that blackhead could spread through a group of turkeys in the absence of caecal worms or other vectors. This contagious nature of histomonosis in turkeys reveals important considerations in its control and prevention. Interestingly, the disease is halted by any barrier in the growing facility, anything that prevents direct contact between birds.

5.5 Chickens as carriers of blackhead disease

Chickens are excellent hosts for caecal worms, and thereby *Histomonas* organisms. Usually infections are confined to the caeca and may pass without notice. As a result, chickens are likely to contaminate growing facilities and litter with worm eggs, which contain the blackhead organism. In some areas, the closeness of chicken farms to turkey farms may contribute to the outbreaks in turkeys. Insects and other vermin are suspected of mechanical dissemination of caecal worm eggs. This has been demonstrated with flies and grasshoppers, but more work is needed to define the risk from such transfer. Barns or houses where chickens have been previously used are likely to be contaminated with caecal worm eggs and should not be used for turkey production. Such contamination is known to persist for several years. Small-flock owners or poultry fanciers should be aware of the risk if turkeys should be kept in the same area where chickens are kept. Poultry producers in the 1930s and 1940s lived by the adage 'don't raise chickens and turkeys together.' In fact, other diseases are also known to be more severe in one bird or the other, as with the mycoplasmas, suggesting strict separation of production areas.

6 *Cochlosoma anatis*

Recent case reports and research suggest that this parasite is a significant pathogen in turkeys and ducks. It is also found in gamebirds such as the chukar partridge, and experimentally in bobwhite quail and chickens. It occurs naturally in a wide variety of wild birds. Usually of low pathogenicity, it causes diarrhoea and stunting of poults and ducklings. Oral or cloacal inoculation of faecal material from infected birds has been used to infect other birds.

6.1 Site of infection

Infections are often confined to the caeca or lower intestine, but occasionally are found throughout the intestinal tract. More severe pathology may be seen when birds are concurrently infected with other parasites, such as *Hexamita* (*Spironucleus*).

6.2 Diagnosis

Diagnosis is by microscopy of fresh live specimens from the affected area (Fig. 7). The ovoid cochlosoma cells are often dense on the mucosa and active.

6.3 Prevention and control

Cysts and pseudocysts have been described, and probably aid the parasite in surviving outside the host for a short time. However, these forms are easily killed by disinfectants, including phenolic or quaternary ammonium types, and formalin. Prevention of infection is mainly by disinfection and quarantine. Wild birds are an obvious source of infection, and infections could be introduced by mechanical transfer from other farms. Flies are thought to be mechanical vectors, but more research is needed. There is no approved treatment for this infection. Historically, roxarsone and nitarsone were partly effective, but these products are no longer available.

7 *Spironucleus meleagridis* (formerly *Hexamita meleagridis*)

Despite changes in the classification of these organisms into other generic groups, the only common name for this parasite is *Hexamita*. These are flagellated protozoa common in turkeys. Typically quite active, they are not easily confused with *Histomonas, Tetratrichomonas* or other flagellates. These parasites are not highly pathogenic, but are thought to contribute to poor performance, culling and reduced weight gain.

8 *Tetratrichomonas gallinae* (formerly *Trichomonas gallinae*)

Like *Hexamita*, this parasite is common in turkeys and other birds. Its pathogenicity is mild, leaving an uncertain position in overall health considerations. Almost all pigeons are infected with this parasite, and are passed on from the adult to the chicks during early feeding. The actual damage caused by trichomonads in turkeys is poorly understood due to lack of controlled experiments with experimental infections. However, it is thought to contribute to poor performance, increased culling and increased secondary bacterial infections.

9 Blood-borne protozoan parasites

Blood parasites are common in wild birds, particularly malaria, *Haemoproteus,* trypanosomiasis and *Leucocytozoon*. Mosquitoes, blackflies and other bloodsucking insects serve as intermediate hosts. However, their importance in poultry is somewhat limited. At one time the turkey industry in South Carolina was threatened by outbreaks of *Leucocytozoon*, transmitted by blackflies. At the time, turkeys were started in brooder houses, then moved outside for finishing. Blackflies (*Simulium* spp.), a bloodsucking fly common around streams in the area, would not commonly go inside buildings. Thus, the birds were only at risk during the 'free-range' part of their growth. *Leucocytozoon* is common in ducks and geese, often resulting in high mortality. Control of this parasite obviously depends on control of blackflies and reducing exposure by keeping the birds indoors.

Birds of all kinds harbour a wide variety of worm parasites. The most important are roundworms (nematodes) and flatworms (tapeworms and trematodes). Some wild birds also have another type, the spiny-headed worms (Acanthocephala). Many worms have complicated life cycles involving intermediate hosts, usually an insect or other arthropod or mollusc, in which they live part of their lives. Others have a direct life cycle, in which the worms in the intestine of a bird shed eggs, which are excreted in the faeces, to be picked up by other birds in the flock. Worm parasites are usually more of a nuisance than a plague, but can detract from the health and profitability of a flock in many ways.

10 Roundworms (Nematoda)

The most important worms in chickens are *Ascaridia galli* (the large intestinal roundworm of chickens) (Fig. 6), *Heterakis gallinarum* (the caecal worm) (Fig. 7) and *Capillaria obsignata* (a thread-like worm of the small intestine). These worms have a short, direct, life cycle. Many other worms use intermediate hosts such as grasshoppers, slugs, fish or crustaceans. For

Figure 6 Nematodes in chickens: These worms (*Ascaridia galli*) were recovered from a few broilers at a processing plant in North Georgia, USA, in late summer. Mature worms reach 12 cm in length. The presence of such worms is seasonal, and may be affected by control programme and overall management (Photo provided by Dr. Robert Beckstead).

Figure 7 Nematodes in chickens: The cecal worm *(Heterakis gallinarum)* is commonly found in layers and breeders, and serves as a reservoir of infection with blackhead disease (histomoniasis) which is contained within its eggs.

obvious reasons, worms with intermediate hosts are more likely to infect free-range or wild birds. Some of the common worms and their hosts are summarized in Table 5.

10.1 Biology of nematodes

Worms begin producing eggs in about 25 days after the egg is ingested (Fig. 8). The bird picks up eggs in the litter or ground in the normal scratching, pecking activities. The egg is crushed in the gizzard and hatches when exposed to digestive acids, releasing infective microscopic larvae, which invade the intestinal epithelium, where they feed and grow. Between growth spurts, they moult, becoming adult after the third moult and start producing eggs. Larvae develop inside the egg in a few days and become infective. The eggs may survive in the soil of a poultry facility or poultry yard for several years, remaining infective when eaten by another bird pecking in the dirt. This is especially important with *Heterakis* (caecal worm) ova, because they carry blackhead disease protozoans. Soil or facilities contaminated with caecal worms can be a danger to turkeys placed in the area for several years.

Nematode worms tend to be chronic and prevalent in layers and breeders, and seasonal in broiler chickens. Diagnosticians report high infection rates in broilers in late summer. Turkeys and other poultry also suffer damaging infections of nematodes.

10.2 Diagnosis of nematode parasites

These parasites shed eggs, which pass out in the faeces. Microscopic examination of faeces is simple and accurate; to detect nematode eggs, take small samples of faeces and mix with saturated salt solution in a small test tube. The eggs will float to the top within a few minutes. Transfer a drop from the top to a microscope slide and examine at 100x. Worm eggs are easy to see. Some are distinctive, such as those of *Capillaria*, which have an operculum (plug) at each end. The *Ascaridia* and *Heterakis* ova are oval shaped and similar. To separate these, one can take caecal droppings (the chocolate-brown material)

Table 5 Nematodes reported from chickens and other poultry in the USA*

Species	Location	Intermediate host	Other definitive hosts
Ascaridia dissimilis	Small intestine	None	Turkey (not in chickens)
Ascaridia galli	Small intestine	None	Turkey, quail
Ascaridia numidae	Small intestine	None	Guinea fowl (not in chickens)
Baylisascaris procyonis	Brain	Raccoon	Raccoon (and other accidental hosts)
Capillaria anatis	Crop, esophagus, proventriculus	None	Turkey, duck, goose, partridge, quail
Capillaria annulata	Esophagus, crop	Earthworm	Turkey, goose, grouse, guinea fowl, pheasant, quail
Capillaria bursata	Small intestine	Earthworm	Turkey, guinea fowl, pheasant, peafowl, quail
Capillaria caudinflata	Small intestine	Earthworm	Turkey, duck, goose, grouse, guinea fowl, partridge, pheasant, pigeon, quail
Capillaria contorta	Proventriculus	Earthworm or none	Turkey, guinea fowl, partridge, pheasant, quail
Capillaria obsignata	Small intestine	None	Turkey, goose, guinea fowl, pigeon, quail
Cheilospirura hamulosa	Proventriculus	Grasshopper, beetle	Turkey, grouse, guinea fowl, partridge, quail
Dispharynx nasuta	Proventriculus	Sowbug	Turkey, goose, grouse, guinea fowl, partridge, pheasant, pigeon, quail
Gongylonema ingluvicola	Gizzard	Grasshopper, beetle	Turkey, partridge, pheasant, quail
Heterakis gallinarum	Ceca	None	Turkey, grouse, guinea fowl, partridge, pheasant, quail
Oxyspirura mansoni	Eye	Cockroach	Turkey, grouse, guinea fowl, pigeon, quail, duck
Strongyloides avium	Ceca	None	Turkey, goose, grouse, quail
Subulura brumpti	Ceca	Earwig, grasshopper, beetle, cockroach	Turkey, dove, duck, grouse, guinea fowl, quail, pheasant
Subulura strongylina	Ceca	Beetle, cockroach, grasshopper	Guinea fowl, quail
Syngamus trachea	Trachea	Beetle, cockroach	Turkey, goose, guinea fowl, pheasant, peafowl, quail
Tetrameres americana	Proventriculus	Grasshopper, cockroach	Turkey, duck, grouse, pigeon, quail
Tetrameres fissispina	Proventriculus	Amphipod, grasshopper, cockroach, earthworm	Turkey, duck, goose, guinea fowl, pigeon, quail
Trichostrongylus tenuis	Ceca	None	Turkey, duck, goose, guinea fowl, pigeon, quail

* Many of these parasites are known throughout the world, as well as others of more local importance.

separately from intestinal droppings. Ova found in intestinal droppings are most likely from *Ascaridia*, while those found in caecal droppings will be *Heterakis*. At necropsy, worms 2–4 inches long in the midgut are ascarids, while smaller worms found in the caeca are *Heterakis*. Capillarid worms are found in the small intestine, and are difficult to see because they are thread-like.

10.3 Pathology of nematodes

Damage done by nematode infections depends almost entirely on the number of worms present, although pathogenicity differs between species. Layers and breeders are more likely to suffer serious infections with worms because they are reared on the floor for at least 16 weeks. This allows time for several generations of parasites. The most severe species infecting chickens is *Capillaria obsignata*. This worm can badly damage the mucosa of the small intestine, resulting in reduced laying performance and loss of flock uniformity. Ascarid worms in turkeys can also cause significant damage to the intestine. Caecal worms are not very pathogenic, but are important because they harbour *Histomonas*, serving as a long-term reservoir and vector for transmission of this important disease.

10.4 Control and treatment of nematode infections

Nematodes are easily killed with drugs called anthelmintics. The most effective are called benzimidazoles. There are several products in this category, including fenbendazole, oxybendazole, cambendazole and mebendazole. These are available in water-soluble formulations. Other products are available, but may be more difficult to use. In particular, piperazine, an older drug used in many formulations, tends to narcotize worms rather than killing them outright. As a result, some of the worms might recover in the cloaca. These worms sometimes migrate back up the intestinal tract, or may enter the oviduct, where they can be encapsulated in hens' eggs before the shell is put on. An ascarid worm in freshly cooked eggs is not a pretty sight.

Older products tended to be effective only against adult worms and not the larvae. Folk remedies for worming abound, but there is little evidence that they are safe and effective.

Commercial poultry, especially layer and breeder replacements, are normally wormed on a regular basis, as needed to keep worm infections at a low level. Even cage layers are likely to have serious worm infections in warm weather, because of the mechanical transmission by flies. Broilers are sometimes wormed as needed during the time of peak exposure (late summer).

10.5 Management and environmental control of nematodes

Worm eggs can be killed by commonly available disinfectants, but earthen-floored houses and farmyards are difficult to clean. The soil around poultry buildings is often contaminated with worm eggs, which are easily tracked inside. In the case of worms utilizing intermediate hosts, it is important to control insects and other invertebrates. This is especially important with free-range poultry or backyard flocks.

11 Tapeworms (Cestoda)

Tapeworms are common in poultry. Even chickens reared indoors are commonly infected with *Raillietina cesticillus*, because of the prevalence of insects in the environment capable of serving as intermediate hosts. Tapeworms are even common in cage layers, because stable flies serve as intermediate hosts and are frequently eaten by the hens. Some common tapeworms in poultry are listed in Table 6.

11.1 Biology of tapeworms

All tapeworms have a life cycle involving intermediate hosts. Common intermediate hosts are flies, beetles and ants. The adult worm living in the intestine is segmented, and as each segment matures it becomes an egg sac, which drops off to be excreted in the faeces (Fig. 8). Insects feeding on the egg sac become infected with a larval stage called a cysticercoid, which is infective to another avian host when it eats the insect. The *Raillietina*

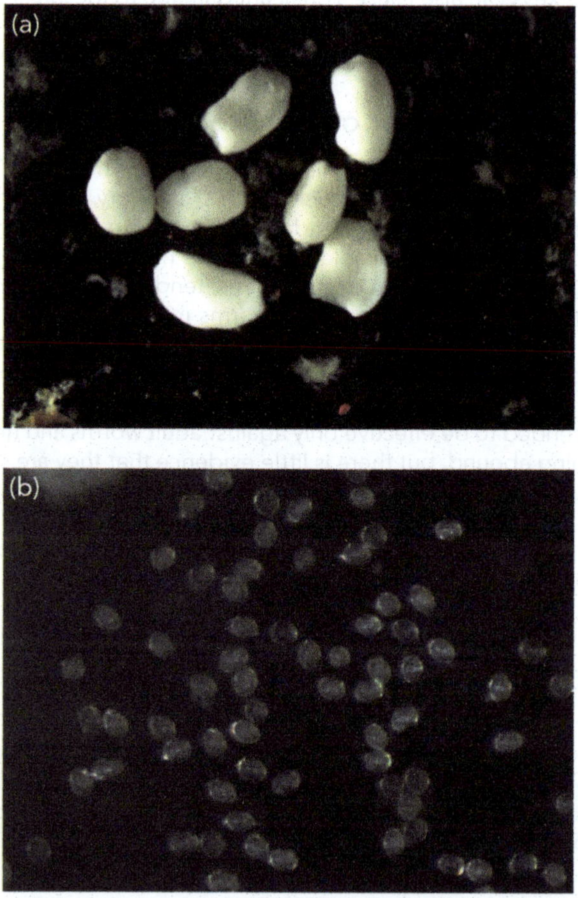

Figure 8 (a and b). Tapeworm proglottids recovered from litter of hens infected with *Raillietina cesticillus*. These have the appearance of grains of rice, and contain hundreds of tiny ova. When beetles or flies consume these eggs, further development takes place into the next stage, called a cysticercoid. When chickens consume the fly or beetle, they become infected with the tapeworm.

spp. of chickens use both flies and beetles as intermediate hosts. Once the insect is consumed by a chicken, the cysticercoid attaches to the intestinal mucosa and grows into another tapeworm.

11.2 Treatment and control of tapeworms

Most tapeworms of poultry cause little harm in terms of weight gain or feed conversion. This is fortunate, as there are no cost-effective treatment drugs approved for use in chickens or other poultry. Control of tapeworms is essentially indirect, by elimination of intermediate hosts. Flies and beetles are the most common intermediate hosts for tapeworms. As these insects are common and difficult to eliminate from poultry facilities and farmyards, it is impossible to eliminate completely the chance of infection.

For obvious reasons, we can expect free-range poultry to host a wider spectrum of worms, as they are in contact with the intermediate hosts. The intermediate hosts listed in Table 6 include houseflies, stable flies, grasshoppers, ants, slugs and several species

Table 6 Tapeworms (cestodes) common in poultry

Species	Normal host*	Intermediate host**	Pathogenicity
Amoebotaenia cuneata	Chicken	Earthworm	Low
Choanotaenia infundibulum	Chicken	Housefly, beetle	Moderate
Davainea meleagridis	Turkey	Unknown	Unknown
Davainea proglottina	Chicken	Slug, snail	Severe
Diorchis nyrocae	Wild and domestic duck	Copepod crustacean	Unknown
Hymenolepis anatina	Wild and domestic duck	Aquatic crustacean	Severe
Hymenolepis cantaniana	Chicken	Beetle	Low
Hymenolepis carioca	Chicken	Stable fly, dung beetle	Unknown
Hymenolepis lanceolata	Goose, duck	Aquatic crustacean	Severe
Hymenolepis megalops	Duck	Aquatic crustacean	Severe
Metroliasthes lucida	Turkey	Grasshopper	Unknown
Raillietina cesticillus	Chicken	Stable fly, beetle	Low
Raillietina echinobothrida	Chicken	Ant	Moderate–severe
Raillietina georgiensis	Wild and domestic turkey	Ant	Unknown
Raillietina magninumida	Guinea fowl	Beetle	Unknown
Raillietina tetragona	Chicken	Ant	Moderate–severe

* Most of these worms occasionally parasitize other domestic and wild birds.
** Most of these parasites are found in birds reared out of doors (free range or open pens), where they have contact with the intermediate hosts.

of aquatic crustaceans (copepods, ostracods, amphipods, etc.). Ducks and geese are commonly reared in contact with these organisms in many countries.

12 Trematodes (flukes) in poultry

Flukes are rarely a problem in poultry, owing to the complicated life cycles involving a snail or other mollusc, and sometimes an insect or other invertebrate. The oviduct fluke of chickens (*Prosthogonimus macrorchis*) is known mostly from Asia, and infects chickens with access to streams harbouring snails. An adult fluke worm lays eggs in the oviduct of chickens, which may fall into water and hatch. The larvae infect snails and develop into another larval form, which is released into the water. These larvae may be consumed directly by another bird, or may encyst on vegetation and be eaten by a bird, completing the life cycle. Infection also may result when a bird eats a snail.

Other flukes are normally parasites of wild birds, with poultry becoming accidentally exposed. For instance, *Echinostoma revolutum* has been known to infect racing pigeons with mortality and severe lesions, apparently because of their eating snails. Most poultry are not exposed to environments where snails would be common.

12.1 Treatment and prevention of trematode infections

There are no approved products for treatment of fluke infections in poultry or other birds. Infections can be prevented by keeping birds away from streams or ponds where snails may be producing infective larvae.

13 Ectoparasites and arthropod pests

The arthropods affecting poultry include flies, ticks, fleas, bedbugs, lice and mites. Beetles and their larvae are serious pests of broilers. Economically, the most significant are northern fowl mites for caged layers and breeder flocks, the lesser mealworm (darkling beetle) for broilers and houseflies for caged layers. Many arthropods serve as intermediate hosts or mechanical vectors for other parasites and diseases.

13.1 Lice (Mallophaga)

These are common in wild birds, but rare in poultry. The chicken body louse (*Menacanthus stramineus*) is the most common species in poultry and may be found all over the body. The shaft louse (*Menopon gallinae*) primarily attach to feather shafts of the breast and thigh. The chicken head louse (*Cuclotogaster heterographa*) is found on the skin of the head and neck areas. This lice can cause extreme irritation and even death in heavy infestations.

13.2 Bedbugs (*Cimex lectularius, Haemotosiphon indorus*)

Poultry houses are ideal for infestations of these pests. These are nocturnal and emerge at night to feed on chickens or other poultry. Infestations may reach enormous proportions and cause losses through irritation and blood loss. Cases are not uncommon, and the bugs can be carried on clothing or other equipment to infest homes.

13.3 Flies

Several species of flies are common pests around poultry production units, including the housefly (*Musca domestica*), stable fly (*Stomoxys calcitrans*), black flies (family Simuliidae), mosquitoes (various species), biting midges (*Culicoides* spp.) and the pigeon fly (*Pseudolynchia canariensis*) and contribute to spread of disease. Flies are more of a pest around cage layer operations because the hens do not have direct contact with piles of droppings where the flies breed. This is important because of the impact on nearby homeowners, who might suffer invasions of flies because of high populations in adjacent poultry operations. Chickens on the floor are efficient at scratching out and eating fly larvae, reducing the potential effects on adjacent properties. Technically, an important hazard of flies is their role as intermediate hosts for parasites. They carry tapeworms, and serve as mechanical vectors for coccidia and worm eggs, accounting for infections in cage layers or broilers, and contamination of nearby farms.

13.4 Beetles

Every poultry facility in the United States is heavily contaminated with beetles, principally the darkling beetle. The darkling beetle causes little direct harm to poultry, and in a few cases, it forms a significant portion of the chicken diet.

The darkling beetle is known to serve as intermediate host for tapeworms (*Raillietina cesticillus*) and may contribute to the survival of bacteria (*Salmonella*, etc.) in the litter. Little is known of the possibility that beetles may fly to neighbouring farms and carry eggs of *Heterakis*, or other parasites, bacterial or viral agents. More research is needed in this area. The application of insecticides is generally effective in reducing populations of beetles, but they are always present. The practice in North America of 'built-up litter', where clean out is practiced only once or twice in a year, contributes to the survival and proliferation of beetles. No currently available treatment programme is capable of elimination of beetles from poultry facilities.

13.5 Mites

Feather mites, northern fowl mites, scaly leg mites and so on (Fig. 9). The most pathogenic mite pathogen, the northern fowl mite, is widespread throughout the poultry world. The mite collects around the posterior of chickens, especially around the vent. Severe irritation results in morbidity and mortality. These mites are best known from layers and breeders. Infestations tend to build up over the course of a flock. Even free-ranging chickens may be affected by these mites. Commonly, it is thought that northern fowl mites can be spread by contact with wild birds, but this is not proven. Some suggest that mites infecting wild birds may even be genetically different than those infecting poultry.

Feather mites and others are common pests of poultry, but cause only minor damage to the growth and productivity of these birds. Anyone who has worked with poultry will attest to the irritating feeling of these mites creeping across the skin.

13.5.1 Control of mite infestations

Poultry facilities should be thoroughly cleaned and disinfected between flocks. Several disinfectants can be used to kill mites and their eggs. Be sure to check before use, to see if particular disinfectants can be used around food-producing animals. For active infestations

Figure 9 Mites: several species are important in poultry, including the northern fowl mite, a variety of feather mites and skin mites. Besides the nuisance factor, they may cause a drop in egg production. Free-range poultry often suffer from the scaly leg mite (shown here), which disrupts the scales of the feet, causes irritation and leads to secondary infections.

on birds, a good dusting with an effective insecticide is effective in greatly reducing the number of parasites. With northern fowl mites, special application to the vent area may be necessary. In the United States, the insecticides used on food-producing animals are carefully regulated by USDA and EPA. Each product is specifically labelled as to how it can be used and under what conditions. Poultry companies are well aware of such limitations, but it is up to individual users to carefully study the specific limitations on the use of each product.

13.6 Fleas

The sticktight flea of poultry (*Echidnophaga gallinacea*) is common on backyard flocks but not on commercial poultry. The sticktight flea is found on a wide variety of birds, and is usually seen attached to the skin around the head or eye. Infestations may reach hundreds or thousands/bird and may interfere with vision.

13.7 Ticks

The fowl tick, *Argas persicus*, is the most important example of this group. Rarely a problem in commercial poultry or even in backyard flocks in the Americas, this tick is more often a problem in the Middle East and warmer climates. These ticks live off the host and move onto the birds to feed at night, making them difficult to notice. If large numbers are present, the birds may die from blood loss before treatment is initiated. Treatment would be the same as for other mites, although it is difficult to kill all the nymphs, larvae and eggs hiding in crevices in the building.

13.8 Arthropods as intermediate hosts

Arthropods are known to transmit a wide variety of diseases and parasites, either as intermediate hosts or as mechanical vectors. In particular, the common stable fly and darkling beetles serve as intermediate hosts for tapeworms (*Raillietina* spp.), and may aid the survival of human foodborne disease bacteria including *Campylobacter* and *Salmonella*. Other arthropods as intermediate hosts for parasites are mentioned previously.

13.9 Control and prevention of ectoparasites

All of the arthropod parasites and pests affecting poultry can be controlled by insecticides or miticides. The application of insecticides is a procedure of dusting or spray application, as well as treatment of the environment. Manufacturers' recommendations are specific in the methods for treatment. Fortunately, the products approved for use on food-producing animals are safe and free of side effects when used as directed. Alternative control methods would include treatment of the environment and the birds with substances such as boric acid (or related borate compounds), diatomaceous earth and other products. In small flocks these products can be successfully applied and may be preferable. However, the effectiveness of such materials is questionable or borderline. Specialists agree that disease outbreaks are sporadic, not occurring in every flock. For this reason, products with marginal effectiveness may appear to be effective most of the time because of low challenge. Routine maintenance of preventive approaches is needed. Ectoparasites may never be a problem in small-flock environments, but require constant vigilance in large-scale production.

Commonly, it is thought that poultry acquire ectoparasites from wild birds, as sparrows and other birds frequently visit commercial poultry production units. However, there is no good evidence for this, and there is much evidence to the contrary. Poultry are most likely infected by residual infective eggs or larvae left over from the previous flock. Thorough clean out and disinfection is important in killing residual arthropods.

14 Conclusion and future trends

We are faced with a shortage of new products for parasite control. Pharmaceutical firms, saddled with enormous costs for development of new products have mostly decided to invest their research dollars elsewhere and have abandoned search for new anti-coccidials and blackhead treatments. Vaccination against coccidiosis is becoming increasingly important with coccidiosis, but this is not possible with blackhead disease or worms.

Clearly, future control of parasites in poultry will have to be maintained with fewer, less-effective control measures. As discussed in previous sections, various avenues for research are possible, including selection of poultry for improved disease resistance, and the use of natural products to stimulate innate and acquired resistance to infection. With increasing importance of free-range rearing of poultry, we need more research on roundworms and tapeworms and their prevention, as this area has been neglected for many years. The interaction of nematode parasites and protozoan parasites is of major importance, as well as the interaction of parasites with bacterial diseases such as necrotic enteritis needs further work. Intense effort and creativity will be needed to find new ways to control parasites if we are to maintain the present level of production and expand to meet the demand for high-quality food by a growing world population.

15 Where to look for further information

More complete information on any of these parasites will be found in recent editions of *Diseases of Poultry* (McDougald 2013a,b).

16 References

Bermudez, Alex J. 2013. Miscellaneous and sporadic protozoal infections. Cochlosoma anatis. Chapter 27, in *Diseases of Poultry*, 13th edition, eds. Swayne, D. E., Glisson, J. R., McDougald, L. R., Nolan, L. K., Suarez, D. L. and Nair, V. Wiley-Blackwell, Ames IA, pp. 1172–82.

Chapman, H. D., Barta, J. R., Blake, D., Gruber, A., Jenkins, M., Smith, N. C., Suo, X. and Tomley, F. M. 2013. A selective review of advances in coccidiosis research. *Advances in Parasitology*, vol. 83. Oxford: Elsevier Ltd, pp. 93–171.

Hess, M. and L. R. McDougald. 2013. Histomoniasis (Blackhead) and other protozoan diseases of the intestinal tract. Chapter 27, in *Diseases of Poultry*, 13th edition, eds. Swayne, D. E., Glisson, J. R., McDougald, L. R., Nolan, L. K., Suarez, D. L. and Nair, V. Wiley-Blackwell, Ames IA, pp. 1172–82.

Hinkle, N. M. and R. M. Corrigan. 2013. External parasites and poultry pests. Chapter 26, in *Diseases of Poultry*, 13th edition, eds. Swayne, D. E., Glisson, J. R., McDougald, L. R., Nolan, L. K., Suarez, D. L. and Nair, V. Wiley-Blackwell, Ames IA, pp. 1099–1116.

McDougald, L. R. 2005. Blackhead disease (histomoniasis) in poultry: A critical review. *Avian Dis.* 49: 462–76.

McDougald, L. R. 2013a. Internal Parasites. Chapter 27, in *Diseases of Poultry*, 13th edition, eds. Swayne, D. E., Glisson, J. R., McDougald, L. R., Nolan, L. K., Suarez, D. L. and Nair, V. Wiley-Blackwell, Ames IA, pp. 1117–46.

McDougald, L. R. 2013b. Cryptosporidosis. Chapter 27, in *Diseases of Poultry*, 13th edition, eds. Swayne, D. E., Glisson, J. R., McDougald, L. R., Nolan, L. K., Suarez, D. L. and Nair, V. Wiley-Blackwell, Ames IA, pp. 1167–71.

McDougald, L. R. and S. H. Fitz-Coy. 2013b. Protozoal Infections: Coccidiosis. Chapter 27, in *Diseases of Poultry*, 13th edition, eds. Swayne, D. E., Glisson, J. R., McDougald, L. R., Nolan, L. K., Suarez, D. L. and Nair, V. Wiley-Blackwell, Ames IA, pp. 1147–66.

McDougald, L. R., Hofacre, C, Mathis, G., Fuller, L., Hargrove, J. L., Greenspan, P. and Hartle, D. K. 2008. Enhancement of resistance to coccidiosis and necrotic enteritis in broiler chickens by dietary muscadine pomace. *Avian Dis.* 52: 646–51.

Reid, W. M. 1962. *Chicken and Turkey Tapeworms: Handbook to Aid in Identification and Control of Tapeworms found in the United States of America*. Georgia Agricultural Experiment Stations, University of Georgia, Athens GA 30602 USA. 71p.

Chapter 5

Disease management in poultry flocks

Peter Groves, University of Sydney, Australia

1 Introduction

This chapter will focus on disease preventative measures, health monitoring and disease investigation techniques, and the management of sick, meat chicken flocks. Poultry operations vary markedly, not only between countries, but also between integrated operations within the same geographic region. Diseases in intensive poultry flocks may occur due to viral, bacterial, fungal, protozoal, metazoan or arthropod infections or infestations. Non-infective metabolic conditions, nutritional deficiencies and toxicities can also present disease or poor performance manifestations. Genetic change in the broiler, which has become exceedingly rapid over the last three to four decades, have made major changes to the birds' physiology and anatomy. These changes have often proved unbalanced and resulted in the emergence of disease (typically cardiovascular and skeletal problems) or impaired health states. Technological changes in environmental control of houses, increased stocking densities, shorter batch cycles due to improved growth rate, increased feed intakes, faster turnarounds between batches, growing birds to larger final body weights, multiple pickups for slaughter within a house and regulatory withdrawal of many medications have complicated disease expression and severity. Development of better and expanded vaccines (especially of the immunosuppressive diseases), vaccination technologies and eradication of some diseases from breeding stock have improved bird health and decreased the reliance on antibiotic medication of earlier decades. With shorter grow out time to final weight (now as young as 32 days), the influence of factors during fertile egg incubation assume far greater effects on the growth and health of the

http://dx.doi.org/10.19103/AS.2016.0011.19

modern broiler. As a result, the occurrence of multifactorial conditions has become much more a feature of the scenario and these issues now dominate veterinary interventions and investigations in intensive poultry. The coverage here will tend to be general in nature.

2 Disease investigation techniques

Disease or flock problems may present as either a localized situation (i.e. one farm or one region showing the problem) or as a widespread issue affecting many flocks, possibly in various geographic locations. The latter will require a different approach to the former in terms of complexity and importance to the operation.

2.1 Initial investigation – routine approach

Investigation of a single problem in bird health or poor performance can be approached in the classical veterinary investigation manner: establish the existence of a problem; gather a sound and comprehensive history from the farmer and the service personnel serving the operation, and whatever records of flock performance are available; examine the birds' environment and management (farm area, house conditions, temperature and humidity, air quality (particularly ammonia level), litter quality, ventilation control, biosecurity, vermin control, bird behaviour and clinical appearance of the flock, condition of the range area if applicable); examine fresh dead and culled birds, and culled 'average' birds if necessary by necropsy; and decide on what laboratory specimens to collect (blood, swabs, tissues, histopathology specimens, feed, water, environmental samples) as appropriate to the condition, including antibiotic sensitivity testing in bacterial cases. These areas are well described in standard poultry health texts[1,2].

A diagnosis can usually be generated in such cases and treatment can be instigated if possible, and control measures can be described to prevent the problem in future batches of birds. In integrated operations, however, a localized disease problem is the exception, and usually a problem will be spread across or beyond the operation into the local industry. Where a widespread problem develops, a more systematic approach is required to better understand the situation.

2.2 Widespread problem – systematic approach

When a problem appears to be spread across multiple farms within an operation or across operations in a location, a systematic approach to understanding the situation is a far more rewarding procedure. An epidemiological approach will yield more useful outcomes than a non-systematic empirical investigation. The latter is much more susceptible to bias[3].

A systematic investigation will have more ability to determine problems that have a multifactorial causation, which are the predominant forms of problems seen in modern poultry operations. Where multiple factors contribute to a presenting problem, simple solutions are rare due to the intricacies of possible interactions between the factors[4] and the frequent presence of confounding factors[5] (i.e. factors that may not have a direct bearing on the condition but are associated with other factors that do).

Diseases in poultry are seldom just the result of the presence of an infectious organism. Although a pathogen may be a 'necessary' cause (i.e. its presence is necessary for the disease to occur), it may only be part of a 'sufficient' cause (i.e. a combination of determinants which will always produce the disease). For example, the necessary cause of necrotic enteritis may be *Clostridium perfringens* but the disease will not occur just because that organism is present in the chickens' intestinal tracts but it requires the interaction of several other factors such as coccidiosis, some immunosuppression and excessive protein in the feed (which may make up a sufficient cause) for disease expression.

The objectives of a systematic epidemiological investigation will be to stop the problem from occurring and to enable the prevention of its reoccurrence[6]. The approach is applicable to both infectious and non-infectious diseases and is equally valuable when poor performance is the problem rather than overt disease. The overall aim is to identify factors that can be considered as 'key determinants'[5,6] of the problem (i.e. those factors which are important in increasing or decreasing the risk of occurrence of the problem, which can be influenced by management practices).

Hancock and Wiske (1998)[6] suggested a useful starting point with an as yet misunderstood flock problem to prepare a path model by drawing together known information about the condition. This is not a mathematical model but a descriptive attempt to describe how some risk factors may relate to the situation. This can form the basis of areas to be evaluated within the overall investigation. An example of a path model developed to aid understanding of the broiler ascites syndrome (pulmonary hypertension syndrome) is illustrated in Fig. 1.

The first requirement in this scenario is to define a 'case'. In poultry, the smallest unit of study possible is often the 'house' or even 'farm' rather than individual animals. A 'case' should be defined to describe a 'unit of study', which has a significant enough level of the disease or depressed performance that is typical of the problem, differing from 'healthier' flocks which can be considered as 'controls'.

The distribution of cases across time and spatial location initially may lead to a closer appreciation of the problem's occurrence. Putative risk factors identified from the path modelling procedure or drawn from other contributing sources should be examined in

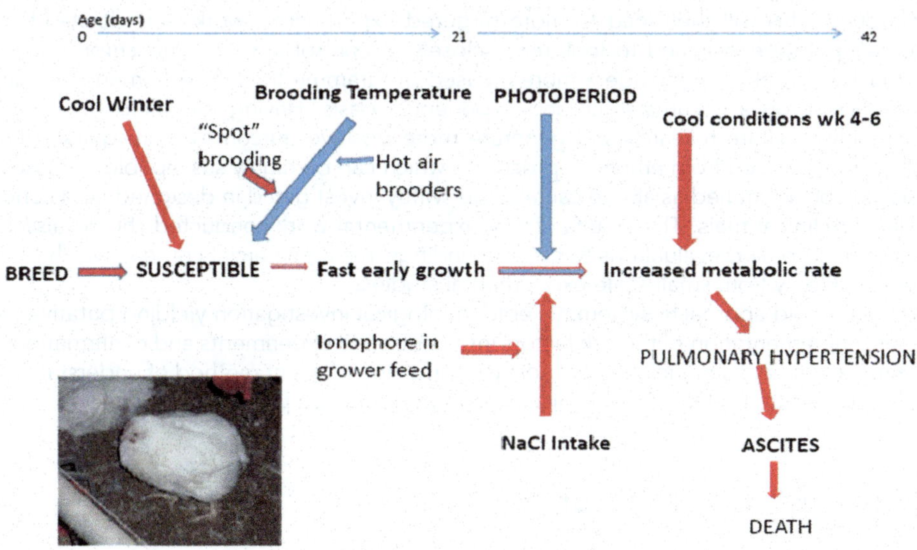

Figure 1 A path model for broiler ascites syndrome. Red arrows represent risk factors and blue arrows represent protective factors (adapted from Groves, 1999[7]).

their association with identified case and control flocks. The use of recorded performance data (see Section 4) can be extremely useful to gain quantitative estimates of the influence of these factors. These may be evaluated using the classical epidemiological contingency table approach (χ^2 analysis) or by *t*-tests or ANOVA if the independent variable is continuous[3]. Bearing in mind the importance of the health and immunological status of the broilers' breeder source (see Section 4.1), the identity of the birds' parents should always be included as a risk factor to be evaluated. Other factors commonly included in the initial investigation include date of placement, age of birds, timing and pattern of mortality or morbidity or performance loss, sex (if farmed separately), breed and natural groupings such as house types, litter type, age of use of litter, source hatchery, feed source and ration stage, flock size, stocking density at certain ages, vaccination history and history of any medications. Other factors need to be included if data is available and can include any factors thought to be associated (e.g. ventilation levels, water consumption, pickup for slaughter ages, proximity to other poultry farms, proximity to main roads, altitude, feeder and drinker types, water quality and source, interaction of staff with other farms, sharing of equipment with other farms, proximity to surface water bodies and presence of wild birds).

A univariate analysis across each factor's association with cases or controls will yield a shorter list of interesting potential risk factors. The analysis should then move to multivariate techniques, searching for confounding interactions that may modify or clarify the picture. Multivariate techniques such as multiple logistic regressions or multiple linear regressions can be powerful in this process. The significant risk or protective factors should have become apparent or developed into suggestions of other areas to investigate. Interactions are particularly important at this stage.

If findings from the systematic approach lead to new knowledge and perspectives, depending on the importance of the problem and the immediate ability to remedy it, the identified risk factors provide opportunities for directed experimental studies to firstly confirm the association between the putative risk factor(s) and the disease occurrence. This will then lead to more targeted experimental work to understand the disease's pathogenesis and to look more closely at control options. This chronology of approach generally will provide a more focused programme than if experimental studies are launched on a circumstantial and experiential basis. Having identified risk factors will greatly facilitate research and generate more positive outcomes more quickly than a project developed from empirical processes which can be highly susceptible to biases.

Risk factors identified as above can also be swiftly investigated in designed field cohort studies or clinical trials. The results of any experimental work conducted should also be positioned into field evaluations to confirm findings in a commercial context, which often differs markedly from small-scale experimental studies.

This threefold approach: systematic epidemiological investigation yielding putative risk factors, implementation of the risk factors into controlled experiments and confirmation of the findings in field situations is the most powerful and efficient method of understanding and improving control of complex disease and performance problems available.

3 Preventative measures

Best practice disease control measures can decrease the risk of infection and may reduce the severity of an endemic infection. The components of disease prevention and control measures include biosecurity, vaccination, medication and health monitoring.

3.1 Biosecurity

Biosecurity is simply defined as 'security from transmission of infectious diseases, parasites and pests'.[8]

Some infectious diseases are capable of being prevented on farms by good biosecurity while others cannot be completely controlled by this means. For example, biosecurity may be effective against diseases where the pathogen does not survive well outside the host (e.g. mycoplasmata, infectious laryngotracheitis virus), requires relatively close contact for transmission (e.g. infectious coryza) and where there are no alternative control measures (e.g. avian influenza (AI) virus where the jurisdiction does not allow vaccination). Situations where biosecurity alone is likely to be ineffective include those where the pathogen has a long survival time in the environment (e.g. Marek's disease virus and fowl cholera), has an airborne route of infection (e.g. infectious bronchitis virus and Marek's disease virus), has an effective and robust vector (e.g. fowl pox virus), is fairly resistant to disinfectants (e.g. coccidiosis and infectious bursal disease virus) or where the consequences of an outbreak are severe (e.g. infectious laryngotracheitis virus). These latter diseases will require added control mechanisms on top of biosecurity (e.g. prophylactic medication or vaccination) to achieve adequate control.

The basics of biosecurity as practised in the chicken meat industry have several components, as listed below.

3.1.1 Hierarchy of operations

Generally, the effort devoted to biosecurity will vary depending on the class of bird and their relative value and importance to the operation. This may also be a determinant in their ability to spread an infection throughout an operation. For example, the introduction of a *Mycoplasma* into an elite breeder flock could lead to vertical transmission throughout several generations of birds through vertical transmission, whereas the introduction of the same organism into a commercial broiler flock would be self-limiting. Hence many integrated operations would apply strict biosecurity to the elite flocks higher in

Table 1 Example of differing emphases on biosecurity applied to different classes of poultry in some countries

Procedure	Breeders and hatcheries	Broilers	Layers
Isolated	Yes, but often on complexes	No	No
All-in, all-out	Yes	Yes	No
Visitor register	Yes	Yes	Larger operations
Shower in	Yes	No	No
PPE	Yes	Yes	Variable
Foot baths and hand washes	Yes	Yes	Variable
Vehicle disinfection	Yes	Yes/No	Mostly No
Remote feed delivery	Yes	No	Mostly No
Youngest to oldest	Only limited visits	Yes	No
Dedicated staff	Yes	Varies	Varies

the hierarchy (nucleus, great grandparent breeders and grandparents) and somewhat less stringent measures to lower stages (hatcheries, parent breeders) and less still to commercial broiler flocks. Differences in biosecurity application to various levels may relate to the limitations imposed on time allowed between visits of personnel (such as service staff and veterinarians) between farms (e.g. a break in contact between a visit to a great grandparent operation of 48 hours, a shorter break before visiting a grandparent flock of 36 hours, an overnight break before visiting a hatchery but would allow multiple visits to broiler farms within the same day).

An example of different levels of biosecurity procedures commonly applied to different classes of poultry is shown in Table 1.

3.1.2 Isolation of premises

Isolation used in this context refers to distance, in space and time, between chicken operations. 'Isolation' in most biosecurity thinking refers to separation of newly introduced animals for a time to observe whether they are healthy. Where commercial poultry flocks introduce new animals it will usually be a group of many thousands and separation for a time would require major facilities and huge labour requirements in moving them a second time. This is highly impractical in a commercial situation. By 'isolation' in this context refers to the separation of farms by distance – not only from other poultry farms but also from main roads which may carry birds being moved and from personnel who may have contact with other birds, poultry or otherwise. Many chicken diseases are capable of airborne spread or can be readily transmitted by fomites, the most important of which would be people and equipment which may be shared between farms. Marek's disease virus, for example, can spread at least 2–3 km through airborne route[9]. Infectious bronchitis virus is also capable of airborne spread over long distances[10]. Birds being trucked to slaughter or replacement pullets that travel past other poultry operations can be a source of infection to flocks close to main roads (e.g. infectious laryngotracheitis virus and *Mycoplasma gallisepticum*).

3.1.3 Single age

Introduction of an 'all-in, all-out' approach to commercial poultry farming improved health dramatically and provided far better disease exclusion capability. Avoidance of multi-age flocks breaks the cycle of many diseases by interrupting the ability of infectious organisms to spread from older, infected flocks into young naïve birds. This is particularly important for pathogens with long carrier states, where recovered individuals may harbour the organism without clinical signs for long periods. Obvious examples are the mycoplasmata, infectious laryngotracheitis and infectious coryza. Large integrated poultry companies have the capacity to place all houses on a large broiler complex within a week. The established flock can also be depopulated within a short period, allowing for terminal hygiene procedures prior to the arrival of the next batch of broilers.

Breeder flocks also should be managed in an 'all-in, all-out' fashion, with all flocks placed within a few weeks span. Once a breeder flock is established as a single age unit it is desirable to maintain it as a closed flock, with no further introductions during its lifetime. This objective is frequently compromised, however, with the need to introduce fresh males in a middle-aged breeder flock to maintain fertility. This practice is a major compromise to biosecurity and risks introduction of a pathogen with the introduced birds, or exposure of the younger, more naïve males to a resident pathogen in the existing flock.

3.1.4 Quarantine and hygiene procedures

There are many practices which can be considered and the level of adoption on a particular farm will depend on the farm's position in the hierarchy, importance and value to the operation and the management philosophy on biosecurity. Practices may include:

- Provision of an entry log for visitors to the farm.
- Vehicle disinfection at the point of entry to the property. Feed trucks and gas delivery trucks may need to enter the site but also have the propensity to have recently visited other poultry farms during their normal business. Thus they represent a frequent risk for the possible introduction of pathogens. Vehicle disinfection facilities may vary from the simple provision of manually operated disinfectant sprays (Fig. 2a) to more sophisticated total vehicle disinfection systems (Fig. 2b and 2c). Restriction of vehicle entry past a certain point on the property, physically separated from the birdhouse area, can also be effective where the vehicle does not need to enter the site (e.g. visitors and service personnel).
- Feed trucks are, by necessity, frequent visitors to poultry farms. The provision of remote feed delivery capacity (e.g. Fig. 3) can remove the need for the vehicle and its driver to enter the poultry holding area.
- Showers and changing facilities should be provided for high-value flocks and even flocks at lower levels on the hierarchy should provide clothing changing facilities. Where amenities for showering are provided, the design allowing staff to disrobe and enter the shower from one side and then emerge in a clean area where on-farm clothing and boots are provided is the most effective set up. Provision to achieve surface sanitation of articles being brought in can also be made, such as ultraviolet light cabinets, with openings from both sides of the change rooms. This can be used to sanitize necessary equipment and personal items (lunch boxes, etc.). It is my personal view that mobile phones should never be brought through onto the farm areas.

Figure 2 (a) Hand-operated vehicle disinfection, (b) automatic vehicle spray and (c) total vehicle spray and wheel wash.

Figure 3 Remote feed delivery system.

- Farms without showers should provide personal protective clothing for any visitor. On commercial broiler farms, this can be cheaply provided through disposable overalls, hairnets and provision of farm boots or the use of strong plastic boot covers.
- The entries to all houses should have a disinfectant footbath and a hand sanitizing device nearby which should be used on entry and exit by all persons. Footbaths need to be maintained and the sanitizer refreshed frequently. An even better approach is to provide dedicated boots for each house.
- Other livestock should not be allowed in the house vicinity, as these, and their faeces, can be sources of *Salmonella* contamination. Pets should also be restricted from the house area and should never be allowed inside a house.

3.1.5 Depopulation

In areas of dense poultry populations it can be advisable to attempt to coordinate flock ages and avoid large variations in broiler age within close proximity. This is essentially trying to make an area run in an 'all-in, all-out' mode. The benefit here is to avoid older flocks, especially those being depopulated, being in proximity to younger flocks. This can be a difficult situation to achieve but the benefits can be major, especially during local outbreaks of infectious disease, such as infectious laryngotracheitis in broiler flocks. This also requires considerable cooperation between different companies and organizations farming in the same locale.

3.1.6 Clean out, composting and re-use of litter

Preferably houses should go through a thorough batch cleaning and disinfection process before new chicks arrive. Difficulties in supply of clean litter materials, appropriate used litter storage and/or land application of used litter often hampers this, however, and it is necessary to re-use multi-batch litter. The latter may also provide some advantages, such as decreased environmental contamination with poultry manure and development of a more valuable fertilizer product. However, the re-use of litter should involve the proper composting (litter pasteurization) procedures for this to be safe and efficient. Good litter pasteurization requires proper litter heaping, monitoring of temperatures achieved during composting and sufficient time for the process to achieve the correct results of litter pasteurization and stabilization of nitrogen into microbial protein which is less volatile as ammonia.

In full clean-out processes, the general procedure is to remove all litter from houses and remove it from the vicinity. The house needs to be swept effectively to remove as much of the litter remnants as possible. The ceilings, walls, curtains, floors, feeders and drinker lines need to be washed – preferably with a heavy-duty detergent. All houses on the farm should reach this stage of preparation before a sanitizer is used (i.e. all litter removed and all houses washed). A sanitizer can then be applied to the clean facilities. An insecticide spray may also be applied after this procedure. The value of a detergent should not be underestimated and where time or economy is limited, use of a detergent should be given preference over a sanitizer. Then fresh clean litter material can be delivered. Some farms may also use a fumigation procedure (using formaldehyde gas, where allowed) once the litter is put in.

3.1.7 Free-range biosecurity issues

Exclusion of contact with wild birds is essential for truly effective biosecurity. Hence houses should be completely wild bird proof. Free-range operations obviously cannot completely exclude wild bird contact possibilities. The major concern here is possible direct or indirect contact with wild waterfowl or their faeces, with low but inherent risk of the introduction of low pathogenic AI virus. Free-range facilities should therefore do as much as possible to discourage the presence of waterfowl in the vicinity of the range areas. There should be no surface water sources (dams, creeks, rivers, drains, etc.) in the vicinity of the range areas. There should be no sources of feed or drinking water outside of the house. Any surface water used for the poultry (for drinking, cooling or irrigation of the range areas) must be effectively sanitized. Grass should be kept mowed short around and in the range areas. Any feed spills around silos or storage areas must be avoided and feed storage areas should be securely fenced off from access by the ranging poultry. Chickens that escape and are seen in the feed storage areas should be caught and culled. They should not be allowed to rejoin the main flock.

In most circumstances the range area is allowed a reasonably lengthy recovery period between batches of meat chickens. This would include the farm downtime while birds are absent and generally the first three weeks of the next flock's life before they are allowed to range. In environments such as Australia, with low humidity, high temperatures and high sunlight intensity, this length of time allows for reasonable natural disinfection of the range area. In more temperate and cooler climates this may need closer assessment with regard to the possibilities of pathogen survival and build up on protected range areas. Survival of large numbers of clostridial spores and coccidial oocysts may present issues depending on environmental conditions. Other possible pathogens could include helminth eggs and possibly histomonads, although these generally require longer lifespans than those afforded to modern meat chickens.

Good biosecurity is expensive. Like insurance, biosecurity is a necessary precautionary expenditure against incidents which we hope will never occur. Its expense often prompts broiler management to compromise and cut corners, particularly when there seems to be little disease operating, which may actually be an unrecognized result of good biosecurity. It is appropriate to consider biosecurity to be equivalent to a necessary insurance process, of similar importance as the financial insurance of physical assets of a business. Losses from disease incursion can be just as costly, and possibly more likely to occur than is, say fire damage to buildings.

In terms of relative effectiveness, isolation is the most effective biosecurity measure possible, and also the most expensive (farm location, access to staff and services). The next most effective measure is single-age (all-in, all-out) operation. Without these two factors, biosecurity relying solely on quarantine and hygiene measures is much more difficult. Reliance on the latter measures makes maintenance of biosecurity very hard and prone to failure.

3.2 Vaccination – breeders versus broilers

Typical vaccination programmes used in various countries for broiler breeders and broilers have been reviewed by Cserep[13]. Breeder vaccinations have the dual justifications of protection of the breeder flock from mortality and production losses and to ensure the provision of high maternally derived antibody for the progeny of the flock. The particular programme will vary depending on the diseases and their virulence and severity active in a particular location.

3.3 Egg-borne infections

Breeder flocks need to be free from or vaccinated against organisms known to be capable of transmission vertically; either within the fertile egg or on the eggshell. This will include the species-specific *Salmonella* serovars of poultry (Pullorum and Gallinarum), mycoplasmata (*M. gallisepticum, M. synoviae* and *M. meleagridis*), avian encephalomyelitis virus and chicken infectious anaemia virus in all regions and various other organisms that may be problematic in certain locations, such as reoviruses in many countries and fowl adenoviruses (e.g. types 8b and 11 in Australia). The paratyphoid serovars of *Salmonella* (particularly Enteritidis and Typhimurium) are also well capable of egg-borne transmission and freedom from these has proven much more difficult to achieve, but breeders supplying broiler chicks and poults should strive towards this goal. The status of the breeder flocks supplying day-old progeny should be known at least.

3.4 Management-related diseases and problems

More detail on management of particular diseases and conditions is provided further in this book, but the role of husbandry in all disease conditions must be recognized and accepted by the farmer in all situations. Although some conditions cannot be prevented at the final farm level, their severity and sequelae may be largely modified by good management conditions.

Brooding conditions can be major contributors to improved bird health. Examples of conditions which may modify an existing or introduced disease include increased severity of aspergillosis with high relative humidity (>80%) in the house, increased severity of infectious bronchitis with chilling and exacerbation of the latter prevalence of broiler ascites with slightly low brooding temperatures and prolonged photoperiods. Provision of correct brooding temperatures as advised by the breeding companies will aid in minimizing the deleterious outcomes of many situations. Correct temperatures of course include heating of the litter substrate to the correct temperature as well as achieving adequate ambient air temperatures, prior to the arrival of day-olds.

Attention to stocking density is crucial to good health, performance and welfare of the flock. Different localities, climates and housing systems will dictate the appropriate level. Overstocking is one of the major contributory factors to health issues in meat birds. A maximum density based on kilograms of bodyweight per square metre of floor space must be established and exceptional attention to not exceeding this must be practised. This is especially so where multiple thin outs are practised, as the first thin out of birds must occur in a timely manner to avoid the density target being over-reached. This can be a logistical challenge in some situations but deserves a strong commitment from management to ensure compliance. As a guide, maximum densities should not be allowed to exceed 34 kg/m^2 in tunnel-ventilated premises and possibly not more than 28 kg/m^2 in open-sided housing, depending on local conditions and house construction.

Nutrient deficiencies are uncommon in most commercial operations due to advanced feed formulation by experienced nutritionists but occasional problems can occur due to mistakes (e.g. premix or other ingredient left out, salt inclusion level incorrect), poor ingredient quality, toxicoses such as from mycotoxins or biogenic amines, contaminants, medication overdoses or *Salmonella* contamination.

Planned medications must be regularly assessed and challenged. Control of coccidiosis or necrotic enteritis may involve in-feed medicaments and adequate rotation or shuttle programmes need to be carefully considered and instituted to avoid or at least delay the development of resistance. Anticoccidial vaccines have undergone continued improvements in recent years with precocial strains providing good immunity without the subsequent need to medicate, but these still produce a mild intestinal infection that does temporarily reduce growth and may cause enough intestinal irritation to promote the occurrence of necrotic enteritis. But as the consumer pressure to remove feed medication from food-producing animals increases worldwide, the need to optimize the use and outcome of coccidial vaccines will continue to grow in importance.

3.4.1 Wet litter

Management of litter condition is another important consideration and this is gaining more prominence due to welfare accreditation programmes in some countries. Wet litter is a serious issue for bird health and is the major contributor to foot pad dermatitis development and cold, wet conditions contribute to severity of respiratory diseases. Contributory factors and causes of wet litter have been recently reviewed[11]. Broiler chickens drink about 1.8 litres of water per kg of feed consumed and 75% of this water ends up in the faeces. Ventilation is the only method available to remove moisture from the house. Ventilation is the major factor involved with litter moisture management and failure to ventilate adequately is the major cause of wet litter development, often closely linked with excessive stocking density. Hence there are more frequent problems seen in the cooler months as there is a compromise between air quality and temperature maintenance. Reduced ventilation leads to increased humidity and litter will become wet once relative humidity exceeds 70%.

Good management of drinkers (avoiding spillage, correct height and water pressure for the age of bird) is also of importance. Maintaining dry friable litter provides advantages: it has a higher gas transfer capacity and hence dries faster and is more resistant to compaction; it will be more alkaline and tend to promote aerobic conditions. Wet litter, however, has slower drying capacity, is easily compacted, promotes anaerobic conditions and tends towards acidity, with the increased risk of subsequent burning of the birds' feet and hocks.

Other factors of excessive moisture in the litter can include physical sources, such as water leaks or excessive condensation; bird health issues causing diarrhoea such as coccidiosis, necrotic enteritis, dysbacteriosis, infectious stunting syndrome, infectious bronchitis, colibacillosis and a myriad of viral infections (infectious bursal disease, Marek's disease, Newcastle disease (ND) and AI; and nutritional factors can be involved. Determining the initial cause of a wet litter problem can be difficult, particularly if the condition has existed for some time, but one approach is to determine the nature of the faeces predominating in the house[12].

Wet litter conditions where the majority of the faeces appear normal and well-formed is suggestive of presence of excessive water or failure to adequately remove moisture: hence ventilation may have been compromised or relative humidity may have been too high; there may have been lack of attention to litter aeration and cultivation or the litter depth or type of litter material may have been inadequate. A lighting regime that affects bird activity may also have contributed to wetter litter conditions as the lowered flock activity may reduce the flocks' ability to cultivate the litter.

Where faeces are observed to be watery (diuresis), some involvement of the kidneys could be suspected[12]. The possibility of renal damage can be investigated in areas including infectious bronchitis virus, presence of mycotoxins (e.g. ochratoxins) in feed and possibly avian nephritis virus infection. A possibility of an electrolyte imbalance (either through feed or from levels in drinking water) must also be considered.

The presence of orange/pink coloured mucous in faeces would indicate intestinal irritation[12] or disease, and examinations for the existence of dysbacteriosis or coccidiosis should be conducted. This can also occur if the birds have been off feed for a few hours.

Observation of undigested feed particles in faeces should alert one to the possibility of some level of infectious stunting syndrome, viral infections, oxidized fats in feed, coccidiosis or necrotic enteritis.

Nutrition-related causes of excessive diarrhoea can involve problems such as excessive electrolyte (sodium, potassium) inclusion in either feed or water (and this can be exacerbated by the use of electrolyte treatment via drinking water which may increase water intake and hence increase faecal moisture). Diarrhoea can also result from feed non-starch polysaccharides which are not controlled by the added enzyme in use. Excessive or imbalanced amino acid levels leading to high blood uric acid levels can stimulate water intake. High fat inclusions in rations may also result in diarrhoea or steatorrhoea. Unintentional presence of toxins (e.g. mycotoxins, rancid fats or biogenic amines) may also lead to excessively fluid droppings. Physical feed factors can also contribute to increase in faecal moisture, such as fine particle size which can cause intestinal irritation, whereas the inclusion of whole grain may reduce it.

3.5 Growth rate and associated problems

Modern commercial meat poultry strains have phenomenal growth rates; continuing improvements in this capacity is achieved each year as a result of genetic selection. Historically this also led to the inadvertent development of some health problems mainly involving the skeletal and cardio-respiratory systems. Leg weakness and ascites syndrome were the main examples. Rapid growth rates have been shown to impinge on bird health in these two problem areas[13]. Breeding companies have made substantial improvements in their breeds more recently and the industry has learnt ways of minimizing problems (e.g. by providing longer scotoperiods and heightened attention to brooding conditions). In some operations, there is a trend towards concentration on achieving higher seven-day weights as a major target. While this is a valid management performance target it should be viewed with some caution. Recent findings have shown that first week growth rates is increased in chickens which have hatched early and experienced a longer sojourn time in the hatcher which is associated with poorer leg health[13], and hence a focus here may need to be carefully considered.

3.6 Dead bird disposal

Removal of mortalities and culled birds is essential and must be done daily, usually the first action of the staff each day. Dead and sick birds are a source of contamination for other birds and their extended presence in a house may lead to problems with botulism. The mortalities must be disposed of so as to minimize the chances of further contamination of the farm itself and so as not to transmit disease to other farms. Environmental concerns often mean it is not possible to bury birds on site. Composting of dead birds is an effective method of dealing with them if done correctly. Incineration is also a good method if done correctly. Birds should not be burned in the open as feather material which may be infective may be spread by the convection currents evoked by the fire. Use of an enclosed incinerator is advisable. Otherwise the birds must be removed. Having a large freezer expedites this process so that mortality, if not excessive, may be stockpiled without massive decomposition and odour. Frozen birds can be removed at convenient intervals for disposal in public landfill or designated sites. Extreme care must be taken if vehicles for dead bird removal travel from farm to farm and thus pick up points for this process must be positioned to minimize the vehicle's entry to each farm site.

3.7 Hatching egg hygiene

The quality of day-old poultry is strongly reliant upon good husbandry and management of their parent breeder flocks. A major consideration here includes good hatching egg hygiene. Contamination of eggshells, especially soon after laying, is a major cause of reduced hatchability and of poor chick quality. This takes on greater significance as breeder flocks age and shell quality declines. Good hygiene practices of nest boxes and communal automatic nests on breeder farms are extremely important. Fumigation of eggs with formaldehyde gas has been a historic procedure that does improve the microbiological quality of eggs used for incubation. This practice has been withdrawn in many regions due to workplace safety requirements and hence a greater attention to nest and egg hygiene is required. Floor eggs should not be incubated, but under market conditions of egg shortages, the temptation is often provided to use lower quality eggs for incubation. If this does occur, poorer quality eggs should be separated and incubated separately and, if possible, kept separately as chicks in the field. More highly contaminated eggs will produce more day-olds with a propensity towards yolk sac infection and it may be beneficial to manage these chicks differently in the field as their first week mortality is usually increased. The use of antibiotics in such flocks is contentious. While this may appear to reduce first week mortality, quite often the effect is only to delay the problem into subsequent weeks and many such flocks experience increased incidences of problems such as femoral head necrosis (bacterial chondronecrosis and osteomyelitis). There is also the danger of provoking the development of multiple antibiotic resistance in bacterial populations in this situation. There are some indications that the use of probiotics in the first days of life for chicks may give some assistance with these problems[14].

3.8 Rodent and insect control

Rodent and vermin control measures form an essential part of quality assurance programmes for most meat chicken producing companies. Adequate positioning of bait stations which are refreshed frequently and monitored for rodent activity outside all houses are requirements of modern poultry production systems.

3.9 Multiple pickup for slaughter

The most desirable approach to house depopulation is to remove the entire flock in a single operation. Some regions, however, practice flock thin outs to reduce flock density at critical stages and meet bird size requirements for their markets while making more efficient use of facilities. This does introduce some disease spread risks as abattoir vehicles and equipment and catching crews move between farms, often during night-time. Where this is necessary it is important to limit the possible spread of contamination as much as possible by measures such as ensuring crews move from youngest to oldest farm and that attention to hygiene of catching equipment and vehicles is of a practically achievable level.

3.10 Free-range issues

The growth in free-range broiler management systems has been notable in several countries in recent years. This adds some complexity to farm management practices. Fortunately, with birds only released to range areas after three weeks of age, there is usually a lengthy downtime for the range areas, allowing for natural decontamination (exposure to sunlight, drying conditions, etc.) and time for regrowth of grass. Predator protection and discouraging of wild birds in the range areas assume more importance. Predation does not contribute markedly to bird losses directly but possible contact with wild animals and birds or their faeces does increase disease introduction risks. There should be no feed or drinking water supplied outside of the house to ensure that extra attractiveness to wild birds is avoided. Provision of shade is essential in range areas as chickens, being descended from jungle fowl, do not appreciate continual full sunlight exposure. If birds spend lengthy periods outside they are not consuming feed as contiguously as they would if fully confined and this can compromise coccidiosis control if it relies on in-feed anticoccidials. Market expectations in many countries also require the non-use of in-feed antibiotics, hence the prevalence of dysbacteriosis and occasionally necrotic enteritis may be higher. Alternative control measures for these intestinal conditions need to be developed for this method of production to be efficient or even sustainable in the long term.

4 Monitoring of poultry health and performance

Poultry companies record and store vast amounts of data on their flocks, primarily for production monitoring and reporting. Data covering flock location/identity, numbers of chicks placed, weekly average bodyweights, crude mortality and culls, feed conversion efficiency and pickup for slaughter information would be commonly kept. Histories of anticoccidial programmes and routine medications should also be documented. Some operations may also record breeder source identities, hatchery information, vaccination details, day-old chick weights and sex (if grown separately). Other more specific records are useful but seldom recorded routinely (e.g. disease-specific mortality or morbidity, lighting programme, farm visitor records, serviceman records of litter quality, ventilation assessments and problem notes). Laboratory records from breeder flocks are also available and taken to reveal the success of vaccination procedures and/or to evaluate the possible exposure to wild pathogens. Routine serological monitoring of broiler flocks, commonly for infectious bursal disease, infectious bronchitis and ND is also sometimes carried out. These resources are of immense value but are often underutilized. They are of use for the monitoring of performance and health. Often the computerized storage format is not easily amenable to formal analysis techniques and considerable editing is necessary to produce a database capable of straightforward statistical manipulation. The time needed to translate data into an efficiently analysable format is, however, completely worthwhile. A baseline for each parameter should be determined[15] for each operation. The knowledge of baseline data, seasonally adjusted, can be a useful technique for early detection of problems or for gaining an understanding of why performance may have improved. There is always a natural level of variability in field data, and assessment of performance must be considered within the framework of this variation to avoid invalid conclusions. A very instructive method of evaluating data on any of the parameters is to display them in a statistical control run chart[16]. Run charts plot data consecutively over time and confine the individual points between control lines calculated at three times the standard deviation (above and below the mean). As these are built up over time the value increases. These provide visual evaluation of trends occurring within an operation. Points plotted between the control lines represent natural variation. A single point outside the control lines would represent a significant outlier and an exception to the general process. Trends can become apparent from a successive run of points on one side of the mean line and can signal that performance is changing from a common source causation (either getting worse or better, depending on the direction). Identified changes can then be more thoroughly analysed against other known factors by standard epidemiological procedures. Any of the above parameters can be examined in this way.

There are systematic health assessment programmes in use in some operations where a sample of birds (often five birds per house) from different ages are collected and examined at post-mortem as a periodic snapshot of gross pathology. These often use a lesion scoring system to provide a record of a semi-quantitative assessment of disease signs occurring across the operation. An example would be the Elanco's Health Tracking System programme, but these carry on from suggested programmes developed by a university veterinary support group[17]. Such systems are extremely powerful and assist in the examination of health and production changes over long periods and provide inter-operation comparisons. These provide magnificent information but need to be more deeply examined, looking for correlations and associations within the recorded factors and to other farming factors to provide even more valuable outcomes.

4.1 Breeder monitoring

Commercial broiler health is heavily dependent upon the health status of their parent flock. Parent breeder flocks may be a source of vertically transmitted organisms and also a source of essential maternally derived antibodies for the young broiler. The latter provision is responsible for an enormous amount of protection for the young chick for up to three weeks of its life, which now represents 30% or more of the broiler's entire lifespan. Breeder flocks are extensively vaccinated with this outcome in mind. The particular diseases where maternal antibody contributes to prevention or management will vary between geographic locations but usually include infectious bursal disease, chicken anaemia virus, fowl adenovirus (e.g. types 8b and 11 in Australia), reoviruses and avian encephalomyelitis. In some situations, maternal antibody can interfere with the quality of vaccinations given to broilers (e.g. Marek's disease and infectious bronchitis) and is also of critical importance in countries where broilers themselves receive live infectious bursal disease vaccines. Serological monitoring of the breeder flocks is imperative to ensure that an adequate transmission of maternal antibody occurs for the diseases endemic to their locality.

Many breeder operations are either free of mycoplasmata or vaccinated against them, and the removal of this source of infection has improved broiler health dramatically. Paratyphoid *Salmonella* serovars, however, remain particular organisms of concern for potential transmission through the human food chain and hence will be monitored by environmental swabbing and culture in breeder flocks.

Hatcheries should provide the broiler farms with health status information on the source flock(s) for their chicks so that appropriate managements can be instigated or varied as necessary.

5 Management of sick birds

5.1 Diagnosis

Arriving at a correct diagnosis is obviously essential. The processes described in Section 2 should be observed. Clinical signs and gross pathology are often sufficient to arrive at a preliminary diagnosis in many conditions. Laboratory and further methods are used frequently and should be applied for confirmation and often for more accurate identification of the strains of the infective organisms involved. The latter investigations are becoming increasingly important in the understanding of the epidemiology of the disease in question and may be very important in the development of future control measures. Appropriate tests for various diseases have been reviewed[18]. In any poultry disease occurrence where there is a relatively large increase in mortality or an unexplained drop in egg production, it is imperative that the possible presence of an emergency disease be excluded. The clinical and gross pathological signs shown in ND and highly pathogenic AI are not sufficiently specific enough to be ruled out. In all such cases, where it is possible, tracheal and cloacal swabs, collected in a viral transport medium, should be submitted to a government laboratory capable of rapidly identifying the presence of these viruses.

Individual sick, lame or injured birds should be euthanized humanely as soon as they are observed. Dead or culled birds should be removed from the house promptly. Setting up 'hospital pens' in large commercial operations is generally inadvisable as very ill birds seldom recover and they act as a source of continued infection. Culling and removal of these animals is appropriate.

5.2 Mass medication methods

The sheer size of modern poultry flocks makes administration of treatment to individual animals logistically and economically unfeasible. Rapid medication of a large flock with a therapeutic product will require administration via drinking water. Poultry houses should each have a water-holding tank that can be used for medication purposes. Some live vaccines are also administered in this way. Dose rate calculation needs to be carefully considered and withdrawal times recorded and observed. Time of administration is important, as it is desirable that the medication is consumed promptly. If administered late in the afternoon, a proportion of the medicated water may sit unconsumed overnight and could be compromised by temperature or time. Administration of vaccines via drinking water requires considerable attention to detail if a successful result is to be obtained. This process usually requires withdrawal of water for an hour or more to ensure the birds are desiring to access the drinkers promptly. Water stabilizers or chlorine inhibitors are usually mixed into the water to receive the vaccine (e.g. skim milk powder and proprietary stabilizers) prior to vaccine suspension. Thorough mixing of the diluted vaccine into the required amount of water for about 2–3 hours of supply is essential. The water containing the vaccine should then be flushed through the drinking lines to ensure an immediate dose is available to the birds and the house should be walked by staff to encourage the birds to move towards the drinkers.

Automatic proportioning devices are also available for adding medication directly into a flowing water line. These may work well for chemical medications but it is difficult to adequately apply an in-water vaccine through this method. Close attention needs to be paid to ensuring that correct dose is achieved across the appropriate time frame required. Where water stabilization of vaccines is required an adequate amount of the stabilizer needed for the entire volume of water to be medicated must be supplied through the proportioner, bearing in mind that it will be difficult to achieve the time of water contact needed to achieve stabilization in some cases.

In-feed medication may be used as a follow-up to water administration of medicines or for longer-term medication. Detailed attention is needed to ensure proper withdrawal times are observed with in-feed medications, however.

5.3 Responsible use of antibiotics

Concerns over the development of antibiotic resistance which may eventually be transferred to the human population are increasing worldwide. Chemotherapeutic products should be under strict veterinary control in all cases. There should be a strong justification for their use to be considered. In commercial chickens, the eradication or control of *Mycoplasma* in breeding flocks, improved biosecurity and better control of immunosuppressive diseases (Marek's disease, infectious bursal disease, chicken anaemia virus) have largely removed the need for therapeutic medication of broiler flocks. Medication for problems such as yolk sac infection is generally unrewarding and should only be used where there are no reasonable alternatives (and management and other practices as described above should take precedence). Withdrawal times of medications need to be strictly observed to avoid meat residues.

The use of in-feed low-level prophylactic antibiotics ('antibiotic growth promoters') has been an historical feature of intensive animal industries. Because of possible antibiotic resistance development concerns, this practice has been banned or is under review in many countries. In many instances, the removal of this prophylaxis resulted in increased clinical disease, especially necrotic enteritis and dysbacteriosis, and actually led to an increase in therapeutic drug use. While arguments rage about the pros and cons of the practice, responsible use of antibiotics is important. Where in-feed prophylactic antibiotic use is allowed, there are obvious measures regarding the choice of such products that could be instituted to avoid or minimize the undesirable outcomes considered. Choice of antibiotics used in this way should be limited to those which are not orally absorbed by the chicken (i.e. only have gut activity), have a limited spectrum of antibacterial activity (e.g. Gram-positive spectrum) and are not chemically related to any antibiotic used in human medicine.

5.4 Communication with the grower/serviceman/manager

Rapid communication between grower and the management of the operation is imperative. The service personnel are usually the first link to a veterinarian and their experience is invaluable in first-line appreciation of the development of a problem that may require veterinary attention. Similarly, it is important for the veterinarian to communicate findings, laboratory results, treatment recommendations and regulatory requirements through the management structure back to the grower. In this way everyone involved is aware of the progress of the situation and mistakes and misunderstandings can be minimized.

5.5 Economic judgement

Final decisions about any health matter including treatment rests with the owners of the birds. In some situations the overwhelming costs may be the deciding factor for the operation in any treatment option. Having said that it is the management's responsibility to make the welfare of the animals and the safety of any food produced from them as the primary consideration, these must override any economic consequences.

6 Emergency disease occurrence

In many countries the outbreak of serious poultry diseases will result in government authority intervention and attempts to eradicate the disease. In commercial meat chicken flocks, the two diseases likely to trigger a disease emergency response are AI[19] and ND[20]. The difficulties with initial recognition and diagnosis of these diseases are covered elsewhere but clinical signs and post-mortem pathology of these diseases may be difficult to distinguish from other serious respiratory diseases including fowl cholera, infectious laryngotracheitis, chlamydophilosis, acute toxicities or management malfunction (e.g. overheating and smothering).

It is in the interests of the local industry, the region and the nation to expedite the detection of outbreaks of emergency diseases as soon as possible. Outbreaks of such diseases need to be contained and eradicated promptly. Whenever confronted with a clinical picture in broilers involving high mortality with respiratory or nervous signs even if other conditions are strongly suspected, steps should be taken to quickly exclude and/or confirm highly pathogenic AI and ND. Swabs collected from trachea and cloaca from a meaningful number of freshly dead or clinically ill birds should be submitted without delay to a government laboratory, preferably transported in a viral transport media. PCR tests are capable of detecting HPAIV and velogenic NDV within 12 hours of arrival at the laboratory.

6.1 Mass destruction techniques

Control of emergency diseases such as AI and ND in many countries is reliant upon eradication which entails the need for rapid mass destruction of poultry flocks. Flocks may also need to be destroyed where there has been a market failure or malfunction, especially so for large broilers that cannot be normally slaughtered and are approaching critical body mass or growing too large in a house, severely compromising their welfare. Methods of killing large numbers of poultry have been reviewed[21,22]. Euthanasia methods that are deemed to be acceptable for poultry include occipitocervical dislocation, decapitation and gassing[21]. CO_2 is the most frequent gas used in mass destruction and this requires the confinement of birds to a contained structure, such as a waste skip bin[23] or a purpose-designed killing cart[21]. The birds need to be caught and manually placed into these devices and the process can cause stress, not only to the birds but also to the people involved. Such procedures are never pleasant and can be a considerable source of distress for owners of the birds and staff associated with the farm. Mass destruction processes are desired to be rapid to interrupt the possible spread of infection. With this end point in mind, welfare needs to be compromised to some extent to enable the destruction. It is important to carry out this procedure in the most humane manner possible in the situation. More recently, fire-suppressing foam has been employed as a method of destroying a large number of birds *in situ* in the house. The birds do not drown with this procedure. The foam kills the birds by asphyxiation within a few minutes of being covered above their heads. The process is not immediate: movement can be appreciated underneath the foam for several minutes. This procedure is rapid and efficient but does not cause immediate death on application. Under the OIE *Terrestrial* Code, the principles needed to be considered for humane destruction refer to the induction of unconsciousness to be under 'the least aversive conditions as possible and should not cause avoidable anxiety, distress or suffering in animals'[21]. In this sense, use of foam does overcome the stressful process of manually herding, catching and carrying of birds to a gassing chamber or the distressing practice of mass neck dislocation within a farm situation. Care needs to be taken in applying the foam to large groups as an advancing foam front may cause birds to panic and flee, leading to packing up of birds. Dividing the flock into smaller groups with portable solid fences may assist in this regard (Fig. 4). The foaming technique is usable where the birds are on the floor but is difficult if the floor is slatted and will not work well in a cage facility.

6.2 Disposal of mass mortality

Masses of potentially contaminated killed birds should be disposed of promptly in a manner that minimizes the chance of further spread of the infection. Effective methods are burial, incineration, rendering and composting (the latter can be completed on site (Fig. 5), avoiding the need to transport dangerous material)[24].

Figure 4 Penned section within a house filled with firefighting foam for mass destruction of poultry.

Figure 5 (a) Composting of mass mortality *in situ*; (b) monitoring of compost temperatures.

7 Future trends and conclusion

Challenges that will remain as issues for developed nations into the near future will include areas of food safety, the move away from the use of antibiotics in production, animal welfare, environmental sustainability and diminishing worldwide resources. Continuing emphasis on endemic diseases and emergency diseases, AI in particular, will also be important. In the developing world maintenance of poultry farming is a priority with local industries experiencing sustainability difficulties[24].

The rapidly increasing world human population will continue to drive the need for food production and poultry meat appears as the most sustainable industry that would be able to meet the growing demand for protein. Poultry production is the most efficient converter of food sources to high-quality protein, produces no liquid waste, uses less water and produces the least amount of greenhouse gas emissions of any of the livestock systems.

7.1 Food safety

High-quality and safe food is the goal of the industry and is demanded more and more by safe food authorities and consumer groups. In some countries the standards required by regulatory authorities extend beyond meat processing right back through the production chain. Enhanced traceability of product has become essential in some jurisdictions and can only be seen to be an increasing requirement in the future[25]. Standardized control procedures such as hazard analysis and critical control point will continue to be important and will be audited even more resolutely. Even though contamination of the final product can be effectively controlled during the slaughter and processing system there is clear desire to decrease the amount of human pathogenic bacteria which come from the farming sector. The major pathogens of interest with poultry meat being *Salmonella* serovars (particularly serovars Enteritidis and Typhimurium) and *Campylobacter* species. Efforts to control these more effectively at the farming level will continue to be driven in the foreseeable future.

7.2 Animal welfare

Consumers, animal welfare interest organizations, major retailers and fast food chains are placing ever-increasing concern on how animals are raised in agriculture. A poor welfare image to the public has been a burden on the poultry industry for many years and activist groups continue to apply pressure in this aspect. Meat chicken growers do have a serious view on the well-being of their birds and genuinely wish to maximize their welfare while still being able to farm sustainably. The clash of views has been volatile in the past and will continue to follow a confrontational path unless the industry takes a proactive role in continually improving and being seen to be improving their systems. The huge increase in the popularity of alternative systems in some countries, notably the move to free-range production systems, has been facilitated and in some cases forced through the increased public interest in animal welfare. As we have noted, moving to alternative production systems is not without the risks of reduced productivity and increased disease prevalence and the industry needs to continually research and evaluate improved methods to actually enhance welfare once moving into alternative systems. This will be a continued area of growth in research expenditure and in communication between industry, government regulators and consumer groups. The introduction of universally accepted standards is becoming an important step towards this end goal.[25,26]

7.3 Environmental sustainability

Chicken meat production is the most environmentally friendly of the livestock meat industries, using less land, more efficient conversion of feed to protein, less water use per kg of meat produced (including that for the feed needed) and lower waste production and the lowest greenhouse gas production.[27] Chicken meat appears as the most likely sustainable source of high-quality protein to support the predicted rise in global human population. Even so the reduction of the industry's environmental footprint remains a challenging goal for sustainability. Issues involving the proximity of urban living to poultry farms continue to cause concern and problems for planning authorities. The main issues here are odour production, visual pollution and noise associated with vehicle movements during collection and transport of birds for slaughter. Better future planning regulations may help decrease these issues in future but they will remain as a problem area for some time yet, particularly as the urban spread encroaches into traditionally rural areas on city fringes. This will encourage the need to move and re-establish poultry enterprises in new regions.

7.4 Diminishing resources

As the human population grows, all resources will come under ever-increasing pressure. Adoption of newer technologies will be essential if the chicken meat industry worldwide is to meet the protein needs of a burgeoning consumption base. One of the major limitations will be sources of phosphorus. The industry has moved to improved utilization of phosphorus sources with the introduction of phytases to make plant phosphorus sources frequently available. Continued improvements will be necessary going forward. Another major challenge is emerging as competition for grains from the ethanol production industry. As the world searches for renewable energy sources, a threat to available food sources emerges and this will need to be carefully balanced into the next two decades.

7.5 Challenges for poultry in the developing world

Perhaps the largest challenge to the sustainability of chicken meat production in developing countries comes from the importation of lower cost meat from other high-producing countries, providing competition with which the local industry cannot compete. The cost of production in developing countries is a major obstacle, the major element being feed costs, which can only be seen to increase in the near future. Returns on their produce is also limited often due to lack of market access[28]. Interest rates and access to finance is also more difficult in the developing world[24]. Access to products to support efficient production, particularly vaccines, in some countries is limited[29]. Inadequate control of diseases and lack of implementation of biosecurity remain as areas where vast improvements in productivity could be made in this region of the world. Water quality is an issue for animal health as much as it is for human health, and its improvement offers massive increases in the standard of living for many regions.

8 Where to look for further information

Sources of information on health investigation include:

Martin, S. W., A. H. Meek and P. Willeberg (1987), *Veterinary Epidemiology: Principles and Methods*. Iowa State University Press, Ames, Iowa, USA.

Pattison, M., P. F. McMullin, J. M. Bradbury and D. J. Alexander (eds) (2008), *Poultry Diseases*, Sixth Edition. Elsevier, Edinburgh, UK.

Saif, Y. M. (ed.) (2008), *Diseases of Poultry*, Twelfth Edition. Blackwell Publishing, Oxford, UK.

Thrusfield, M. (ed.) (2017), *Veterinary Epidemiology*, Third Edition. Blackwell Publishing, Oxford, UK.

9 References

1 Bermudez, A. J. 2008. Principles of disease prevention: Diagnosis and control introduction. In *Diseases of Poultry*, 12th Ed., Y. M. Saif (Ed.), pp. 3–46. Blackwell Publishing Ltd.: Oxford, UK.

2 Alcorn, M. J. 2008. How to carry out a field investigation. In *Poultry Diseases*, 6th Ed., M. Pattison, P. F. McMullin, J. M. Bradbury and D. J. Alexander (Eds), pp. 14–38. Elsevier: Edinburgh, UK.
3 Lessard, P. 1988. The characterization of disease outbreaks. *Vet. Clin. North Am. Food Anim. Pract.*, 4:17–31.
4 Hueston, W. D. 1988. Evaluating risk factors in disease outbreaks. *Vet. Clin. North Am. Food Anim. Pract.*, 4:79–96.
5 Willeberg, P. 1979. The analysis and interpretation of epidemiological data. *Proceedings of the Second International Symposium of Veterinary Epidemiology and Economics*, pp. 185–90. Canberra, Australia.
6 Hancock, D. D. and Wikse, S. E. 1988. Investigation planning and data gathering. *Vet. Clin. North. Am. Food Anim. Pract.*, 4:1–15.
7 Groves, P. J. 1999. The ascites syndrome in broiler chickens: identification of epidemiological risk factors associated with the syndrome and examination of their nature. PhD Thesis, The University of Sydney, Sydney, Australia.
8 Blood, D. C. and Studdert, V. P. 1999. In *Saunders Comprehensive Veterinary Dictionary*, 2nd Ed., p. 132. WB Saunders: London, UK.
9 Groves, P. J., Walkden-Brown, S. W., Islam, A. F. M. F., Reynolds, P. S., King, M. L. and Sharpe, S. M. 2008. An epidemiological survey of MDV in Australian broiler flocks. *Proceedings 8th International Marek's Disease Symposium*, p. 26. James Cook University: Townsville, Queensland, Australia.
10 Cavanagh, D. and Naqi, S. A. 2003. Infectious Bronchitis. In *Diseases of Poultry*, 11th Ed.,Y. M. Saif (Ed), p. 107. Iowa State Press: Ames, Iowa, USA.
11 Dunlop, M. W., Moss, A. F., Groves, P. J., Wilkinson, S. J., Stuetz, R. M. and Selle, P. H. 2016. The multidimensional causal factors of 'wet litter' in chicken-meat production. *Sci. Total Environ.*, 562:766–76.
12 LaVorgna, D. and Vancraeynest, D. 2104. Evaluating wet droppings as a useful guide to bird health. *Int. Poult. Prod.*, 22:11–13.
13 Julian, R. J. 2005. Production and growth related disorders and other metabolic diseases of poultry–a review. *Vet. J.*, 169:350–69.
14 Wideman, R. 2016. Probiotics as an alternative to antibiotics for treating lameness due to bacterial infections in broilers. *Proceedings APSS*, 27:200–5. University of Sydney: Sydney, Australia.
15 Keirs, R. W. 1993. Flock health monitoring programs. *Proceedings XIII Latin American Poultry Congress*, pp. 29–36. Santo Domingo, Dominican Republic, October.
16 Tague, N. R. 2005. *The Quality Toolbox*, 2nd Ed., pp. 155–8. Milwaukie, USA: ASQ Society for Quality. Available at http://asq.org/learn-about-quality/data-collection-analysis-tools/overview/control-chart.html.
17 Keirs, R. W., Magee, D., Purchase, H. G., Underwood, R., Boyle, C. R. and Freund, J. 1991. A new system for broiler flock-health monitoiring. *Prevent. Vet. Med.*, 11:95–103.
18 Morrow, C. 2008. Laboratory investigation to support health programmes and disease diagnosis. In *Poultry Diseases*, 6th Ed., M. Pattison, P. F. McMullin, J. M. Bradbury, D. J. Alexander (Eds), pp. 39–47. Elsevier: Edinburgh, UK.
19 Animal Health Australia. 2011. Disease strategy: Avian influenza (version 3.4). *Australian Veterinary Emergency Plan (AUSVETPLAN)*, Edition 3, Primary Industries Ministerial Council: Canberra, ACT, Australia. Available at http://www.animalhealthaustralia.com.au Validated 18 August 2016.
20 Animal Health Australia. 2014. Disease strategy: Newcastle disease (version 3.3). *Australian Veterinary Emergency Plan (AUSVETPLAN), Edition* 3, Agriculture Ministers' Forum: Canberra, ACT, Australia. Available at http://www.animalhealthaustralia.com.au Validated 18 August 2016.
21 Thornber, P. M., Rubira, R. J. and Styles, D. K. 2014. Humane killing of animals for disease control purposes. *Rev. Sci. Tech. Off. Int. Epiz.* 33:303–10.
22 Tahir, R., Anjum, A. A., Muhammad, K., Rasool, A. and Khan, F. 2015. Avian influenza and its mass depopulation strategies in infected poultry birds. *Taiwan Vet. J.*, 41:51–7.

23 Animal Health Australia. 2015. Operational Procedures Manual: Destruction of animals (version 3.2). *Australian Veterinary Emergency Plan (AUSVETPLAN)*, Edition 3, Agricultural Ministers' Forum: Canberra, ACT, Australia. Available at http://www.animalhealthaustralia.com.au/ Validated 18 August 2016.
24 Kusi, L. Y., Agbeblewu, S., Anim, I. K. and Nyarku, K. M. 2015. The challenges and prospects of the commercial poultry industry in Ghana: a synthesis of literature. *Int. J. Manag. Sci.*, 5:476–89.
25 Penz-Jr., A. M. and Bruno, D. G, 2011. Challenges facing the global poultry industry until 2020. *Proceedings of the Australian Poultry Science Symposium*, 22:49–55, Sydney, Australia.
26 Australian Chicken Meat Federation, 2013. Bird Welfare. Available at http://www.chicken.org.au/page.php?id=44 Accessed 14 January 2017.
27 de Vries, M. and de Boer, I. M. J. 2010. Comparing environmental impacts for livestock products: a review of life cycle assessments. *Livest. Sci.*, 128:1–11. doi:10.1016/j.livsci.2009.11.007.
28 Karthikeyan, S. 2012. Problems faced by small scale poultry farmers in developing countries. *The Poultry Site.* Available at http://thepoultryguide.com/problems-of-small-scale-poultry-farmers/Accessed 14 January 2017.
29 Anonymous, 2008. Plenty of poultry farming challenges. *Poultry News*, The Poultry Site available at http://www.thepoultrysite.com/poultrynews/15627/plenty-of-poultry-farming-challenges/ Accessed 14/01/2017.

Chapter 6

Understanding and boosting poultry immune systems

Rami A. Dalloul, Virginia Tech, USA

1 Introduction

In the perpetual battle against common poultry pathogens and other environmental antigens, the role of the immune system is an important and ongoing topic of research. With the current trend of reduced reliance on preventative drugs and emphasis on more natural alternatives, the proper functioning of the host defences becomes even more significant. Careful priming and activation of targeted immune responses not only leads to better protection against challenges, but also helps in optimizing the birds' performance through proper allocation of nutrients towards maintenance and growth. Modulation of immune responses is best exemplified by routine vaccination to build specific immunity against common pathogens and prevent disease development. However, stressors such as subclinical infections that often go undetected pose the most significant threat to profit margins of the poultry industry. Low pathogenic loads, especially in the gastrointestinal tract, frequently lead to diminished performance as they compromise the epithelial lining. By priming local immune elements at mucosal sites (digestive and respiratory), subclinical infections could be averted or suppressed, thus enhancing growth and general health. Among the most commonly used modulators are feed and/or water supplements such as probiotics and prebiotics where much of the recent research focus has been invested. Therefore, this chapter will first review the various elements of the host defence system, and then will provide specific examples on how certain supplements can modulate the immune system and its responses to specific challenges that are common in commercial production.

http://dx.doi.org/10.19103/AS.2016.0011.21

2 Overview of the avian defence system

The structure of the avian defence system, including the immunity components, is similar to its mammalian counterpart regarding organization and mechanisms of action, but with few avian-specific differences. The second edition of avian immunology (Schat et al., 2014) provides complete and detailed description of structure and function of this system. The avian defences and immune system, like in all vertebrates, are multi-layered and can be broken down into two key components: innate and adaptive.

2.1 Innate defences

Innate defences comprise various physical and chemical elements including the innate branch of the immune system. These are generally characterized by being non-specific defence mechanisms that are present and ready to be mobilized on the day the bird hatches. This system acts as the first line of defence, providing immediate protection against impending foreign antigens and environmental challenges.

2.1.1 Physical and chemical barriers

The epithelial layers, body secretions and mucous membranes of the body constitute physical barriers that are generally impermeable to most infectious agents. Most pathogens gain access to the host through mucosal surfaces of the respiratory and gastrointestinal tracts, which are in direct contact with the external environment. If a pathogen is inhaled, ciliated epithelium and mucus found within the respiratory tract act as filters to assist in keeping the airways clear (Juul-Madsen et al., 2014). If a pathogen enters via the digestive tract, a thick mucus layer, produced by goblet cells, will block the pathogen from penetrating the host's cells. Additionally, the gastrointestinal tract is lined with a single layer of columnar intestinal epithelial cells (IECs), which regulate nutrient and water uptake and act as a physical barrier separating the potentially harmful contents of the intestinal lumen from the underlying tissues (Oswald, 2006). Due to their protective function, IECs have developed several mechanisms to reduce the risk of pathogen invasion in the surrounding tissues. Such mechanisms include the inhibition of bacterial passage and colonization along the luminal surface of the epithelial layer through production of secretory proteins, such as host defence and antimicrobial peptides, as well as the interaction with components of the underlying immune system (Pitman and Blumberg, 2000). The gut maintains a dense microbial population capable of preventing colonization by invading organisms through competitive exclusion, as well as producing soluble factors capable of inhibiting the growth and development of pathogens in the gut. The lower pH found within the gastrointestinal tract also serves as a powerful chemical defence against ingested pathogens (Juul-Madsen et al., 2014). In addition, higher body temperature found in avian species is capable of prohibiting many infectious agents from causing disease (Butcher and Miles, 2001). Except in situations where these physical and chemical barriers are compromised, entry of the majority of pathogens is prohibited, preventing access to the host.

2.1.2 Cellular barriers

If pathogens are capable of breaching the physical and chemical barriers of the innate defence system, there are cellular checks in place to provide protection. The most widely known cells of the innate immune system are phagocytic cells that are capable of ingesting and destroying antigens. These cells include macrophages, heterophils (the avian equivalent to mammalian neutrophils) and dendritic cells. Natural killer (NK) cells constitute another major cellular component of the innate immune system. NK cells are cytotoxic lymphocytes that do not require activation in order to destroy cells deemed to be 'non-self', and thus are not restricted by the major histocompatibility complex (MHC), unlike B and T lymphocytes. These cells of the innate system are derived from the lymphocyte lineage and are usually characterized by a large granular morphology in the cytoplasm. NK cells function by releasing these granules from their cytoplasm, which contain proteins such as perforin and proteases in order to kill virus-infected and tumour cells (Juul-Madsen et al., 2014).

The fundamental concept behind eliciting a proper immune response is the ability to determine self from non-self. Due to lack of specificity, the innate immune system utilizes a unique mechanism for detecting invading pathogens involving recognition of molecular structures present in foreign organisms. The cells of the innate immune system are usually triggered when conserved microbial motifs known as microbe-associated molecular patterns (MAMPs) are recognized by pattern recognition receptors (PRRs) located on the surface of the immune cell. These structures (MAMPs) are critical for pathogen survival and can be found in microbial membranes, cell walls, proteins, and nucleic acids (Jenkins et al., 2007). Following activation and phagocytosis, the phagocyte will present a processed fragment of the pathogen to members of the adaptive immune system, mainly B and T lymphocytes, which will stimulate a response to the pathogen. As such, these innate immune cells also function as antigen-presenting cells (APCs). Recognition of pathogens by the innate immune system triggers both immediate innate defences as well as the activation of the adaptive immune response. Dendritic cells and macrophages play a key role in detecting and processing antigens, and then dictating the differentiation of naïve lymphocytes into appropriate effector cells in order to defeat specific types of pathogens (Lee and Iwasaki, 2007). Initial encounter of pathogens with the innate system leads to destruction of pathogens, as well as initiation of a cascade of events which include recruitment of various immune components, and induction and modulation of the adaptive immune system.

2.2 Adaptive immune system

For pathogens that cannot be controlled by innate barriers and cells, the adaptive immune system acts as the next line of defence and provides protection against re-offending pathogens. Adaptive immunity is more complex than innate immunity and offers antigen-specific protection to the host. After the antigen is processed by cells of the innate immune system, it is recognized by B and T lymphocytes, the key components of the adaptive immune system. Adaptive immunity also has a 'memory' feature, which allows future responses against a specific antigen to be quicker and more robust.

The adaptive response is classified as humoral or cell-mediated immunity (CMI). The immune system will utilize the humoral branch, CMI, or a combination of the two in order to clear the offending pathogen, depending on the antigen characteristics. Primary lymphoid organs such as the thymus and bursa of Fabricius are responsible for lymphocyte production and development. Once developed, lymphocytes travel to the secondary lymphoid organs such as the spleen and mucosa-associated lymphoid tissues, where they come into contact with potential pathogens and other antigens and differentiate into effector cells (Dalloul and Lillehoj, 2006).

2.2.1 Humoral immunity

The humoral immune system is primarily mediated by B lymphocytes. B cells produce specific immunoglobulins (Ig), or antibodies, when stimulated by microbial exposure or other antigens. Antibodies defend the host against these invaders by three mechanisms: 1) Opsonization: antibodies will bind to receptors and coat the surface of the pathogen for it to be more readily and efficiently engulfed by phagocytes; 2) Neutralization: antibodies will interact with epitopes on an infectious agent and inhibit its ability to infect the host; and 3) Complement Activation: the complement system is activated when antibodies bind to the surface of invading pathogens, which aid in phagocytosis (Juul-Madsen et al., 2014).

Unlike mammals who carry five main types of immunoglobulins, only three main classes are identified in birds: IgM, IgY (also known as IgG), and IgA. IgM is found on the surface of B cells and is the first antibody found in circulation during a primary immune response. IgY (or IgG) is the most abundant immunoglobulin found in avian as well as mammalian blood, and is the primary antibody produced during a secondary immune response. IgA plays a critical role in mucosal immunity where it is secreted. Secretory IgA exists as a dimer and is most concentrated in mucosal surfaces, tears and saliva (Härtle et al., 2014).

2.2.2 Cell-mediated immunity

CMI is characterized and controlled by T lymphocytes. As found in mammals, chicken T lymphocytes can be categorized into CD4+ (helper T cells or T_H cells) and CD8+ (cytotoxic T cells or T_C cells) subpopulations (Viertlboeck and Göbel, 2014). Helper T cells are activated by recognition of a class II MHC coupled with a processed antigen on an APC. Once activated, T_H cells divide and produce a variety of cytokines in order to activate B and T lymphocytes as well as other immune cells. Proliferating T_H cells will differentiate into one of two major subtypes known as Type 1 and Type 2 helper T cells (T_H1 and T_H2, respectively), depending on the cytokine profile present in the environment. The T_H1 cells are primarily responsible for producing cytokines that promote inflammation and activation of macrophages, and both B and T lymphocytes. The cytokines produced also inhibit the function of T_H2 cells, which is part of the response to generate immunity to intracellular pathogens. Cytokines secreted by T_H2 cells stimulate B lymphocyte proliferation and antibody production while inhibiting T_H1 cell function to enhance immunity to extracellular pathogens (Kaiser and Stäheli, 2014). A third and more recently discovered T_H cell lineage has been identified as T_H17 due to the production of its signature pro-inflammatory cytokine IL-17. Cytotoxic T (T_C) cells, on the other hand, are responsible for the recognition and lysis of cells infected with endogenous pathogens in association with a class I MHC molecule (Dalloul and Lillehoj, 2006).

2.2.3 Immune response genes

Numerous genes are involved in directing both innate and adaptive immunity by encoding for production of various elements of the immune system including antigen receptors, antimicrobial peptides, chemokines, cytokines and other factors. Few elements are briefly described below and discussed later in this chapter; for a detailed list and description of such gene elements, the readers are referred to Kaiser and Stäheli (2014). When exposed to antigens or chemotactic agents, macrophages will begin to produce inducible nitric oxide synthase (iNOS). This enzyme leads to the production of nitric oxide, which will subsequently react with superoxide anions to generate toxic derivatives, allowing macrophages to proficiently kill numerous types of pathogens (Kaspers and Kaiser, 2014). Trefoil factor (TFF)-2 is a stable secretory protein expressed in gastrointestinal mucosa responsible for protecting the epithelial layer from insults, stabilizing the mucus layer and promoting the healing of the epithelium (Jiang et al., 2011).

Interferon (IFN)-γ is a vital pro-inflammatory cytokine that plays a central role in regulating the innate and adaptive immune responses, and is responsible for promoting T_H1 cell differentiation, suppressing T_H2 cell activity and enhancing innate immune cell activation and function (Kaiser and Stäheli, 2014). Expression of lipopolysaccharide-induced tumour necrosis factor-α (LITAF) is principally in the spleen of chickens as well as in intestinal intraepithelial lymphocytes. LITAF is a transcription factor that mediates the expression of members of the tumour necrosis factor (TNF) ligand superfamily (Hong et al., 2006b).

Members of the interleukin family take part in many aspects of immunity. Interleukin (IL)-1β is produced by activated macrophages and plays a role in inflammatory reactions and host defence (Kaiser and Stäheli, 2014). Produced by macrophages and dendritic cells after activation, IL-12 is responsible for driving inflammatory T_H1 responses (Kaiser and Stäheli, 2014). IL-12 is also capable of inducing IFN-γ production from T_H1 and NK cells. IL-1β, TNF-α and IL-12 facilitate innate immunity and associated inflammation (Elgert, 2009); however, an avian TNF-α gene has yet to be identified. IL-13 is a T_H2 cytokine that is predominant in the cytokine response initiated by an extracellular pathogen (Kaiser and Stäheli, 2014). IL-13 increases B cell proliferation and inhibits activation of macrophages through antagonizing IFN-γ mediated activities (Elgert, 2009). Transforming growth factor (TGF)-β plays an important role in immunoregulation (Kaiser and Stäheli, 2014), a cytokine with anti-inflammatory and immunosuppressive activities thought to prevent potential damage to the host from an excessive immune response (Elgert, 2009).

2.3 Gut-associated lymphoid tissues (GALT)

The primary function of the gastrointestinal tract is to digest and absorb nutrients in order to meet metabolic demands for normal growth and development, but it also acts as a vital barrier preventing the entry of several antigens and potentially harmful pathogens from the external environment (Kogut and Swaggerty, 2012; Smith et al., 2014). The gut is a major site of development, residence and portal of entry for pathogenic microorganisms. As such, any disruption of gut physiology is easily capable of resulting in substantial clinical consequences. Therefore, an effective immune capability in the gut is essential to cope with the plethora of potentially pathogenic microorganisms present (Smith et al., 2014). The GALT make up the largest component of the mucosa-associated lymphoid tissues (MALT) and are a significant source of immune cells that monitor and protect the mucosal layers of the intestine. The GALT are continuously being exposed to food antigens, microbiota and ingested pathogens (Yun et al., 2000). Protection of the gut is achieved through use of both the innate and adaptive immune systems.

The mucosal layer of the gut is comprised of the epithelium and the lamina propria. The epithelial layer is marked by the presence of mostly T lymphocytes, while the lamina propria is populated by a variety of immune cells including antibody-producing B lymphocytes (Dalloul and Lillehoj, 2006). Unlike the mammalian GALT, chickens do not possess lymph nodes; instead, they have scattered lymphoid aggregates as well as organized lymphoid structures, such as the bursa of Fabricius, caecal tonsils, Meckel's diverticulum and Peyer's patches (PP). The epithelium of the PP, as well as the other lymphoid structures of the GALT, contains areas occupied by phagocytic antigen sampling cells known as microfold

(M) cells (Muir et al., 2000). These M cells are responsible for taking up antigens from the lumen and delivering them to APCs. Upon encountering the antigen in the PP, B and T cells will mount their specific immune response in order to combat the invading pathogen (Smith et al., 2014).

In healthy birds, the immune system provides a layered defence mechanism with increasing specificity against particular pathogens. While costly, such responses tend to be measured and targeted to vaccines or other immunization methods. However, poultry are faced daily with several disease challenges and environmental stressors and their immune system must cope with the extensive array of antigens associated with such threats. More so are the mucosal sites particularly at the gastrointestinal level, where it is constantly exposed to foreign antigens. The defence system therefore must deal with this plethora of diverse menaces costing the host precious resources that would otherwise be directed towards growth and production. Among the constant threats to the industry are enteric diseases including coccidiosis, which also can predispose flocks to necrotic enteritis. The healthier and better-prepared birds respond more efficiently to challenges, and much evidence points to the significance of a healthy gut immune system. As traditional control strategies (e.g. use of antibiotic growth promoters) are questioned, alternatives are tested and employed to curtail the negative impacts of enteric diseases. Some of the case studies and results are presented below in the context of enteric pathogens, with a particular focus on the recent use of probiotics during coccidiosis challenges. To this end, an introduction to this impactful disease, the causative agents and the host immune responses is warranted with reference to a number of resources that the readers can further explore.

3 Coccidiosis

Coccidiosis is an economically devastating parasitic disease of the poultry industry caused by the development and reproduction of several species of the *Eimeria* protozoan within the intestine, resulting in an estimated cost of $3 billion annually worldwide (Dalloul and Lillehoj, 2006). The chicken is host to seven species of *Eimeria*, including the most common *E. acervulina*, *E. maxima* and *E. tenella*, each of which infects a specific area of the gut and invades the IECs, resulting in varying levels of tissue damage and morbidity (McDonald and Shirley, 2009). Depending on the species, magnitude and site of infection, coccidiosis can result in a limited enteritis resulting in fluid loss and malabsorption of nutrients (typically due to *E. acervulina* and *E. mitis*), inflammation of the intestinal wall with pinpoint haemorrhages and sloughing of epithelia (as seen with *E. brunetti* and *E. maxima*), or a complete villus destruction, leading to extensive haemorrhage and death (encountered with *E. necatrix* and *E. tenella*) (Chapman, 2014). The disruption of the intestinal epithelial layer naturally leads to the diminished ability of the intestine to absorb nutrients, resulting in reduced performance and higher susceptibility to other diseases, such as necrotic enteritis (Yegani and Korver, 2008).

Eimeria sp. parasites present a unique challenge to avian immunity as they possess a complex multi-stage life cycle, each expressing numerous antigens to which the host immune system generates specific responses. Birds become infected through the ingestion of sporulated *Eimeria* oocysts from contaminated feed, water or litter (Li et al., 2005). As the oocyst wall is disrupted in the gizzard, sporocysts are released and broken down with the aid of trypsin and bile in the duodenum, releasing sporozoites. The sporozoites then invade the IECs where they develop into schizonts containing many offspring called merozoites. This developmental stage is capable of breaking out of the epithelial cells, invading other cells and asexually replicating further. Due to this cyclic reproduction, many of the epithelial cells are destroyed. Once asexual reproduction ceases, the later generation merozoites develop into sexual male and female gametocytes, identified as microgametes and macrogametes, respectively. The microgamete subsequently fertilizes the macrogamete to produce a zygote. The zygote matures into an oocyst, ruptures the intestinal cell, and is passed with the faeces (Blake and Tomley, 2014). Once outside the host, the oocysts will sporulate when environmental conditions such as temperature, humidity and oxygen become conducive to growth and once again become infective (Williams, 2002). The entire lifecycle from ingestion to release may take 4–6 days to complete (McDougald, 1998; Allen and Fetterer, 2002). With each stage, the bird's gut immune system responds differently based on the antigens presented eventually leading to memory that protects the bird against homologous reinfection. Further, the bird must cope with every species via generating separate sets of immune memory cells.

3.1 Coccidiosis prevention and treatment

Currently, most poultry producers rely on prophylactic measures such as anticoccidial drugs and vaccines to prevent coccidiosis in their flocks. Anticoccidial drugs, commonly referred to as coccidiostats, are administered in the feed at low doses during the grow-out period, with a withdrawal period prior to market age and weight. There are two classes of coccidiostats: chemical and ionophore. Chemical anticoccidials have specific modes of action against parasite metabolism, while ionophore anticoccidials work by disrupting osmotic balance and altering ion transport (McDougald, 1998). However, *Eimeria* species have developed resistance to both chemical and ionophore drugs over time (Stringfellow et al., 2011). These means of control are considered effective still only because parasite growth is suppressed sufficiently to allow natural immunity to develop (Blake and Tomley, 2014). Due to drug resistance and consumer concerns regarding drug usage, the practice of live vaccines to control coccidiosis has greatly increased. Vaccines provide an alternative for disease protection, capable of limited efficacy as they induce specific protective immunity by exposing the chicken's immune system to *Eimeria* antigens (Williams, 2002; Dalloul and Lillehoj, 2005; Stringfellow et al., 2011). Immunity is subsequently boosted and maintained by multiple re-infections caused by oocysts present in the litter due to shedding and ingestion (Williams, 2002; Chapman et al., 2005). Li and colleagues (2005) found that birds receiving a coccidia vaccine without therapeutic levels of an ionophore anticoccidial in the feed had less severe lesion scores in the upper and middle intestinal segments when compared to non-vaccinated, medicated birds following challenge with three *Eimeria* species. A second study found that birds given one of three different coccidia vaccine doses had less severe lesion scores than the positive non-vaccinated control (Li et al., 2005). Fewer and less severe lesion scores are indicative of less damage to the intestinal epithelium, leading to infected birds having a greater chance of recovery from disease. However, one drawback to live vaccines is that immunity to avian coccidia is strongly species-specific; therefore, the bird will only develop immunity to the species of *Eimeria* present in the vaccine (Williams, 2002; Dalloul and Lillehoj, 2006). This specificity mandates that the vaccine contain the species of *Eimeria* known to be prevalent in a particular region and therefore more likely to cause an outbreak (Dalloul and Lillehoj, 2005). The necessary early administration of vaccines (typically during the first 1–7 days of life) becomes a second disadvantage regarding live vaccines. Early exposure of the chick's immune system to antigens results in immunity developing at a younger age, minimizing risk of exposure while the chick is unprotected. However, administration of live oocysts in a vaccine at a young age can result in a low level infection, which can cause an early reduction in growth and may increase the chick's susceptibility to secondary infections, such as necrotic enteritis (Dalloul and Lillehoj, 2005; Li et al., 2005; Stringfellow et al., 2011). Even though these methods are generally considered to be successful, due to the issues related to the use of anticoccidial drugs and vaccines, as well as the impending ban on animal feed additives, research has recently focused on more 'natural' means of controlling and managing coccidiosis.

3.2 Host immune response to *Eimeria*

The complex life cycle of coccidia elicits a number of immunological responses involving both the innate and adaptive immune systems, with each response being specific to the *Eimeria* species involved. Prior to the activation of an adaptive immune response, the innate immune system of a naïve host will attempt to prohibit the *Eimeria* infection through various pathways such as competitive exclusion by commensal microbiota, gastric secretions, phagocytosis, and complement components. The GALT play an invaluable role in protecting the host and bridging the innate and adaptive immune responses. The GALT provide protection to the host by performing three main functions: processing and presenting antigens, producing intestinal antibodies by activating the humoral immune system, and activating CMI (Yun et al., 2000; Dalloul and Lillehoj, 2005).

3.2.1 Humoral immune response

The intestine is considered to be the largest immunological organ containing approximately 70–80% of the total immunoglobulin-producing cells, with IgA and IgM being the predominant isotypes found within intestinal secretions (Yun et al., 2000). Chickens infected with *Eimeria* produce increased levels of parasite-specific antibodies in response to the challenge, confirming that *Eimeria* parasites promote activation of the humoral immune system (Lillehoj and Lillehoj, 2000). However, the role of the humoral immune response in protecting the bird is still not well understood. This is partly attributed to studies using hormonally and chemically bursectomized chickens that were resistant to reinfection, thus suggesting that antibodies play a lesser role in protecting the host by potentially reducing, not eliminating, the pathogen (Lillehoj and Trout, 1996; Dalloul and Lillehoj, 2006).

3.2.2 Cell-mediated immune response

Cell-mediated immune responses are thought to be the most effective against intracellular parasites such as *Eimeria*. Following an exposure to *Eimeria*, proliferation and infiltration of T lymphocytes, especially CD8+ T lymphocytes, are increased at the site of infection. Elimination of CD8+ T lymphocytes results in aggravated severity of coccidiosis and greater oocyst shedding, demonstrating their importance in providing immunity to the disease. Suppression of T lymphocyte function also resulted in impaired immunity (Dalloul and Lillehoj, 2006). Higher levels of CD8+ T lymphocytes were also found to be related to reduced oocyst shedding (Bessay et al., 1996). Furthermore, in chickens, splenocytes and peripheral blood lymphocytes from immune birds can transfer resistance to infection to naïve birds. Though the role of the CD8+ T lymphocytes in controlling *Eimeria* infections is evident, it has been suggested that CD4+ T lymphocytes are important during primary infections while CD8+ cells are essential during secondary infections. Chickens treated with anti-CD4 antibodies shed more oocysts following a primary infection of *E. tenella* (Trout and Lillehoj, 1996). Single infections of *E. maxima* or *E. tenella* have led to an increase of CD4+ T cells in the duodenum and caeca, respectively (Cornelissen et al., 2009). Hong et al. (2006c) reported a rise in CD4+ and CD8+ cells following an initial exposure to *E. maxima*. These results suggest that the different subsets of T lymphocytes work together in order to clear initial infections of *Eimeria* and provide enhanced protection against future encounters.

The response of T lymphocytes against an *Eimeria* infection is predominantly controlled and regulated by cytokines. Following primary *E. maxima* infection, mRNAs of the T_H1 cytokines IFN-γ, IL-1ß, IL-6, IL-12, IL-15, IL-17 and IL-18 are upregulated, as well as those of T_H2 cytokines IL-4, IL-10, IL-13 and granulocyte macrophage colony-stimulating factor (GM-CSF) (Hong et al., 2006c). In a similar study, an initial infection of *E. acervulina* resulted in the upregulation of IFN-γ, IL-2, IL-12, IL-15, IL-16, IL-18, and GM-CSF and downregulation of IL-4 and IL-13 mRNAs (Hong et al., 2006a). Similarly, Cornelissen et al. (2009) reported that *E. acervulina* or *E. tenella* infection upregulated mRNA expression of the T_H1 cytokines IL-2, IL-18 and IFN-γ. However, they observed an increase in gene expression of T_H2 cytokines IL-4 and IL-10 as well as the chemokine IL-8 (Cornelissen et al., 2009). These data further confirm the importance of both the T_H1 and T_H2 subsets in controlling *Eimeria* infections.

4 Probiotics and poultry

Prophylactic use of anticoccidial feed additives is currently the most common method of controlling coccidiosis. However, due to increasing concerns regarding drug use and high costs of vaccines, alternative control methods have taken the forefront in the research community. The use of an immunomodulator to manipulate the immune system is currently the most promising alternative. Probiotic supplementation is one option currently being explored as a means of reducing the amount and severity of enteric diseases in poultry and subsequent contamination of poultry products for human consumption (Rolfe, 2000; Patterson and Burkholder, 2003; Eckert et al., 2010). A probiotic is defined as 'a live microbial feed supplement, which beneficially affects the host animal by improving intestinal balance' (Fuller, 1989). The use of probiotics is based on the understanding that a well-balanced gut microbiota is crucial for optimal animal health and performance. Alterations in the microbial profile can influence all aspects of the gastrointestinal tract, including development, physiology, immunology and resistance to enteric infections, through interactions of the microorganisms with the intestinal lining and lymphoid tissues.

4.1 Modes of action

Probiotics are used to help maintain a healthy microbial balance within the intestine to promote gut integrity and immune health, and potentially prevent enteric disease. As the gut microbiota begins to establish within hours after the chick hatches, the earlier the introduction of non-pathogenic microorganisms, the more effective their establishment in the digestive tract (Timmerman et al., 2006; Torok et al., 2007; Cox and Dalloul, 2015). This is accomplished through three main mechanisms: competitive exclusion, bacterial antagonism, and stimulation of the immune system (Ohimain and Ofongo, 2012). Competitive exclusion is the idea that probiotic strains have the ability to inhibit establishment of pathogenic bacteria through competition for space, attachment sites and available nutrients by colonizing the intestinal tract themselves. Probiotic bacteria may also produce antimicrobial substances such as volatile fatty acids, bacteriocins and low pH that limit the growth and/or survival of pathogenic microbes (Hume, 2011). Bacteriocins are a small class of secreted peptides or proteins produced by bacteria that kill closely related bacterial strains by forming pores in their cellular membranes or hindering essential enzymes. Furthermore, many probiotics, predominantly those containing *Lactobacillus* species, lower the environmental pH through their production of lactic acid, which can also impede the growth of acid-sensitive microbes (Travers et al., 2011). One of the most important functions of probiotics is to stimulate the immune system against invading microorganisms as discussed below.

4.2 Performance and intestinal development

Key to being effective immunomodulators, probiotic supplementation needs to help performance that is the bottom line for producers (Kabir, 2009). Due to the reduction in the sub-therapeutic use of antibiotics and chemotherapeutic agents, poultry producers and researchers have been actively searching for a viable alternative that offers the same growth-promoting benefits while thwarting pathogenic threats. The status of the intestinal tract plays a major role in influencing the performance of poultry (Getachew, 2016). The microbiota present in the gut is an essential component of a healthy intestinal tract (Pan and Wu, 2014). The gut microbial profile can be manipulated through the use of probiotics in order to create conditions favourable to enhancing performance. Probiotics have been shown to result in improved body weight gain and feed conversion ratios in chickens (Kabir et al., 2004; Khaksefidi and Ghoorchi, 2006; Nayebpor et al., 2007; Talebi et al., 2008; Ignatova et al., 2009; Sen et al., 2012). Probiotic supplementation has also been shown to have growth-promoting effects comparable to those of avilamycin, further promoting it as a viable antibiotic alternative (Mountzouris et al., 2007). Despite the plethora of data demonstrating the positive effects of probiotics on performance, several researchers have reported no significant enhancements due to probiotic supplementation (Rahimi et al., 2011; Wolfenden et al., 2011; Getachew, 2016). These differences could be due to a variety of factors that can alter the efficacy of a probiotic such as strain(s) of bacteria utilized, composition and viability of the probiotic bacteria, and the preparation methods. Further, other factors may include probiotic dosage, method and/or frequency of application, overall diet, condition and age of the birds, potential drug interactions, as well as environmental stress factors such as temperature and stocking density (Patterson and Burkeholder, 2003; Mountzouris et al., 2007; Cox and Dalloul, 2015). More recently, administration of a commercial (lactobacilli- and bifidobacteria-based) probiotic as early as day 18 of embryonic age was shown not to impact hatchability and could enhance early performance of broilers during the first week post-hatch (Pender et al., 2017). This early application further demonstrated that such effects may be dose-dependent and the numbers of inoculated bacteria (colony-forming units) would need to be optimized.

Augmented performance due to probiotic supplementation could be attributed to enhanced gut development. The use of probiotics has been shown to increase the length of intestinal villi and decrease the depth of crypts in poultry (Samanya and Yamauchi, 2002; Marković et al., 2009). The increase in villus length suggests a greater surface area with increased absorptive capacity. The intestinal crypt is the site of continuous proliferation of enterocytes in order to replace cells lost at the villus tip due to normal sloughing or inflammation from insults (Awad et al., 2009). The depth of the intestinal crypt is directly correlated with epithelial cell turnover. Increased cellular turnover requires a substantial amount of energy that might otherwise be utilized towards growth. Thus, longer villi and shallower crypts are related to decreased cell replacement, longer enterocyte lifespan, and improved performance (Marković et al., 2009). Probiotics may also offer protection against damage to intestinal morphology caused by deleterious agents. For example, Awad and colleagues (2006) evaluated the effects of probiotics in broiler chicks fed diets contaminated with deoxynivalenol (DON), which altered the small intestine morphology by decreasing villus height and width in the duodenum and jejunum. Probiotic supplementation ameliorated the negative effects caused by the mycotoxins by diminishing villus atrophy (Awad et al., 2006).

4.3 Probiotics and innate immunity

Characterized by non-specific defence mechanisms, the innate immune system acts as the first line of defence to provide immediate protection against an impending foreign challenge. Several studies have provided evidence that specific strains of probiotics are able to stimulate various aspects of innate immunity. The avian heterophil is considered to be equivalent to the mammalian neutrophil, and is therefore the first line of cellular defence against a broad spectrum of microbial pathogens. These innate immune cells are highly phagocytic and utilize bactericidal mechanisms such as oxidative burst and degranulation to destroy phagocytized pathogens (Harmon, 1998). Farnell and colleagues (2006) found that heterophils isolated from broiler chicks fed the probiotic bacteria *Bacillus subtilis*, *Lactococcus lactis lactis* or *Lactobacillus acidophilus* elicited an increase in oxidative burst and degranulation. Similar results were seen in a later study where birds treated with a commercial probiotic (containing *Enterococcus faecium*, *Pediococcus acidilactici*, *Bifidobacterium animalis*, and *Lactobacillus reuteri*) had greater levels of heterophil oxidative burst (Stringfellow et al., 2011). Macrophages are phagocytic immune cells that play a key role in detecting and processing antigens, as well as dictating the differentiation of naïve lymphocytes into appropriate effector cells in order to defeat specific types of pathogens (Qureshi et al., 2000; Pan and Yu, 2014). Non-challenged broiler chicks given a commercially available *Lactobacillus*-based probiotic had an increased number of macrophages in the ileum and caeca, but those birds exposed to *Salmonella* Enteritidis had decreased numbers of macrophages in the ileum when given probiotics one hour after challenge. The reduction of macrophage numbers within the infected birds could be due to a decrease in the bacterial load caused by the addition of probiotics, potentially mediated by competitive exclusion. Furthermore, abdominal exudate cells from *Salmonella*-challenged birds had greater phagocytic capacity when treated with probiotics (Higgins et al., 2007a).

The upregulation of host innate immunity, including enhanced antimicrobial defences and phagocytosis, may result in the reduction and possible elimination of pathogenic invaders. Recently, administration of a probiotic to day-18 broiler embryos was shown to act as an immunomodulator by altering the expression of several immune-related genes within the ileum and caecal tonsils. For example, in addition to an initial upregulation of inducible nitric oxide synthase on day of hatch (3 days post inoculation), in ovo probiotic supplementation was associated with downregulated expression of Toll-like receptor-2 and -4, inducible nitric oxide synthase (iNOS), trefoil factor-2 (TFF-2) and mucin-2 (Muc-2), all innate immunity markers (Pender at al., 2017).

4.4 Probiotics and adaptive immunity

The adaptive immune system acts as a second line of defence and provides protection against re-offending pathogens. Recent reports demonstrate the importance of probiotics in potentiating the adaptive immune response in chickens. Probiotic supplementation increases antibody titres to sheep red blood cells, as well as many important disease agents such as Newcastle disease virus and infectious bursal disease virus (Kabir et al., 2004; Haghighi et al., 2005; Khaksefidi and Ghoorchi, 2006; Nayebpor et al., 2007; Apata, 2008; Karimi Torshizi et al., 2010). Furthermore, Haghighi et al. (2006) determined that female broilers orally gavaged with probiotics had enhanced production of natural antibodies. Probiotic-treated birds had increased levels of IgA reactive to tetanus toxoid, alpha-toxin, and bovine serum albumin, as well as increased levels of IgG against tetanus toxoid in the intestine. The administration of probiotics also augmented serum levels of IgG and IgM antibodies reactive to tetanus toxoid and alpha-toxin (Haghighi et al., 2006). In a study to compare the efficacy of the antibiotic avilamycin and probiotics on the humoral immune response, Mountzouris et al. (2009) found that both treatments resulted in reduced levels of plasma IgA and IgG and intestinal IgA against *Salmonella* Enteritidis when compared to the challenged control. In addition, the treated levels were not different from those of the negative controls. The reduction in antibody levels could be a result of enhanced clearance and accelerated recovery caused by their experimental treatments. It has also been shown that *Eimeria acervulina* infected birds fed probiotics produced more *Eimeria*-specific antibodies (Lee et al., 2007a,b).

Probiotics magnified the numbers of intestinal intraepithelial lymphocytes expressing the cell surface markers CD3, CD4 and CD8 (Dalloul et al., 2003; Noujaim et al., 2008). Supplementing probiotics through drinking water enhanced cellular immune response to a 1-chloro-2,4-dinitrobenzene (DNCB) challenge, while administration in feed or water improved cellular immunity to a phytohaemagglutinin-M (PHA-M) injection, demonstrated through increased skin thickness (Karimi Torshizi et al., 2010). *Lactobacillus*- and *Bacillus*-based probiotics can modulate the levels of several cytokines including pro-inflammatory cytokines (IL-1ß, IL-6, IL-17a, IL-18), Th1 cytokines (IFN-γ, IL-2, IL-12), and Th2 cytokines (IL-4, IL-10, IL-13). However, discrepancies have been noted due to differences in the probiotic strains used (Dalloul et al., 2005; Brisbin et al., 2010; Lee et al., 2010a; Pender et al., 2017). As such, generalization of specific effects on immune responses is cautioned and researchers are advised to limit interpretation of findings to within the scope of reported studies.

4.5 Probiotics and host defence against pathogens

The immune enhancing capabilities of probiotics have resulted in the clearance of several economically important pathogens such as *Eimeria* spp., *Salmonella* spp., *Escherichia coli* and *Clostridium perfringens*, further asserting their potential use as antibiotic alternatives (Dalloul and Lillehoj, 2005; Knap et al., 2010; Ritzi et al., 2014, 2016; Pender at al., 2016, 2017). The use of probiotics is based on the understanding that microbiota in the gut are an essential component involved in resistance to enteric infections. Probiotics aid in protection against a variety of economically important enteric diseases (Dalloul and

Lillehoj, 2005). Chicks fed diets supplemented with probiotics and subsequently infected with *E. acervulina* or *E. tenella* had significantly reduced oocyst shedding (Dalloul et al., 2003, 2005; Lee et al., 2007a,b), suggesting reduced susceptibility to infection. Further, *Bacillus*-based probiotics reduced lesion severity in *E. maxima* challenged broilers (Lee et al., 2010b). More recent work revealed a differential effect of the application method (feed vs. water) on performance and response of broiler chickens to an *Eimeria* challenge (Ritzi et al., 2014). Data indicated that water supplementation resulted in lower duodenal and jejunal lesion scores, indicating a healthier intestine and enhanced resistance to *Eimeria* sp. Also, birds that were intermittently given a water-based probiotic shed fewer oocysts in the faeces, although this was not a trend for all of the probiotic treatment groups (Ritzi et al., 2014). In a subsequent study, the authors reported that a combination of probiotics and a coccidiosis vaccine could enhance performance and provide an additional protective effect (reduced lesion scores) against a mixed *Eimeria* challenge (Ritzi et al., 2016). Pender et al. (2016) also reported reduced disease severity in birds inoculated with a water-soluble probiotic at embryonic day 18 followed by a coccidia challenge post hatch.

The addition of probiotics was also efficacious in reducing *Salmonella* colonization in the caeca, liver and spleen of broiler chicks (Revolledo et al., 2009). Furthermore, researchers have observed a reduction in lesion score severity, mortality, and numbers of *Clostridium perfringens* due to probiotic treatment during experimentally induced cases of necrotic enteritis (McReynolds et al., 2009). Numerous studies have found that probiotic supplementation leads to significant reductions in numbers of other intracellular pathogens present in the intestine, such as *Salmonella* Enteritidis and *Campylobacter jejuni* (Higgins et al., 2007b; Ghareeb et al., 2012). Such effect of probiotic supplementation could prove to be exceptionally beneficial to the bird, as some microorganisms such as *Clostridium* and *Salmonella* may exacerbate *Eimeria* infections and vice versa (Chapman et al., 2002). Ultimately, the reduction in the presence of intracellular pathogens is indicative of a healthier intestine, with minimal damage done to the epithelium. An intact intestinal epithelium serves as the vital barrier preventing entry of potential pathogens and results in proper nutrient absorption and utilization, leading to optimal health and performance of the bird. This handful of reported findings, which by no means is inclusive, suggests that maintaining a proper balance of intestinal microbiota through the use of probiotics could enhance the host immune response, and prove to be effective in preventing and potentially treating enteric infections as a suitable alternative to the use of chemotherapeutic agents. Similar to other 'health-promoting' products, probiotic formulations are unique to their development purposes and therefore should not be considered a commodity. The source and selection of species as well as strains for a particular formulation may convey specific functional traits that are different among probiotic products, thus rendering it difficult to obtain comparable outcomes in the field.

5 Conclusion and future trends

The poultry immune system is under constant pressure reacting to a myriad of challenges in commercial settings. Such responses are often costly yet rarely accounted for in production costs in terms of energy and nutrients that the host defences require for proper functioning and protection against insults on the farm. Mucosal pathogens present a particularly serious threat, especially at the gut level where digestion and absorption are key to the bird's performance. The small intestine is the largest lymphoid organ as well as a vital component of the digestive system. The integrity of the digestive tract is essential for protecting the host against enteric diseases such as coccidiosis. At such mucosal sites, the host immune system continuously evolves to deal with enteric and environmental antigens, resulting in either clearance of those threats or subclinical infections and stress. Consequently, the host regularly redirects significant resources to generating immune responses, maintaining the mucosal layer, replenishing enterocytes, etc. Additional pressures by the ban of growth-promoting doses of antibiotics and marketing of 'no antibiotics ever' poultry products have created an impetus for finding alternatives capable of maintaining animal health without impeding performance. Therefore, the role of the immune system has become even more critical and perhaps accounting for its proper development and nutritional demands needs to be more seriously addressed. Positive responses in immune stimulation or priming have been frequently reported by supplying feed or water additives. This is especially the case for probiotics, prebiotics, and classical nutrients such as amino acids and vitamins. Modulating the bird's immune responses based on particular field situations and needs may prove effective. Ample evidence shows the usefulness of probiotics as an alternative with their potential to modulate the immune system, and reduce the rate and severity of enteric infections while maintaining optimal performance in poultry.

The findings of various studies with probiotics, including those reported herein, support the protective effects provided by probiotic administration, both in feed and water applications, against enteric pathogens. Supplementation of probiotics can enhance performance in poultry and provide protection against several economically important diseases. Manipulation of the intestinal microbiota through the use of probiotics can also promote the development of the intestinal tract and boost its local immune system. Through numerous studies, the safety and efficacy of probiotic administration in challenge settings have been demonstrated. However, the underlying mechanisms of probiotic activity are multi-faceted with effects on both the innate and the subsequent adaptive immune responses. With such interventions, the role of commensal microbiota must be taken into consideration with a caution against the assumption that the addition of relatively few probiotic bacteria (i.e. probiotics) would have the metabolic capacity to modify that of the larger population of indigenous microbiota. Any resulting impact is expected to be relative and dependent on the status of commensal bacteria. Combining probiotics with prebiotics may even be more efficacious at stimulating defences and maintaining gut integrity, whereby beneficial organisms colonize mucosal linings and prebiotics provide substrates otherwise not used by the host. This interaction between various immunomodulators is being investigated by researchers in both industry and academia with encouraging outcomes. Pre-hatch immune interventions have also shown promise, yet the immune system and its response to modulatory and nutritional manipulations remain poorly understood. In this context, another caution is warranted in terms of avoiding 'over-stimulating' the local immune response, which may result in excessive inflammation and potential damage to the gut epithelium. In other words, excessive immunomodulation could lead to stronger than needed innate responses while adequate priming of local innate defences would yield better clearance of pathogenic elements. Therefore, 'positive immunomodulation' should result in a shortened response to challenges with limited collateral damage to epithelial integrity and reduced resources that would be reallocated to repair those tissues. Further, adequate innate immunity will help direct the appropriate adaptive response as needed, ending up with specific immunological memory to each challenge, which would swiftly deal with subsequent exposure to the same pathogens. With this in mind and based on the specific product(s) formulations and particular challenge models, including the genetics and environments of experimental or field birds, one should be careful measuring the proper parameters at the appropriate time points. This aspect is critical in the context of immune responses that tend to be fluctuant and transient, especially when poultry researchers have to face the reality of the limited immunological toolbox at their disposal. Therefore, better integrated and larger scale studies are needed to further elucidate the complex and intricate relationships between the host immune system and various well-defined modulators, which would culminate in consistent outcome and benefits to scientists and producers.

6 Where to look for further information

For further information related to the poultry immune system, readers are referred to *Avian Immunology* (2nd ed) edited by K. A. Schat, B. Kaspers, and P. Kaiser (Elsevier Ltd., London, UK). Other review articles cited throughout this chapter are also useful resources on further aspects not detailed hereto.

7 References

Allen, P. C. and R. H. Fetterer. 2002. Recent advances in biology and immunobiology of *Eimeria* species and in diagnosis and control of infection with these coccidian parasites of poultry. *Clin. Microbiol. Rev.* 15:58–65.

Apata, D. F. 2008. Growth performance, nutrient digestibility and immune response of broiler chicks fed diets supplemented with a culture of *Lactobacillus bulgaricus*. *J. Sci. Food Agric.* 88:1253–8.

Awad, W. A., J. Bohm, E. Razzazi-Fazeli, K. Ghareeb and J. Zentek. 2006. Effect of addition of a probiotic microorganism to broiler diets contaminated with deoxynivalenol on performance and histological alterations of intestinal villi of broiler chickens. *Poult. Sci.* 85:974–9.

Awad, W. A., K. Ghareeb, S. Abdel-Raheem and J. Bohm. 2009. Effects of dietary inclusion of probiotic and symbiotic on growth performance, organ weights, and intestinal histomorphology of broiler chickens. *Poult. Sci.* 88:49–56.

Bessay, M., Y. Le Vern, D. Kerboeuf, P. Yvore and P. Quere. 1996. Changes in intestinal intra-epithelial and systemic T-cell subpopulations after an *Eimeria* infection in chickens: comparative study between *E. acervulina* and *E. tenella*. *Vet. Res.* 27:503–14.

Blake, D. P. and F. M. Tomley. 2014. Securing poultry production from the ever-present *Eimeria* challenge. *Trends Parasitol.* 30:12–19.

Brisbin, J. T., J. Gong, P. Parvizi and S. Sharif. 2010. Effects of *Lactobacilli* on cytokine expression by chicken spleen and cecal tonsil cells. *Clin. Vaccine Immunol.* 17:1337–43.

Butcher, G. D. and R. D. Miles. 2001. *The Avian Immune System*. University of Florida Cooperative Extension Service Institute of Food and Agriculture Sciences EDIS: Gainesville, FL.

Chapman, H. D. 2014. Milestones in avian coccidiosis research: a review. *Poult. Sci.* 93:501–11.

Chapman, H. D., T. E. Cherry, H. D. Danforth, G. Richards, M. W. Shirley and R. B. Williams. 2002. Sustainable coccidiosis control in poultry production: the role of live vaccines. *Int. J. Parasitol.* 32:617–29.

Chapman, H. D., B. Roberts, M. W. Shirley and R. B. Williams. 2005. Guidelines for evaluating the efficacy and safety of live anticoccidial vaccines, and obtaining approval for their use in chickens and turkeys. *Avian Pathol.* 34:279–90.

Cornelissen, J. B., W. J. Swinkels, W. A. Boersma and J. M. Rebel. 2009. Host response to simultaneous infections with *Eimeria acervulina, maxima* and *tenella*: a cumulation of single responses. *Vet. Parasitol.* 162:58–66.

Cox, C. M. and R. A. Dalloul. 2015. Immunomodulatory role of probiotics in poultry and potential in ovo application. *Benef. Microbes* 6:45–52.

Dalloul, R. A. and H. S. Lillehoj. 2005. Recent advances in immunomodulation and vaccination strategies against coccidiosis. *Avian Dis.* 49:1–8.

Dalloul, R. A. and Lillehoj, H. S. 2006. Poultry coccidiosis: recent advancements in control measures and vaccine development. *Expert Rev. Vaccines* 5:143–63.

Dalloul, R. A., H. S. Lillehoj, T. A. Shellem and J. A. Doerr. 2003. Enhanced mucosal immunity against *Eimeria acervulina* in broilers fed a *Lactobacillus*-based probiotic. *Poult. Sci.* 82:62–6.

Dalloul, R. A., H. S. Lillehoj, N. M. Tamim, T. A. Shellem and J. A. Doerr. 2005. Induction of local protective immunity to *Eimeria acervulina* by a *Lactobacillus*-based probiotic. *Comp. Immunol. Microbiol. Infect. Dis.* 28:351–61.

Eckert, N. H., J. T. Lee, D. Hyatt, S. M. Stevens, S. Anderson, P. N. Anderson, R. Beltran, G. Schatzmayr, M. Monhl and D. J. Caldwell. 2010. Influence of probiotic administration by feed or water on growth parameters of broilers reared on medicated and nonmedicated diets. *J. Appl. Poult. Res.* 19:59–67.

Elgert, K. D. 2009. Cytokines, pp. 285–319. In K. D. Elgert (Ed.), *Immunology: Understanding the Immune System*. John Wiley & Sons, Inc.: Hoboken, NJ.

Farnell, M. B., A. M. Donoghue, F. Solis de los Santos, P. J. Blore, B. M. Hargis, G. Tellez and D. J. Donoghue. 2006. Upregulation of oxidative burst and degranulation in chicken heterophils stimulated with probiotic bacteria. *Poult. Sci.* 85:1900–6.

Fuller, R. 1989. Probiotics in man and animals. *J. Appl. Bacteriol.* 66:365–78.

Getachew, T. 2016. A review on the effects of probiotic supplementation in poultry performance and cholesterol levels of egg and meat. *J. World Poult. Res.* 6:31–6.

Ghareeb, K., W. A. Awad, M. Mohnl, R. Porta, M. Biarnés, J. Böhm and G. Schatzmayr. 2012. Evaluating the efficacy of an avian-specific probiotic to reduce the colonization of *Campylobacter jejuni* in broiler chickens. *Poult. Sci.* 91:1825–32.

Haghighi, H. R., J. Gong, C. L. Gyles, M. A. Hayes, B. Sanei, P. Parvizi, H. Gisavi, J. R. Chambers and S. Sharif. 2005. Modulation of antibody-mediated immune response by probiotics in chickens. *Clin. Diagn. Lab. Immunol.* 12:1387–92.

Haghighi, H. R., J. Gong, C. L. Gyles, M. A. Hayes, H. Zhou, B. Sanei, J. R. Chambers and S. Sharif. 2006. Probiotics stimulate production of natural antibodies in chickens. *Clin. Vaccine Immunol.* 13:975–80.

Harmon, B. G. 1998. Avian heterophils in inflammation and disease resistance. *Poult. Sci.* 77:972–7.

Härtle, S., K. E. Magor, T. W. Göbel, F. Davison and B. Kaspers. 2014. Structure and evolution of avian immunoglobulins, pp. 103–20. In K. A. Schat, B. Kaspers and P. Kaiser (Eds), *Avian Immunology* (2 ed). Elsevier Ltd.: London, UK.

Higgins, S. E., G. F. Erf, J. P. Higgins, S. N. Henderson, A. D. Wolfenden, G. Gaona-Ramirez, and B. M. Hargis. 2007a. Effect of probiotic treatment in broiler chicks on intestinal macrophage numbers and phagocytosis of *Salmonella* Enteritidis by abdominal exudate cells. *Poult. Sci.* 86:2315–21.

Higgins, J. P., S. E. Higgins, J. L. Vicente, A. D. Wolfenden, G. Tellez and B. M. Hargis. 2007b. Temporal effects of lactic acid bacteria probiotic culture on *Salmonella* in neonatal broilers. *Poult. Sci.* 86:1662–6.

Hong, Y. H., H. S. Lillehoj, S. H. Lee, R. A. Dalloul and E. P. Lillehoj. 2006a. Analysis of chicken cytokine and chemokine gene expression following *Eimeria acervulina* and *Eimeria tenella* infections. *Vet. Immunol. Immunopathol.* 114:209–23.

Hong, Y. H., H. S. Lillehoj, S. H. Lee, D. W. Park and E. P. Lillehoj. 2006b. Molecular cloning and characterization of chicken lipopolysaccharide-induced TNF-alpha factor (LITAF). *Dev. Comp. Immunol.* 30:919–29.

Hong, Y. H., H. S. Lillehoj, E. P. Lillehoj and S. H. Lee. 2006c. Changes in immune-related gene expression and intestinal lymphocyte subpopulations following *Eimeria maxima* infection of chickens. *Vet. Immunol. Immunopathol.* 114:259–72.

Hume, M. E. 2011. Historic perspective: prebiotics, probiotics, and other alternatives to antibiotics. *Poult. Sci.* 90:2663–9.

Ignatova, M., V. Sredkova and V. Marasheva. 2009. Effect of dietary inclusion of probiotic on chickens performance and some blood indices. *Biotech. Anim. Husbandry* 25:1079–85.

Jenkins, K. A., A. G. D. Bean and J. W. Lowenthal. 2007. Avian genomics and the innate immune response to viruses. *Cytogenet. Genome Res.* 117:207–12.

Jiang, Z., A. C. Lossie and T. J. Applegate. 2011. Evolution of trefoil factor(s): genetic and spatio-temporal expression of trefoil factor 2 in the chicken (*Gallus gallus domesticus*). *PLoS One* 6:e22691.

Juul-Madsen, H. R., B. Viertlboeck, S. Härtle, A. L. Smith and T. W. Göbel. 2014. Innate immune responses, pp. 121–47. In K. A. Schat, B. Kaspers and P. Kaiser (Eds), *Avian Immunology* (2nd ed). Elsevier Ltd.: London, UK.

Kabir, S. M. L., M. M. Rahman, M. B. Rahman, M. M. Rahman and S. U. Ahmed. 2004. The dynamics of probiotics on growth performance and immune response in broilers. *Int. J. Poult. Sci.* 3:361–4.

Kabir, S. M. L. 2009. The role of probiotics in the poultry industry. *Int. J. Mol. Sci.* 10:3531–46.

Kaiser, P. and P. Stäheli. 2014. Avian cytokines and chemokines, pp. 189–204. In K. A. Schat, B. Kaspers and P. Kaiser (Eds), *Avian Immunology* (2nd ed). Elsevier Ltd.: London, UK.

Karimi Torshizi, M. A., A. R. Moghaddam, S. Rahimi and N. Mojgani. 2010. Assessing the effect of administering probiotics in water or as a feed supplement on broiler performance and immune response. *Br. Poult. Sci.* 51:178–84.

Kaspers, B. and P. Kaiser. 2014. Avian antigen presenting cells, pp. 169–88. In K. A. Schat, B. Kaspers and P. Kaiser (Eds), *Avian Immunology* (2nd ed). Elsevier Ltd.: London, UK.

Khaksefidi, A. and T. Ghoorchi. 2006. Effect of probiotic on performance and immunocompetence in broiler chicks. *J. Poult. Sci.* 43:296–300.

Knap, I., B. Lund, A. B. Kehlet, C. Hofacre and G. Mathis. 2010. *Bacillus licheniformis* prevents necrotic enteritis in broiler chickens. *Avian Dis.* 54:931–5.

Kogut, M. H. and C. L. Swaggerty. 2012. Effects of prebiotics and probiotics on the host immune response, pp. 61–72. In T. R. Callaway and S. C. Ricke (Eds), *Direct-Fed Microbials and Prebiotics for Animals: Science and Mechanisms of Action*. Springer Science and Business Media, LLC.

Lee, H. K. and A. Iwasaki. 2007. Innate control of adaptive immunity: dendritic cells and beyond. *Semin. Immunol.* 19:48–55.

Lee, K. W., S. H. Lee, H. S. Lillehoj, G. X. Li, S. I. Jang, U. S. Babu, M. S. Park, D. K. Kim, E. P. Lillehoj, A. P. Neumann, T. G. Rehberger and G. R. Seragusa. 2010a. Effects of direct-fed microbials on growth performance, gut morphometry, and immune characteristics in broiler chickens. *Poult. Sci.* 89:203–16.

Lee, K. W., H. S. Lillehoj, S. I. Jang, G. Li, S. H. Lee, E. P. Lillehoj and G. R. Siragusa. 2010b. Effect of *Bacillus*-based direct fed microbials on *Eimeria maxima* infection in broiler chickens. *Comp. Immunol. Microbiol. Infect. Dis.* 33:105–10.

Lee, S., H. S. Lillehoj, D. W. Park, Y. H. Hong and J. J. Lin. 2007a. Effects of *Pediococcus*- and *Saccharomyces*-based probiotic (MitoMax) on coccidiosis in broiler chickens. *Comp. Immunol. Microbiol. Infect. Dis.* 30:261–8.

Lee, S. H., H. S. Lillehoj, R. A. Dalloul, D. W. Park, Y. H. Hong and J. J. Lin. 2007b. Influence of *Pediococcus*-based probiotic on coccidiosis in broiler chickens. *Poult. Sci.* 86:63–6.

Li, G. Q., S. Kanu, S. M. Xiao and F. Y. Xiang. 2005. Responses of chickens vaccinated with a live attenuated multi-valent ionophore-tolerant *Eimeria* vaccine. *Vet. Parasitol.* 129:179–86.

Lillehoj, H. S. and E. P. Lillehoj. 2000. Avian coccidiosis. A review of acquired intestinal immunity and vaccination strategies. *Avian Dis.* 44:408–25.

Lillehoj, H. S. and J. M. Trout. 1996. Avian gut-associated lymphoid tissues and intestinal immune responses to *Eimeria* parasites. *Clin. Microbiol. Rev.* 9:349–60.

Marković, R., D. Šefer, M. Krstić and B. Petrujkić. 2009. Effect of different growth promoters on broiler performance and gut morphology. *Arch. Med. Vet.* 41:163–9.

McDonald, V. and M. W. Shirley. 2009. Past and future: vaccination against *Eimeria*. *Parasitology* 136:1477–89.

McDougald, L. R. 1998. Intestinal protozoa important to poultry. *Poult. Sci.* 77:1156–8.
McReynolds, J., C. Waneck, J. Byrd, K. Genovese, S. Duke and D. Nisbet. 2009. Efficacy of multistrain direct-fed microbial and phytogenic products in reducing necrotic enteritis in commercial broilers. *Poult. Sci.* 88:2075–80.
Mountzouris, K. C., C. Balaskas, I. Xanthakos, A. Tzivinikou and K. Fegeros. 2009. Effects of a multi-species probiotic on biomarkers of competitive exclusion efficacy in broilers challenged with *Salmonella* Enteritidis. *Br. Poult. Sci.* 50:467–78.
Mountzouris, K. C., P. Tsirtsikos, E. Kalamara, S. Nitsch, G. Schatzmayr and K. Fegeros. 2007. Evaluation of the efficacy of a probiotic containing *Lactobacillus, Bifidobacterium, Enterococcus,* and *Pediococcus* strains in promoting broiler performance and modulating cecal microflora composition and metabolic activities. *Poult. Sci.* 86:309–17.
Muir, W. I., W. L. Bryden and A. J. Husband. 2000. Immunity, vaccination and the avian intestinal tract. *Dev. Comp. Immunol.* 24:325–42.
Nayebpor, M., P. Farhoman and A. Hashemi. 2007. Effects of different levels of direct fed microbials (Primalac) on growth performance and humoral immune response in broiler chickens. *J. Anim. Vet. Adv.* 6:1308–13.
Noujaim, J. C., R. L. Andreatti Filho, E. T. Lima, A. S. Okamoto, R. L. Amorim and R. T. Neto. 2008. Detection of T lymphocytes in intestine of broiler chicks treated with *Lactobacillus* spp. and challenged with *Salmonella enterica* serovar Enteritidis. *Poult. Sci.* 87:927–33.
Ohimain, E. I. and R. T. S. Ofongo. 2012. The effect of probiotic and prebiotic feed supplementation on chicken health and gut microflora: a review. *Int. J. Anim. Vet. Adv.* 4:135–43.
Oswald, I. P. 2006. Role of intestinal epithelial cells in the innate immune defence of the pig intestine. *Vet. Res.* 37:359–68.
Pan, D. and Z. Yu. 2014. Intestinal microbiome of poultry and its interaction with host and diet. *Gut Microbes* 5:108–19.
Patterson, J. A. and K. M. Burkholder. 2003. Application of prebiotics and probiotics in poultry production. *Poult. Sci.* 82:627–31.
Pender, C. M., S. Kim, T. D. Potter, M. M. Ritzi, M. Young and R. A. Dalloul. 2016. Effects of in ovo supplementation of probiotics on performance and immunocompetence of broiler chicks to an *Eimeria* challenge. *Benef. Microbes* 7:699–705.
Pender, C. M., S. Kim, T. D. Potter, M. M. Ritzi, M. Young and R. A. Dalloul. 2017. In ovo supplementation of probiotics and its effects on performance and immune-related gene expression in broiler chicks. *Poult. Sci.* 96:1052–62.
Pitman, R. S. and R. S. Blumberg. 2000. First line of defense: the role of the intestinal epithelium as an active component of the mucosal immune system. *J. Gastroenterol.* 35:805–14.
Qureshi, M. A., C. L. Heggen and I. Hussain. 2000. Avian macrophage: effector functions in health and disease. *Dev. Comp. Immunol.* 24:103–19.
Rahimi, S., S. Kathariou, J. L. Grimes and R. M. Siletzky. 2011. Effect of direct-fed microbials on performance and *Clostridium perfringens* colonization of turkey poults. *Poult. Sci.* 90:2656–62.
Revolledo, L., C. S. Ferreira and A. J. Ferreria. 2009. Prevention of *Salmonella* Typhimurium colonization and organ invasion by combination treatment in broiler chicks. *Poult. Sci.* 88:734–43.
Ritzi, M. M., W. Abdelrahman, M. Mohnl and R. A. Dalloul. 2014. Effects of probiotics and application methods on performance and response of broiler chickens to an *Eimeria* challenge. *Poult. Sci.* 93:2772–8.
Ritzi, M. M., W. Abdelrahman, K. van-Heerden, M. Mohnl, N. W. Barrett, and R. A. Dalloul. 2016. Combination of probiotics and coccidiosis vaccine enhances protection against an *Eimeria* challenge. *Vet. Res.* 47:111.
Rolfe, R. D. 2000. The role of probiotic cultures in the control of gastrointestinal health. *J. Nutr.* 130:396S–402S.
Samanya, M. and K. Yamauchi. 2002. Histological alterations of intestinal villi in chickens fed dried *Bacillus subtilis* var. *natto. Comp. Biochem. Physiol. A. Mol. Integr. Physiol.* 133:95–104.
Schat, K. A., B. Kaspers and P. Kaiser. 2014. *Avian Immunology* (2nd ed). Elsevier Ltd.: London, UK.

Sen, S., S. L. Ingale, Y. W. Kim, J. S. Kim, K. H. Kim, J. D. Lohakare, E. K. Kim, H. S. Kim, M. H. Ryu, I. K. Kwon and B. J. Chae. 2012. Effect of supplementation of *Bacillus subtilis* LS 1–2 to broiler diets on growth performance, nutrient retention, caecal microbiology and small intestinal morphology. *Res. Vet. Sci.* 93:264–8.

Smith, A. L., C. Powers and R. Beal. 2014. The avian enteric immune system in health and disease, pp. 227–50. In K. A. Schat, B. Kaspers and P. Kaiser (Eds), *Avian Immunology* (2nd ed). Elsevier Ltd.: London, UK.

Stringfellow, K., D. Caldwell, J. Lee, M. Mohnl, R. Beltran, G. Schatzmayr, S. Fitz-Coy, C. Broussard and M. Farnell. 2011. Evaluation of probiotic administration on the immune response of coccidiosis-vaccinated broilers. *Poult. Sci.* 90:1652–8.

Talebi, A., B. Amerzadeh, B. Mokhtari and H. Gahri. 2008. Effects of a multi-strain probiotic (Primalac) on performance and antibody responses to Newcastle disease virus and infectious bursal disease virus vaccination in broiler chickens. *Avian Pathol.* 37:509–12.

Timmerman, H. M., A. Veldman, E. van den Elsen, F. M. Rombouts and A. C. Beynen. 2006. Mortality and growth performance of broilers given drinking water supplemented with chicken-specific probiotics. *Poult. Sci.* 85:1383–8.

Torok, V. A., K. Ophel-Keller, R. J. Hughes, R. Forder, M. Ali and R. Macalpine. 2007. Environment and age: impact on poultry gut microflora. *Aust. Poult. Sci. Symp.* 19:149–52.

Travers, M. A., I. Florent, L. Kohl and P. Grellier. 2011. Probiotics for the control of parasites: an overview. *J. Parasitol. Res.* 2011:610769.

Trout, J. M. and H. S. Lillehoj. 1996. T lymphocyte roles during *Eimeria acervulina* and *Eimeria tenella* infections. *Vet. Immunol. Immunopathol.* 53:163–72.

Viertlboeck, B. and T. W. Göbel. 2014. Avian T cells: antigen recognition and lineages, pp. 91–102. In K. A. Schat, B. Kaspers and P. Kaiser (Eds), *Avian Immunology* (2nd ed). Elsevier Ltd.: London, UK.

Williams, R. B. 2002. Anticoccidial vaccines for broiler chickens: pathways to success. *Avian Pathol.* 31:317–53.

Wolfenden, R. E., N. R. Pumford, M. J. Morgan, S. Shivaramaiah, A. D. Wolfenden, C. M. Pixley, J. Green, G. Tellez and B. M. Hargis. 2011. Evaluation of selected direct-fed microbial candidates on live performance and *Salmonella* reduction in commercial turkey brooding houses. *Poult. Sci.* 90:2627–31.

Yegani, M. and D. R. Korver. 2008. Factors affecting intestinal health in poultry. *Poult. Sci.* 87:2052–63.

Yun, C. H., H. S. Lillehoj and E. P. Lillehoj. 2000. Intestinal immune responses to coccidiosis. *Dev. Comp. Immunol.* 24:303–24.

Chapter 7

Competitive exclusion (CE) treatment to control pathogens in poultry

Carita Schneitz, Orion Pharma, Finland; and Martin Wierup, Swedish University of Agricultural Sciences (SLU), Sweden

1 Introduction

The gastrointestinal (GI) tract of a healthy adult bird harbours microbial communities that play important roles in maintaining the health and well-being of the host. The mature gut flora competes effectively with any invading organisms that may be harmful to avian or human health and can prevent them from colonizing the digestive tract. Microbial colonization of the GI tract begins soon after hatching (Mead, 2000; Revolledo et al., 2006), but it has been shown that introduction of bacteria into the GI tract can happen already in the pre-hatch state (Klasing, 1998). With coprophagous birds like the domestic fowl or turkey, the transfer of microorganisms is particularly efficient when newly hatched chicks are reared with the adults. This is not possible in industrialized production where chickens are hatched and reared under hygienic conditions. In consequence, the development of the protective microflora is considerably delayed, which leaves the newly hatched chicken exposed to a variety of enteric disease agents (Pivnick and Nurmi, 1982).

http://dx.doi.org/10.19103/AS.2016.0011.22

The high susceptibility of the young bird and its lack of competitive GI microflora were highlighted by Nurmi and Rantala in 1973. The Finnish broiler industry suffered from a severe outbreak of *Salmonella* when the majority of broiler flocks had become infected. At the same time the incidence of human cases caused by the same serotype increased dramatically. The problem with *Salmonella*-infected poultry flocks could not be solved by hygienic measures or by antibiotics, so research was started to find a different approach. In 1952 Milner and Schaffer had infected chicks with *S. typhimurium* and observed that the resistance of young chickens against *Salmonella* increases by age, but the researchers had no explanation for this. The theory of Nurmi was that newly hatched chickens are sensitive to infection by *Salmonella* and other enteropathogens because of delayed establishment of normal intestinal microflora due to lack of contact with the brooding hen and nest environment. Nurmi and his research group were the first to demonstrate that administration of intestinal contents of adult birds to newly hatched chickens prevented them from colonization by *Salmonella*. Their results have later been verified in numerous studies around the world (Snoeyenbos et al., 1978, 1979; Barnes et al., 1980a; Rigby and Pettit, 1980; Stavric et al., 1985; Corrier et al., 1991; Stern, 1994; Hofacre et al., 2000).

The Nurmi approach has been fully accepted and in many countries it is applied as a tool to control *Salmonella* in poultry and is referred to as the 'Nurmi concept' or 'competitive exclusion (CE)' (Pivnick and Nurmi, 1982). Although 'competitive exclusion' means a process, it has been adopted as a term to describe the protective effect of the natural GI microflora in limiting the colonization of pathogenic bacteria.

The CE concept was originally designed to prevent the colonization of food-borne *Salmonellae* in the gut of young chickens but has with time been extended to also cover other bacteria potentially pathogenic to human and poultry like *Escherichia coli*, *Campylobacter*, *Clostridium perfringens* and *Listeria*.

In this chapter, the background of the CE concept, the nature of the protective mechanisms, experiences regarding the applicability of CE in poultry management as well as factors that may affect the efficacy of the treatment, are reviewed. In addition, strategies for safe usage and licensing of CE products are discussed.

2 The CE principle

2.1 History

The first observations that normal intestinal microflora exerts an inhibitory effect on invading pathogens originate from the 1920s (Meynell and Subbaiah, 1963). Meynell (1963) observed that streptomycin destroyed the mechanism responsible for elimination of *Salmonellae* from the gut of adult mice. He concluded that *S. typhimurium* given orally to a mouse 'failed to multiply in the normal mouse gut because of the combined inhibitory effects of a low oxidation-reduction potential and short-chain fatty acids produced by the normal gut flora'. Van der Waaij (1971) who studied the susceptibility of antibiotic-treated mice to oral challenge with *Escherichia coli, Klebsiella pneumoniae* and *Pseudomonas aeruginosa* and the repopulation of the mouse gut with the protective flora, came to the conclusion that 'colonization resistance or CR', a term synonymous to CE, was due to the presence of those anaerobic bacteria that survived the antibiotic treatment and repopulated the intestinal tract.

Milner and Schaffer (1952) had observed that the resistance of young chicks to *Salmonella* colonization increases with age, so that 2-week-old birds were more difficult to infect, even with relatively high doses of *Salmonella*, but there was no explanation of the phenomenon at that time. The answer was provided by a Finnish research group two decenniums later.

The broiler industry in Finland suffered from a widespread outbreak of *Salmonella infantis* infection in the beginning of the 1980s. More than 80% of the broiler flocks were found to be *Salmonella* infected and at the same time the incidence of human cases from this serotype increased considerably (Nurmi et al., 1992). To solve the problem of *Salmonella* infection in poultry, research was started at the former National Veterinary and Food Research Institute. Nurmi and his research group were the first to demonstrate experimentally that oral administration of intestinal contents from healthy, adult birds to newly hatched chicks prevented them from becoming colonized by *Salmonella* (Nurmi and Rantala, 1973). In addition, the protective microflora could be cultured for administration by a relatively simple method (Rantala and Nurmi, 1973). Competitive exclusion was used as a term for the first time by Greenberg in 1969 and in relation to poultry by Lloyd in 1974.

2.2 Development of the chicken gastrointestinal microflora

Bacterial species colonizing the chicken GI tract appear in communities and are dependent on each other and numerous environmental factors. Modern molecular biology, especially DNA sequencing technologies, have given the researches the ability to explore these communities at an unprecedented level (van der Wielen et al., 2002; Yeoman et al., 2012; Pan and Yu, 2014). Employing new methods, the chicken intestinal flora has been analysed for variations of the GI microflora among individuals in the same flock (van der Wielen et al., 2002), in response to anticoccidial and growth promoter treatments (Pedroso et al., 2006; Danzeisen et al., 2011), different genetic lines of chickens (Lumpkins et al., 2010) and interaction with host and diet (Pan and Yu, 2014). With regard to the composition of the chicken GI microflora, Apajalahti et al. (2004) found that 90% of the bacteria in the chicken GI tract represent previously unknown species and, in fact, Yeoman et al. (2012) claimed that the majority of all microbial species colonizing the chicken GI tract have not been cultivated.

Although, knowledge regarding the composition of the chicken GI microflora has changed tremendously along with the new technologies, the early cultivation-based studies give a good overall picture of the development of the chicken GI microflora, which normally changes with age. Within a few hours of hatching, coliforms, enterococci, and sometimes clostridia and bacilli have been found in the ceca and scattered through the rest of the GI tract. Lactobacilli become established by about the third day after the start of feeding (Barnes, 1972, 1977; Barnes et al., 1980b; Revolledo et al., 2006). Non-sporing anaerobic bacteria gradually outnumber the facultative anaerobes, but the flora continues to increase in complexity for several weeks (Barnes and Impey, 1970; Barnes et al., 1972b; Mead and Adams, 1975). While studying the chicken GI microflora between 2 and 6.5 weeks of life, Barnes et al. (1972) found that lactobacilli dominated in the small intestine and clostridia occurred regularly at lower levels, with the exception of *Clostridium perfringens*, which was isolated only occasionally. In the ceca, the total bacterial count was 10^{11} per gram and there was an observable succession of different bacterial types during the growth of the bird.

Some management practices allow rearing of chicks on used litter and these have been shown to harbour both less *Salmonellae* (Corrier et al., 1992) and generic *Clostridium perfringens* (Wei et al., 2013) than chicks reared on new litter. This may indicate a faster development of a protective GI microflora and could be of importance in maintaining a healthy, pathogen-free adult microflora from flock to flock.

2.3 Mechanisms of CE

The mechanism of competitive exclusion is still uncertain. Most probably it is a complex process with an integrated involvement of several factors. Simplified, there are four mechanisms by which one or more bacterial species may inhibit the growth or reduce the numbers of other bacterial species. These are, competition for bacterial receptor sites, creation of a restrictive physiological environment (like production of short-chain fatty acids), elaboration of antibiotic-like substances (e.g. bacteriocins produced by lactobacilli) and depletion of or competition for essential substrates (like fermentable carbohydrates) (Rolfe, 1991; Fons et al., 2000). However, their relative importance in the protective process is not known. There is only one generally accepted reliability and that is that protection depends upon the oral administration of viable bacteria (Mead, 2000). Further, as suggested by Rantala (1974), the protective microflora must be cultured under anaerobic conditions to be fully effective, which means involvement of strictly anaerobic bacteria in the exclusion process. On the other hand, it has also been shown that aerobically incubated cultures give protection equal to that obtained from anaerobic incubation against *S. typhimurium*, *S. infantis*, *S. agona* and *S. enteritidis* in newly hatched chicks (de Oliveira et al., 2000; Filho et al., 2003). This finding does not exclude the presence of strictly anaerobic bacteria because static broth cultures develop very low redox potentials due to the many facultative anaerobes in the inoculum (Barnes et al., 1980b).

The capacity to adhere to an epithelial surface is an important property for some microorganisms that colonize the GI tract of animals (Fuller and Turvey, 1971; Savage, 1983; Rolfe, 1991; Schneitz, 1993). Adhesion is also the first essential step in an infection process (Criado et al., 1989). Compared to other animals the length of the GI tract of a chicken is short. This is why the average passage time of feed is only 3.5 h (Hughes, 2008). To be able to colonize the GI tract, bacteria must either adhere to the gut mucosal layer or grow fast (Pan and Yu, 2014). Soerjadi et al. (1982b) demonstrated adherence of both *S. typhimurium* and native gut flora to the cecal mucosa in germ-free chickens using scanning electron microscopy.

According to Hume et al. (1998b), chicks challenged four hours after CE treatment are relatively well protected against *Salmonella* colonization. Further, Seuna (1979) showed that treatment given just one hour before administration of *Salmonella* conferred considerable protection against the challenge organism. A similar observation was made by Mead et al. (1989b) when studying the effect of CE treatment on the transmission of *S. enteritidis* in chick transport boxes. The importance of adherent bacteria for the protection was also supported by Stavric et al. (1987) who demonstrated that all bacteria relevant to protection were represented in the adhering microflora already 6 to 8 h after administration. Because chickens have been shown to be relatively well protected against an oral *Salmonella* challenge just few hours after CE treatment, the protection is thought to be primarily physical (Donoghue et al., 2006), which means that the protection is obviously provided by adherent bacteria in the treatment material.

Van der Wielen et al. (2000) showed that *Bacteroides* and *Eubacterium* spp. were established at stable levels in the cecal contents of chicks after 14 days and that propionic and butyric acids, metabolites typically produced by strictly anaerobic bacteria, were detected in 12- to 15-day-old birds. Simultaneously, a decrease in counts of Enterobacteriaceae was observed. When pure cultures of these organisms were grown in the presence of volatile fatty acids, growth rates declined as the acid concentrations were increased. Hume et al. (1998b) has also shown that chicks challenged only four hours after CE treatment are relatively well protected against *Salmonella* colonization and that significant increases in cecal propionic acid concentrations were observed within one day of the treatment. Correlation between the cecal propionic acid concentrations in relation to resistance against *Salmonella* colonization have also been shown by other researchers (Martin et al., 2000; Schneitz et al., 1998, 2016). But, as stated by Nisbet et al. (1996), it is unlikely that the propionic acid alone is responsible for the decrease in *Salmonella* colonization, but is more likely that other factors like competition for limiting nutrients are involved that result in propionic acid formation. However, the appearance of propionic acid in the cecal contents of CE-treated chicks indicates at least the presence of strictly anaerobic bacteria.

In addition, there are many other factors that may be involved in the CE process, for example, local immunity, pH and Eh (redox potential), peristalsis, diet and body temperature of the host, as well as inhibitory substances like H_2S and deconjugated bile acids (Meynell, 1963; Savage, 1983; Barnes et al., 1979; Bailey, 1987; Mead and Barrow, 1990; Corrier et al., 1995a,b; Klasing, 1998; Fons et al., 2000; Mead, 2002).

3 Development of treatment materials: undefined CE cultures

The initial results by Nurmi and Rantala (1973) have been verified around the world and scientists have also been encouraged to identify the bacterial species or combinations of species being responsible for the CE effect. These numerous experiments are largely based on either (1) the original contents from the GI tract, here called 'undefined treatment materials' or (2) combinations of pure cultures of different bacterial species and strains, here called 'defined treatment materials'. The overall purpose has been to develop and produce treatment materials with an optimal CE effect, to be used in practice to prevent *Salmonella* infections and possibly the colonization of other potential pathogens in chickens. This section will cover undefined treatment materials, while Section 4 will discuss defined treatment materials.

3.1 Experimental CE cultures

In their first work, Nurmi and Rantala (1973) used diluted contents of the crop and intestinal tract of adult chickens to protect newly hatched chickens against a challenge with *S. infantis*. A similar degree of protection was obtained using the third daily subculture from an anaerobic broth culture of intestinal contents (Rantala and Nurmi, 1973). The results of their studies have been confirmed over years by several research groups around the world (Snoeyenbos et al., 1978, 1979; Barnes et al., 1980a; Rigby and Pettit, 1980; Stavric et al., 1985; Corrier et al., 1991; Stern, 1994; Hofacre et al., 2000).

A modified VL (Viande Levure) broth is probably the most often used medium in the production of effective CE cultures (Barnes and Impey, 1971). The advantage of a VL broth culture was that it could be sub-cultured many times without losing its efficacy (Snoeyenbos et al., 1978; Mead and Impey, 1986). Later, VL broth was also used to produce effective continuous-flow cultures and to develop one of the commercial CE products (Corrier et al., 1995a; Nisbet et al., 1995). The benefit with these two culturing methods is that non-bacterial pathogens such as viruses and protozoa, which are unable to grow in bacteriological culture media, are likely to be diluted out.

Lloyd et al. (1977) observed an increased resistance of newly hatched chicks and turkey poults to colonization with *S. typhimurium* following oral treatment with filtered cecal material from an adult bird. Treatment material from other areas of the alimentary tract was less protective. Snoeyenbos et al. (1979) used faeces from a specific-pathogen-free (SPF) flock to minimize the risk of transferring any pathogens present from donor birds to the recipients. The gut microflora of the SPF flock originated from a group of conventional birds, specially selected as one offering good protection. This approach may solve the problem of insufficient protection caused by the delayed development of the intestinal microflora in SPF chickens (Coloe et al., 1984; Impey et al., 1984; Mead and Impey, 1986).

An innovative CE preparation in the format, known as vermicompost, was produced by feeding the earthworm *Eisenia foetida* on a mixture of fresh chicken faeces and vegetable matter, the faeces being obtained from an adult SPF bird (Spencer and Garcia, 1995). This approach was seen as a means of producing large quantities of protective material easily and inexpensively. Treatment material produced accordingly, when sprinkled on the chicken feed, significantly reduced the cecal *Salmonella* colonization.

3.2 Commercial CE products

To date five commercial CE products, Aviguard, AviFree, Broilact, Mucosal Starter Culture (MSC) and Preempt (or CF-3 or DeLoach 29), have been developed. They are all undefined cultures from the cecal contents and/or cecal wall of domestic fowl. MSC may no longer be produced commercially but is included here because of extensive and valuable scientific work.

Broilact was the first commercial CE product and was developed by Orion Corporation in Finland under the leadership of Professor Esko Nurmi. It was launched in Finland and Sweden in 1987. Competition for bacterial receptor sites on the intestinal mucosa is one of the suggested mechanisms of CE and Broilact is based on this hypothesis (Nurmi et al., 1987). Broilact is a freeze-dried, well-characterized product. Spore-forming bacteria were excluded from the basic inoculum and there is only one Gram-negative facultative anaerobic rod in the product, a well-characterized *Escherichia coli* which was chosen for its sensitivity to all tested antibiotics. Since 1991, all batches of Broilact originate from the washed cecal walls of a single, 33-week-old grandparent breeder hen reared in Swedish quarantine which applies extensive testing of the birds to ensure freedom of *Salmonella* (Wierup, 1995). Continued use of the same basic inoculum ensures that each production batch is comparable in composition, quality and safety.

Aviguard is a freeze-dried partially characterized product from whole cecal contents from adult, specific-pathogen-free (SPF) chickens. It was developed by Life-Care Products Ltd (now Microbial Developments Ltd), UK, and launched in 1995. Since 2007 it has been marketed by Schering-Plough Animal Health.

AviFree was developed by Alltech Ltd and launched in 1996. There is very little information concerning the composition of AviFree. But evidently, it is a freeze-dried undefined culture from whole cecal contents of adult chickens.

Preempt (or CF-3 or DeLoach 29) (MS Bioscience, Dundee, Illinois, USA) is a partially characterized mixed-culture preparation developed by Corrier et al. (1995b) using a continuous-flow culture system and low pH of the culture medium to select for certain facultative and obligate anaerobes. The starting material was a homogenate of cecal tissue and contents obtained from ten-week-old broiler chickens and the researchers reported to have isolated 29 different bacteria from the product (Corrier et al., 1995a; Nisbet et al., 1995). Milk Specialities Co. developed the commercial version of CF-3 (Burns, 1995).

The product named MSC (Mucosal Starter Culture), which was identical to MCE (Mucosal Competitive Exclusion), was developed by Stern et al. (1995). The preparation was derived from scrapings of washed ceca or cecal sections and incubated under anaerobic conditions in an appropriate culture medium (Stern, 1990).

4 Development of treatment materials: defined CE cultures

Substantial efforts have been laid down to identify a defined CE culture with an equal efficiency to prevent the colonization of *Salmonella* as observed for undefined CE cultures described above. According to Stavric (1992), the efficacy of a defined treatment material depends on the complexity of the bacterial mixture included. The most effective pure-culture preparations consisted of tens of different bacterial species or strains representing several genera (Impey et al., 1982, 1984; Stavric et al., 1985, 1987, 1991; Gleeson et al., 1989).

Protection against *Salmonella* was obtained with a mixture of 23 bacterial strains, including lactobacilli and facultative and obligate anaerobes isolated from cecal material (Barnes et al., 1980b). While Goren et al. (1984a) reported that a mixture of 295 obligate anaerobes failed to provide any protection against *Salmonella*, Impey et al. (1982, 1984) demonstrated protective activity equivalent to that of cecal cultures with mixtures of 48 and 65 organisms, respectively, in which the bacterial counts of *Bacteroides* dominated. Stavric et al. (1985) succeeded in protecting chicks against *Salmonella* using a mixture of 50 different bacterial strains. They also demonstrated that the protective effect was dose dependent and less protection was evident when the challenge dose of *Salmonella* was increased beyond 10^4 cfu per chick or when the number of different pure cultures in the mixture was reduced. The proportion of strains from individual genera in the study by Stavric et al. (1985) differed from that used by Impey et al. (1982).

Neither of the two defined culture mixtures of chicken origin by Impey et al. (1984) provided any protection against *Salmonella* colonization in turkey poults. Important observations were also reported by Gleeson et al. (1989) and Stavric et al. (1991), who found that mixtures of defined cultures prepared from stored isolates gradually became ineffective with time, the most stable mixtures containing 28 or 50 different bacteria. Loss of efficacy upon subculture of a defined culture preparation consisting of different *Lactobacillus* species was reported by Schneitz et al. (1993). Some of the protective ability of the 28 and 50 pure-culture mixtures was restored after passaging the mixtures in newly hatched chickens (Stavric, 1992). Nisbet et al. (1996), on the other hand, claimed that growing the organisms as a mixture under conditions of continuous-flow culture is a way of stabilizing their protective activity.

The anti-*Salmonella* activity of lactobacilli has been studied extensively, but the results are very conflicting (Soerjadi et al., 1981b; Weinack et al., 1985a; Hofacre et al., 2000; Hinton et al., 1991a; Stavric, 1992; Shin et al., 2002; La Ragione et al., 2004; Higgins et al., 2007; Zhang et al., 2007b; Menconi et al., 2011; Schneitz et al., 2016b). It has even been suggested that lactobacilli, which may help to confer protection against *Salmonella* when added together with the other organisms of a poultry cecal suspension, can increase *Salmonella* shedding of the chicks when administered alone (Barnes et al., 1980b).

Bielke et al. (2003) tested the anti-colonization efficacy of 24 facultative anaerobic bacteria against *S. enteritidis* in turkey poults. The bacteria were chosen for their ability to inhibit the growth of the challenge organism *in vitro*. Treatment-related protection ranged from 0 to 100% in three trials. For some reason, the greatest protection was gained with the lowest concentrations of the protective microflora in each trial. The efficacy of a defined CE culture, consisting of seven members of the family Enterobacteriaceae and two *Lactobacillus* spp., against *S. enteritidis* were tested in newly hatched chickens by Wolfenden et al. (2007). One experiment consisted of three treatment groups that received different doses of the treatment material which was first delivered by oral gavage followed by drinking water application. The chicks were challenged by oral intubation 2 d later and the trial was terminated 5 d post challenge. The percentage of *Salmonella*-positive chicks in the treated groups ranged from 20 to 55% when it was 88% in the control group. The highest percentage of *Salmonella*-positive chicks was found in the group that received the highest dose of the treatment material resembling the results obtained by Bielke et al. (2003).

Attempts have also been made to develop defined cultures for prevention of colonization of *Campylobacter jejuni* in day-old chickens. Both aerobic and anaerobic cultures of *Citrobacter diversus, Klebsiella pneumoniae* and *Escherichia coli* isolated from *Campylobacter*-free hens and shown earlier to be inhibitory to *Campylobacter jejuni in vitro*, with or without simultaneous administration of sugars, was tested by Schoeni and Wong (1994). Prevention, elimination or direct challenge of *Campylobacter* infection was simulated in a chicken assay by applying treatment materials before, after or at the same time as that of the *Campylobacter* challenge. All treatments reduced *C. jejuni* colonization, also the carbohydrates alone.

Mead et al. (1996) tested the efficacy of ten aerobic isolates from adult *Campylobacter*-free hens, nine strains of *E. coli* and one *Enterococcus faecium* and cecal mucus collected and handled anaerobically from the same source. The ten strains were chosen for their ability to inhibit *C. jejuni in vitro*. In the same way coliforms from pigs showing anti-*Campylobacter* activity were chosen to be tested in the chicken assay described by Mead et al. (1989a). According to the results, none of the pure-culture combinations tested showed any efficacy against *Campylobacter*. Only the anaerobic preparations of cecal mucus provided clear protection. Zhang et al. (2007a) isolated from donor birds 636 bacterial strains that were inhibitory to six *Campylobacter jejuni* strains *in vitro*. Of these, 194 strains were strongly inhibitory to *C. jejuni*. Robyn et al. (2013) tested the efficacy of an *Enterococcus faecalis* isolate in a chicken assay. The strain was chosen because of its efficacy against *Campylobacter in vitro*. As with Mead et al. (1996), no inhibition was observed in the *in vivo* experiments.

5 Evaluation and administration of CE treatment materials

5.1 Experimental evaluation of the efficacy of treatment materials

The efficiency of the CE concept to prevent intestinal colonization of *Salmonella* in day-old chickens has been verified in numerous experimental studies (reviewed by Pivnick and Nurmi, 1982; Schleifer, 1985; Mead and Impey, 1986; Stavric and D'Aoust, 1993; Mead, 2000; Revolledo et al., 2006; Schneitz and Mead, 2010). The chicken assay included application of the treatment material to newly hatched chickens, followed by a marked challenge strain of *Salmonella* a few days later. After a certain period of time, the experiment was terminated and the birds were euthanized and tested for the challenge organism, primarily in the GI tract.

In small-scale experiments the chicks have been housed in cardboard boxes, cages or isolators with wire mesh floors, or in wire-walled or solid-walled pens on concrete floors covered with litter, mostly wood shavings or peat (Martin et al., 2000; Schneitz et al., 1991). Three methods have been applied for administering the treatment material: dosing of individual birds, via drinking water or by spraying. For the challenge, individual dosing by oral gavage was most frequently applied, although seeder-bird technique and challenge via feed or water have also been used (Weinack et al., 1979; Linton et al., 1985; Hinton et al., 1991b; Schneitz et al., 1991; Schneitz, 1992; Martin et al., 2000). Most often the trial has lasted for one week if the chicks have been treated and challenged individually. If the treatment material has been given by spraying or via drinking water, and/or if seeder-bird technique has been employed, the rearing period has usually been 10–14 days (Schneitz et al., 1990, 1991; Schneitz, 1992; Wolfenden et al., 2007).

Both the rearing conditions and the method of administration may affect the results as well as the time period between treatment and challenge and the length of the rearing period itself. In addition, chicks kept in cages with wire floors may be better protected against a secondary challenge from their environment, compared to chicks reared on litter. Nevertheless, trials involving individual CE treatment by oral gavage preceding the challenge with *Salmonella* are the most reliable way to determine the efficacy of a CE preparation because spraying small groups of chicks effectively is difficult and drinking water application may lead to uneven spread of the treatment material. Exact individual dosing of the treatment material is also impossible if the two latter methods are employed.

Mead (1989a) described a recommended assay to standardize the methods used to evaluate the efficacy of CE treatment. Chicks are treated orally as newly hatched, challenged 24 h later orally with *Salmonella* and examined 5 d after challenge to determine both the proportion of positive birds in treated and control groups and the levels of *Salmonella* carriage in infected individuals. The efficacy of the treatment is determined by the calculation of an Infection Factor (IF) value, which is the geometric mean of the number of *Salmonellae* per gram of cecal contents for all chicks in a particular group, and Protection Factor (PF), which is obtained by dividing the IF value for the control group by that for the treated group (Pivnick et al., 1985). Pivnick et al. (1985) suggested a PF value of 4.0 as a minimum for a treatment material to be effective against *Salmonella* in field conditions. Later, Schneitz and Hakkinen (1998) suggested using the difference (ΔIF) between the

IF values when presenting the results of a chicken assay trial, because calculating the PF value is not very informative as an index of efficacy, as it can be misleading if the infection rate of the control chicks is very low. For instance, if the IF value of the control group is 6.0 and that of the treated group 0.6, the PF value is 10. However, the same PF value is achieved if the IF value of the control group is 2.0 and that of the treated group is 0.2, but the differences between the mean *Salmonella* counts of the control and treated groups are 5.4 and 1.8 $\log_{10}$ units, respectively. Statistical evaluation of the results of a chicken assay comprising large numbers of bacteria may also be misleading and is not, on the other hand needed, if the differences between the CE-treated and control groups are manifest (Schneitz, 2016b).

Propionic acid concentrations in the cecal contents of CE-treated chicks have also been suggested as a biological indicator for early establishment of a microbial ecosystem inhibitory to *Salmonella* in chicks (Nisbet et al., 1996b). Hume et al. (1998b) found that chicks challenged only four hours after CE treatment were relatively well protected against *Salmonella* colonization and that significant increases in cecal propionic acid concentrations were observed within 1 d of the treatment. Martin et al. (2000) treated the chicks on day one, challenged them with *S. typhimurium* on d 3 and terminated the trial on d 10. There was an indisputable correlation between increasing cecal propionic acid concentrations and decreasing *Salmonella* colonization of the ceca. A clear correlation between the cecal propionic acid concentrations in 30-day-old chicks in relation to resistance against *S. infantis* in one-week-old chicks was observed by Schneitz et al. (2016b). Telg and Caldwell (2009) tested the efficacy of same treatment material as Martin et al. (2000) either alone or in combination with fructooligosaccharide for protection against *S. typhimurium* in broiler chickens. However, protection provided by the treatment material was less prominent in this study, and in spite of significant increase in the propionic acid concentrations compared to the control, the treatment material showed efficacy only in one out of three trials, which was similar to the previous study by Martin (2000).

5.2 Administration of CE treatment

The original administration method of CE preparations to day-old chickens in the field was via the first drinking water, a method that is still used when CE is applied to older birds, as a means to restore the normal GI microflora following treatment with antibiotics. In Finland, the effect of CE treatment given in the drinking water was tested during a three-year period on farms suffering from a prolonged *Salmonella* contamination. A total of 400 flocks were given 48 h-broth cultures of modified VL medium (Barnes and Impey, 1971) inoculated with cecal contents of healthy adult hen, and another 192 flocks were left without treatment. At the time of slaughter, 6.5% of the treated flocks were *Salmonella*-positive, while the corresponding figure for the untreated flocks was 21% (Hirn et al., 1992). The drinking water method and similar CE cultures were also used successfully in Sweden particularly during a period when *Salmonella* was spread by feed (Wierup et al., 1988, 1992). The same method has also been tested beneficially in small-scale studies (Schneitz et al., 1990; Martin et al., 2000). However, drinking water application is not always reliable. Chicks may hatch over a two-day period and the youngest often refuse to drink during the first hours on the farm. Thus, the most oxygen-sensitive anaerobic organisms in the CE preparation die over time and the product becomes ineffective before all the chicks have consumed an adequate dose (Seuna et al.,1978; Schneitz et al., 1991).

For the best possible result, CE treatment should precede the *Salmonella* challenge (Seuna, 1979). However, chicks may be exposed to *Salmonella* during transportation to the farm and even earlier if there is hatchery contamination or vertical transmission from infected breeders (Al-Zenki et al., 2009). While CE treatment cannot be expected to stop vertical transmission of *Salmonella* via the egg, cases of early chick infection following horizontal transmission could be prevented by CE treatment already in the hatchery. From this point of view, a spray application is the most suitable method for dosing chicks under field conditions. The idea of using an aerosol generated by a spraying method for administering a CE preparation was first introduced by Pivnick and Nurmi (1982). However, to avoid clinical disorders due to inhalation, the droplets sprayed on the chicks must be coarse so that the chicks can peck them from each other. Goren et al. (1984b, 1988) developed such a spraying method to treat chicks, either in the hatchers or in the transport boxes. Spraying in the hatchery followed by drinking water administration on the farm has also been used to ensure maximum efficacy of the treatment (Blankenship et al., 1993). Manual spray application employing a hand-held garden sprayer (Schneitz et al., 1990) preceded automated spraying devices that offer a much more rapid and evenly distributed treatment to each tray of chicks (Schneitz, 1992; Chen et al., 1998). Spray application does not have any adverse effects on the health or performance of the birds during grow-out (Corrier et al., 1995a). In fact, spraying is an oral treatment just like the drinking water application.

The fact that chicks may become infected vertically has urged researchers to look for a method of administration that would enable the birds to be treated prior to hatch (Cox et al., 1990, 1991). An *in ovo* method was therefore developed by Cox and Bailey (1993). A CE preparation was introduced into either the air cell or amnion of the egg a few days before hatching. However, that administration method generally resulted in depressed hatchability, and more specifically embryos were killed when the CE preparation was injected into the amnion (Cox et al., 1992; Cox and Bailey, 1993). Similar results were obtained with Broilact by Meijerhof and Hulet (1997). Nevertheless, it is possible to develop effective CE preparations that do not affect hatchability by excluding strongly proteolytic and gas-forming organisms; also the timing of injection is crucial (Schneitz, unpublished data).

6 Pathogen and host specificity of CE treatment

6.1 *Salmonella*

The CE concept applies to all serotypes of *Salmonella* that are capable of intestinal colonization in the chicken (Soerjadi et al., 1982b; Wierup et al., 1988; Impey and Mead, 1989; Hinton et al., 1991a; Nisbet et al., 1993; Cameron et al., 1997; Ghazikhanian et al., 1997; Newman and Spring, 1996; Schneitz and Hakkinen, 1998a; Hofacre et al., 2000; Ferreira et al., 2003; Al-Zenki et al., 2009; Revolledo et al., 2009; Telg and Caldwell, 2009). CE preparations have been shown to give good protection against both cecal colonization and invasion of inner organs by *S. enteritidis* PT4 or *S. typhimurium* (Cameron and Carter, 1992; Corrier et al., 1995b; Mead et al.,1989b; Nuotio et al., 1992; Bolder et al., 1992; Schneitz, 1992). CE treatment has also been shown to prevent horizontal transmission of *S. enteritidis* PT4 in chicken transport boxes, prohibit horizontal spread and decrease mortality caused by the host-specific *S. gallinarum*, and decrease horizontal spread of

Salmonella in those commercial broiler flocks that have been infected by *Salmonella* already in the hatchery, as presented below.

The GI tract of a newly hatched chicken is not necessarily sterile. When being laid, the outside of the egg becomes contaminated by organisms from the parent bird and, as the egg cools, these organisms may be sucked through the pores of the shell onto the underlying membrane. In the same way bacteria in the air at the hatchery can enter the egg. From this point the organisms may infect the developing embryo where they can multiply without killing it. This type of event is particularly hazardous if a breeder flock is infected by *Salmonella* or any other pathogens that are capable of multiplying in the embryo and infecting other chicks when the birds hatch.

The capability of CE treatment to prevent horizontal transmission of *Salmonella enteritidis* PT4 in transport boxes simulating 3 d transport conditions of newly hatched chickens was shown by Mead et al. (1989b). Four trials were carried out, two in England and two in Finland. The *Salmonella* challenge was introduced by means of seeder birds. All chicks, both seeders and contacts, were given CE treatment by oral gavage, to seeder chicks some 20 min after they had been challenged with *Salmonella*. After a further 20 min the seeders and contact chicks were placed together in cardboard transport boxes. Equal numbers of similarly boxed control chicks that had received sterile water instead of CE treatment were likewise placed in contact with seeder chicks. At 72 h, the trial was terminated and the chicks were examined for *Salmonella*. In the English trials, the challenge organism had spread to ca 50% of control chicks when only 15% of birds given CE treatment acquired *Salmonella* infection. In the Finnish trials, the challenge organism had spread to ca 70% of control chicks and to ca 13% of the birds given CE treatment. The levels of *S. enteritidis* in all contact birds in the CE treated group in both countries were lower than those of controls.

CE treatment has also been shown to prohibit horizontal spread and decrease mortality caused by *S. gallinarum* during the first 10–12 d post-hatch (Nisbet et al., 1998). Seeder bird technique was also employed in this study. Seeder chicks that were challenged as newly hatched with *S. gallinarum* and provided no CE treatment averaged 80% mortality during the first 12 d post hatch across four replicates. Untreated contact chicks that were mixed with the seeder chicks averaged 54% mortality, and the CE-treated contact chicks (provided CE treatment on the day-of-hatch) mixed with the seeders averaged 9% mortality during the same time period. *S. gallinarum* which causes fowl typhoid has been eradicated from developed countries by vaccination. However, the results of this study may be of importance to the poultry industry in geographic areas where fowl typhoid is still prevalent in poultry production.

Further, there is evidence that CE treatment is also beneficial for those commercial broiler flocks that have been infected by *Salmonella* already in the hatchery. According to the results of a field study carried out in France, 13% of the 34 CE treated flocks were *Salmonella* infected as newly hatched, but only 6% were still carriers at 45 days of age. Of the 34 flocks that were left untreated, 25% were positive as newly hatched and 42% at day 45. In addition, contamination of neck-skin samples taken at the processing plant was significantly reduced by the treatment (Palmu and Camelin, 1997). Another field study, conducted in the Netherlands, confirmed that treatment of *Salmonella*-infected chicks with a CE preparation can reduce the subsequent level of infection in the GI tract of the birds (Bolder et al., 1995b). Reynolds et al. (1997) found that treatment of 11 infected breeder flocks with enrofloxacin, followed by CE treatment, resulted in a long-term reduction in *Salmonella* in two trials and in a short-term reduction in another five trials.

Although there is evidence that levels of *Salmonella* infection in commercial broiler flocks can be reduced by the use of CE, treatment must precede infection for maximum benefit (Seuna, 1979).

6.2 *Campylobacter*

Contradictory results have been presented regarding the efficacy of CE treatment against *Campylobacter* (Mead, 2002). Some protection against an experimental *Campylobacter* challenge was provided by diluted faecal suspensions (Soerjadi et al., 1982a; Soerjadi-Liem et al., 1984a) and clear protection by anaerobic preparations of diluted cecal mucus from adult *Campylobacter*-free chickens was shown by Mead et al. (1996).

Later Mead (2002) studied the development of the chicken's natural resistance against *Campylobacter* on a farm where the broiler flocks were consistently *Campylobacter*-free. Cecal mucus samples from different flocks were taken when the birds were 3, 8, 14, 21 and 37 days old. Using the Mead model (Mead et al., 1989a) the efficacy of the mucus samples was tested in day-old chicks. The cecal mucus flora from up to two-week-old chickens was relatively ineffective against *C. jejuni*, in contrast to that from 21- and 37-day-old birds, which gave excellent protection against the challenge organism. The results revealed that the protective flora developed slowly in chickens, which is well in agreement with the fact that the cecal microflora increases in complexity when the bird grows older. According to Stern (1994), anaerobic cultures of cecal mucus from an appropriate donor bird gave variable protection of chicks against a subsequent *Campylobacter* challenge, and a clear loss of protective capability was seen after storage at −80°C for three months.

In a two-week study comprising three trials, Broilact reduced both the proportion of colonized birds and the level of intestinal carriage of *Campylobacter* (Hakkinen and Schneitz, 1999). In a recent five-week study by Schneitz and Hakkinen (2016a), the first two weeks gave almost exactly the same results as those of the preceding study, but in the third week, the *Campylobacter* numbers in the Broilact-treated chicks also started to increase so that at the end of the rearing period there was on average a 1.4 $\log_{10}$ units difference between the treated and control groups. At the end of a five-week rearing period, a similar result was obtained by Aho et al. (1992) when testing the combined effect of Broilact and *Campylobacter*-like bacteria (K bacteria). The results of a field study comprising 2.4 million Broilact-treated chicks and 2.3 million control chicks on different farms scattered around the Netherlands showed that broilers treated with Broilact had a significantly lower *Campylobacter* incidence in their cecal contents compared to the controls (Bolder et al., 1995).

All studies presented here have lasted till the maximum two-week period apart from the five-week studies by Aho (1992) and Schneitz and Hakkinen (2016a) and the field study by Bolder et al. (1995). The results of one- and two-week studies as well as the two first weeks in the five-week studies have been very promising. However, the results reveal that the level of *Campylobacter* colonization started to increase only after the second week of the rearing period along with the litter humidity. This may indicate that the Mead model with a trial period of one or two weeks is less suitable for testing the efficacy of a CE preparation against *Campylobacter*.

6.3 *Escherichia coli* including antibiotic-resistant strains

CE treatment has also been shown to protect newly hatched chicks against chicken and human pathogenic *E. coli* (Soerjadi et al., 1981a; Weinack et al., 1981, 1982, 1984;

Stavric et al., 1992; Hakkinen and Schneitz, 1996; Hofacre et al., 2002). Hakkinen and Schneitz (1996) tested the efficacy of Broilact against the colonization of ceca by a chicken pathogenic *E. coli* and by the human pathogen O157:H7. The treatment reduced the numbers of both *E. coli* strains approximately by 4 $\log_{10}$ units. Later, Nuotio et al. (2013) found that Broilact was able to reduce the numbers of ESBL- and pAmpC-producing *E. coli* in the chicken cecal contents. A similar result was obtained with Aviguard against an antibiotic-resistant poultry-pathogenic *E. coli* (Hofacre et al., 2002). As with vertical *Salmonella* infections, *E. coli* infections originating from the breeder flock are difficult to control by CE. However, treatment in the hatcher could diminish horizontal transmission of the pathogens among the hatchlings.

Recently, a remarkable observation regarding the possible effect of CE treatment against antibiotic-resistant *E. coli* was done in Sweden and Finland. The occurrence of ESBL- and pAmpC-producing *E. coli* among broiler flocks has been monitored in Sweden since 2010. Among the 200 broiler flocks examined in 2010, the proportion of ESBL or pAmpC-positive flocks was 34%, whereas in 2011 the corresponding proportion was 54% among 100 batches (SWARM, 2011, 2012). The same year in Finland 352 broiler flocks were examined for ESBL/pAmpC types using the same methodology as in Sweden. Only three of them (0.85%) yielded pAmpC *E. coli*, whereas ESBL types were not encountered. CE treatment has been used in Finland since 1976 but has not been widely used in Sweden. The markedly different prevalence of ESBL- or pAmpC-producing *E. coli* at slaughter in broilers produced in the two countries suggest a potential role for CE, especially when the production systems in Finland and Sweden are rather similar and at the time of the study the grandparent stock was obtained from the same primary breeder in Scotland (Nuotio et al., 2013).

6.4 *Clostridium perfringens*

Dietary factors that have an impact on the microbial ecology of the GI tract include the amount of fibre, the viscosity of the fibre and fats that are refractory to digestion (Klasing, 1998). For example, grains like barley and rye, which are high in non-starch polysaccharides, are associated with increased incidence of necrotic enteritis (NE) (Hofshagen and Kaldhusdal, 1992; Kaldhusdal and Hofshagen, 1992; Riddle et al., 1992), the causative agents of which are *Clostridium perfringens* types A and C (Ficken and Wages, 1997). Flocks with acute clinical disease show marked symptoms and mortality when a subclinical form has been associated with increased feed conversion ratio and retarded growth rate (Kaldhudal and Hofshagen, 1992).

Although not tested so intensively, CE treatment has been shown to reduce levels of cecal *C. perfringens* and decrease mortality due to NE. Reduced levels of cecal *C. perfringens* was obtained by Barnes et al. (1980b) and Snoeyenbos et al. (1983) by use of CE treatment. In a pilot-scale study, Elwinger et al. (1998) found that the use of Broilact was associated with reduced incidence of NE as well as cecal carriage of *C. perfringens* in broiler chickens. Hofacre et al. (1998) challenged broiler chickens experimentally with *C. perfringens* and found that Aviguard reduced gross intestinal lesions and improved feed efficiency in their disease model. In another study, employing similar challenge model, Hofacre et al. (2003) found that chicks treated with AviFree had significantly less mortalities due to NE than the controls. MSC (Mucosal Starter Culture) was also shown to reduce both the *C. perfringens* counts and the percentage of *C. perfringens*-positive broilers that were raised on a wire floor and offered a rye supplemented feed (Craven et al., 1999). Further, the efficacy of Broilact against NE was tested in Norway on a farm having recurrent *C. perfringens*-associated health problems. The study consisted of four broiler houses, with one treated and one untreated flock per house. Treated birds were sprayed with Broilact upon arrival at the farm. All flocks were offered feed containing the ionophorous anticoccidial agent narasin. The feed did not contain antimicrobial growth promoters. Broilact reduced mortality, the frequency of birds with intestinal NE lesions and the median *C. perfringens* counts during weeks 3 and 4 of the rearing period as well as condemnations at slaughter, but the differences were not significant (Kaldhusdal et al., 2001).

6.5 Other pathogenic bacteria

Cox et al. (1997) found *Listeria monocytogenes* on 6% of the egg fragments and 1% of the chicken transport pads. On the processing plant, only 1 of 115 whole bird rinses and none of the 115 ceca examined were contaminated with *L. monocytogenes*. However, after processing, 25.7% of the 105 carcasses examined were contaminated, which indicates contamination during processing. Very little is known about the ways by which *L. monocytogenes* enters poultry operations but evidently it comes with the chickens and resides on the processing plant. Hume et al. (1998a) showed that CE treatment significantly reduced cecal colonization by *L. monocytogenes* in chicks, although, according to another study, most chicks became *Listeria* negative within nine days without any treatment (Husu et al., 1990). In addition, CE treatment has also been shown to inhibit colonization of *Yersinia enterocolitica* in newly hatched chickens (Soerjadi-Liem et al., 1984b).

6.6 Host specificity of CE treatment

The results of numerous chicken assay studies show that treatment materials originating from adult birds protect newly hatched chickens of the same species against *Salmonella* colonization and the protection seems to be independent of breed, strain or sex of the birds. However, there are differences among donor birds, with respect to protective capacity (Schneitz, unpublished results). This is not surprising when, according to van der Wielen et al. (2002), each chicken within a flock has its own unique dominant intestinal bacterial community indicating that host-specific factors are important in the establishment of the intestinal bacterial community. Further, it has been shown that different genetic lines of broilers have varying rates of intestinal development, which may affect the bacterial community (Lumpkins et al., 2010).

Chickens can be protected to some extent by the microflora of a few other avian species (Snoeyenbos et al., 1979), but material from animals, such as horse or cow, proved to be ineffective (Rantala and Nurmi, 1973). Coliforms from pigs showing anti-*Campylobacter* activity *in vitro* were ineffective against *Campylobacter* when tested in chickens (Mead, 1996). Native chicken and turkey microfloras were shown to provide reciprocal protection in chicks and turkey poults (Weinack et al., 1982; Impey et al., 1984; Schneitz and Nuotio,1992). Cox et al. (2001) demonstrated that CE cultures generated from mucosal scrapings of adult turkeys effectively controlled *Salmonella* in turkey poults during brooding. Hofacre et al. (2000) showed that fresh turkey cecal material was significantly more protective in turkeys than Aviguard. Bamba et al. (1997) demonstrated that levels of *S. typhimurium* decreased significantly in artificially challenged Japanese quails that had been treated on day one with chicken cecal contents. Pheasant chicks were successfully protected against *S. infantis* by Broilact (Schneitz and Renney, 2003).

7 Factors affecting the efficacy of CE treatment

7.1 Antimicrobial-associated factors

Several growth-promoting agents, either alone or together with anticoccidials, have been examined for their effects on *Salmonella* and *Campylobacter* shedding in chickens and turkey poults. According to Smith and Tucker (1978, 1980) and Barrow et al. (1984), chickens fed with diets containing avoparcin remained infected for longer periods and shed higher numbers of *Salmonellae* than those given diets without avoparcin. On the other hand, trials conducted by Smith and Green (1980a) and Gustafson and Kobland (1984) showed no such disadvantage. However, it should be noted here that the possible antimicrobial effect on the intestinal colonization of *Salmonella* in a field situation may be influenced by the *Salmonella* status of the production, as well as the *Salmonella* control applied, as highlighted in Section 9. The studies by Smith and Green (1980) and Gustafson and Kobland (1984) showing no avoparcin-associated negative effect on *Salmonella* colonization were performed in chickens reared on used litter. Further, Cox et al. (2003) suggested that naturally occurring *Campylobacter* in commercial turkey poults were unaffected by flavomycin, virginiamycin or monensin added in the feed, but *Salmonella* populations were significantly decreased by the antimicrobial feed additives.

The synergy or antagonism of antimicrobial feed additives on the protective effect of CE treatment has also been evaluated. Humbert et al. (1991) showed that chicks receiving avoparcin continuously in the feed, following CE treatment, had significantly more *Salmonellae* in their ceca than birds given feed containing no antimicrobials. Chicks given a diet supplemented with flavomycin had levels similar to those of control birds with no dietary antimicrobials, whereas the chicks fed on a diet supplemented with bacitracin or virginiamycin exhibited the lowest level of *Salmonella* carriage. Bailey et al. (1988) demonstrated that a combination of nicarbazin and bacitracin interfered with the protective effect of CE treatment. In contrast, treatment of newly hatched chicks for five days with either furazolidone (400 ppm in the feed) or trimethoprim-methoxasole sulphate (0.02 g trimethoprim and 0.1 g sulphamethoxasol per litre in drinking water) did not affect the protective capability of Broilact because the birds were repopulated by the beneficial bacteria from the environment (Bolder and Palmu, 1995).

The effect of *in ovo* administration of gentamicin or ceftiofur on the efficacy of CE treatment has also been a matter of concern. According to McReynolds et al. (2000), both antibiotics could cause a marked depression in levels of cecal propionate that was used to indicate whether or not a microflora had become established from treatment of chicks with Preempt. However, it has also been shown that gentamicin, at a rate of 0.4 mg per egg and administered *in ovo* on day 18, had no adverse effect on the efficacy of MSC (Bailey and Line, 2001).

7.2 Disease- and stress-associated factors

Other factors that may reduce the efficacy of CE treatment include disease and stress. Administration of *Eimeria tenella*, at a level known to produce cecal coccidiosis, to

CE-treated chicks with low numbers of *Salmonellae* in their intestines, resulted in the shedding of large numbers of *Salmonellae* for more than two weeks (Lafont et al., 1983). Weinack et al. (1984) determined the effect of aerosols of *Mycoplasma gallisepticum* and/or infectious bronchitis virus (MG-IBV) on chicks given CE treatment at one day of age and challenged at three days of age with *E. coli* or *S. typhimurium*. The chicks were exposed at 8 to 24 days of age by aerosol to a virulent strain of MG and/or a commercial IBV vaccine which increased the number of birds shedding the challenge organisms.

Commercial hatchery environments can be contaminated with *Salmonella* and the parent stock may be carriers (Cox et al., 1990, 1991), both of which are likely to reduce the efficacy of CE treatment (Bailey et al., 1998).

Goren et al. (1984a) noticed that starving chicks for the first 24 hours of life also has a negative effect on the protective microflora, whereas Snoeyenbos et al. (1985) showed that in older birds, the protective flora is more difficult to hamper. With the day-old chick, physiological stress induced by high or low environmental temperatures or removal of feed and water either interfered with the colonization of protective organisms or reduced the protection provided by these organisms; however, there was no obvious effect at two weeks of age (Weinack et al., 1985b).

Subjecting market-age broilers to feed-withdrawal and induced molting of White Leghorn layers have also been shown to increase both the numbers of *Salmonellae* in the gastrointestinal tract and the proportion of infected individuals (Holt and Porter, 1993; Holt, 1995; Macri et al., 1997; Ramirez et al., 1997; Corrier et al., 1999; Kubena et al., 2005).

8 Field experience of CE

The appropriate evaluation of the preventive effect of CE on the colonization of *Salmonella* in chickens requires insight into the management of poultry production, in particular, including the prevention and control of *Salmonella* generally and the way it is applied at the actual production site. Such a basic knowledge is also required for other potential pathogens considered. Below, the field experiences of the prevention of *Salmonella*, the initial and major target for the CE, are summarized.

8.1 Principles for prevention and control of *Salmonella*

Worldwide, *Salmonella* is one of the most common food-borne zoonoses and the cause of substantial economic losses (Majowicz et al., 2010). It emerged as a major human health problem following the industrialization of animal production, and the poultry production was early recognized as a major source for human infections. Efforts were made to control *Salmonella*, which required new approaches, in particular, when the traditional veterinary inspection at slaughter did not disclose *Salmonella*-infected animals or contaminated carcasses because *Salmonella*-infected animals usually do not show any clinical signs of disease or lesions on the carcass. Apart from the preventive measures at slaughter and consumer levels, it is found necessary to prevent the infection already at the production, also referred to as the pre-harvest control (WHO 1980). Methods for the control at farm level are well known and basically include biosecurity to prevent the introduction of *Salmonella*, monitoring to detect when infection in birds or contamination of their environment occurred, and if so, undertake actions, including cleaning and disinfection, to eliminate the pathogen. The implementation of these principles in fact has successfully resulted in what can be considered as a *Salmonella*-free poultry production in the Nordic countries with a zero tolerance policy for *Salmonella* (*Salmonella* low-prevalence status; EFSA, 2006) where the bacteria are rarely detected in the monitoring of all flocks before slaughter (EFSA and ECDC, 2015). In many other countries a similar situation can exist for the breeding herds, while at the production level of the breeding pyramid, a substantial proportion of the flocks are often contaminated by *Salmonella* (*Salmonella* medium- or high-prevalence status). However, also in low-prevalence countries, *Salmonella* can never be considered as eradicated from the poultry production because there is always a risk for contamination from wild animals. Particularly, in all countries, there is a continuous risk for the introduction of *Salmonella* from the feed (Wierup, 2013).

8.2 Field experience

Reliable field studies are difficult to arrange because artificial challenge cannot be used and uncontrolled natural challenge from unknown sources may occur in both experimental and control flocks, in particular, on farms in medium- or high-prevalence countries where *Salmonellae* are constantly present. In addition, the outcome of the trials can be affected by numerous environmental factors. Nevertheless, the efficacy of CE treatment against *Salmonella* has also been demonstrated under field conditions (Snoeyenbos et al., 1978; Wierup et al., 1988, 1992; Blankenship et al., 1993; Bolder et al., 1995b; Corrier et al., 1995a; Deruyttere et al., 1997; Palmu and Camelin, 1997) as well as its ability to reduce levels of cecal *Clostridium perfringens*, one of the causative factors in necrotic enteritis (Kaldhusdal et al., 2001).

Finland apparently has the longest experience of CE because the concept has been employed since 1976 (Raevuori et al., 1978). The treatment materials initially were 48 h broth cultures of modified VL medium (Barnes, 1971) inoculated with cecal contents of healthy adult hen. Broilact substituted the VL culture in 1987 (Nurmi et al., 1992), and today practically all commercially produced broilers and turkeys in Finland are treated as newly hatched. It has been suggested that the consistent use of CE treatment has contributed significantly to the low incidence of *Salmonella*-contaminated poultry flocks and carcasses (Hirn et al., 1992) as well as to the low incidence of *Campylobacter* in broilers in Finland (Aho and Hirn, 1988).

The CE concept is an official part of the Finnish *Salmonella* Control Program which covers all serotypes of *Salmonella*, not only *S. enteritidis* and *S. typhimurium*, like in many other countries. In addition to comprehensive control of the breeding flocks and hatcheries, all broiler flocks have to be tested negative for *Salmonella*. Additionally, carcass samples are taken regularly from the processing plant (Maijala et al., 2005).

In Sweden, CE treatment has been used sporadically since 1981 and an epidemiological study conducted by Wierup et al. (1988), during a period when *Salmonella* was being spread from contaminated feed, also supported the view that CE treatment can be effective under field conditions. During 1981–90, 179 flocks, involving 3.82 million chicks, were treated and only one of these flocks was found subsequently to be infected with *Salmonella* (Wierup et al., 1992). As in Finland, the incidence of *Salmonella* in Swedish poultry is very low and a similar *Salmonella* control is applied (EFSA and ECDC, 2015).

9 Strategies for use, safety and licensing

9.1 Strategies for the use of CE

In all countries there is a continuous risk for poultry productions to be exposed to *Salmonella* from the environment or the feed (EFSA, 2008). Therefore, measures are always required to prevent and control the exposure of *Salmonella* (Stenberg-Lewerin et al., 2005). The protection of the chickens directly after hatch is of crucial importance. This is a critical period because the establishment of a protective intestinal microflora, effective against *Salmonella*, is delayed in chickens hatched in a brooder instead of naturally in a nest under the mother hen. CE compensates for the consequences of the industrialized poultry production by the establishment of a protective intestinal microflora in day-old chickens.

Methods for the administration are described above (Section 5). In addition, the possible influence of antimicrobials on the efficiency of CE (Section 8) should be considered. It is well known that in particular oral use of antibiotics to treat bacterial infections in both animals and humans can adversely affect the intestinal microflora of the host. Therefore, it is obvious that the continuous use of growth-promoting antibiotics, which has coincided with the appearance of bacteria with multiple drug resistance (Ginns et al., 1996; Khan et al., 2005; Simjee et al., 2007), may also interfere with the effect of CE. On the other hand, the concept may be used to 'normalize' the intestinal flora when antimicrobials have temporarily been used to treat infections (Johnson, 1992; Reynolds et al., 1997). Also in humans, the transplantation of faecal microflora has recently been found to be an effective measure to treat *Clostridium difficile* infection, a well-known serious side effect of certain antibiotics (Rao and Safdar, 2016).

The use of CE can be considered as the method of choice to prevent an early exposure to chickens by *Salmonella* from being further spread horizontally in a flock or vertically in case of breeding animals. This is also the case for day-old chickens in a moderate to heavily *Salmonella*-contaminated environment, although the chickens in such a situation are likely to later become infected by from other sources. In summary CE is a valuable tool, additional to those required to decrease the prevalence and ultimately possibly eliminate *Salmonella* from a poultry production site. These general considerations for *Salmonella* may well be applied for the use of CE against other potential pathogens.

9.2 Safety and licensing of CE

The safety of a CE product is of crucial importance. The product must be safe both for chickens and consumers. Regarding undefined CE products, special requirement should be placed on the donor bird. The donor bird should originate from a flock or production that for long has been under appropriate surveillance over the health status, particularly for *Salmonella*, but also for other specific poultry pathogens. The flock should never have been treated with antimicrobials. Reliable data in this respect are in practice only available from well-controlled breeding flocks and naturally raised SPF flocks. Further, the donor bird must undergo thorough *ante*- and *post-mortem* examination and all inocula thereof, as well as the final product, must be examined carefully for both human and avian pathogens in laboratories certified by licensing authorities. Also, good laboratory and manufacturing practices must be employed throughout the production process (Nurmi and Nuotio, 1994).

In case of legal requirements for a CE product, a logical approach would be to ensure freedom from specified pathogens instead of a demand for an identification of all microbes present in the product, which may not even be possible when the majority of the bacteria in the chicken GI tract represent previously unknown species and have never been cultivated (Yeoman et al., 2012; Apajalahti et al., 2004). However, the new sequencing technologies would make it possible to describe properties even of less abundant organisms in an undefined CE product in more detail. Further, in terms of the possible presence of antimicrobial and pathogenicity traits, such can also easily be revealed, as has been done from the GI tract of commercial broiler chickens (Danzeisen et al., 2011; Yeoman et al., 2012). Consequently, for the purpose of assessment and management of the possible risk involved in the approval of an undefined CE product, a comparison should be made to the health requirements applied for importation of live breeding animals that harbour the same unknown intestinal microflora as an undefined CE product.

The licensing/registration of CE products has probably limited their use. To overcome associated problems WHO has suggested a special category for the licensing (WHO, 1994). The category could be termed 'normal gut flora' and was suggested to be described as follows: 'In relation to the avian intestinal tract, "normal gut flora" is an undefined preparation of live obligate and facultatively anaerobic bacteria, originating from normal, healthy, adult individuals of an avian species, which is free from specific pathogenic microorganisms and is quality controlled. The purpose of such a preparation is to compensate for any deficiencies in the composition of the normal intestinal microbiota that relate to the natural control of undesirable microorganisms and arise from modern systems of poultry production.'

As described earlier, the use of CE products has since long been approved in Finland and in the Scandinavian countries and also in some other EU member states on the basis of their national legislation. In addition, CE products have been used to some extent in poultry-producing countries in South America, Africa and the Far East.

10 Conclusions

The CE or Nurmi concept is a well-documented tool for effective mitigation of the risk for *Salmonella* infections in day-old chickens. Because there is a continuous risk for exposure to *Salmonella* in all countries, the use of CE can be the crucial measure required to prevent the exposure to newly hatched chickens from being further spread horizontally in a flock or vertically in case of breeding animals. However, CE is no silver bullet for the control of *Salmonella* but should be seen as a valuable additional tool to those major efforts that are always required to control and decrease prevalence and possibly eliminate the pathogen from a production site.

The CE compensates for the delayed development of the protective intestinal microflora in chickens hatched under hygienic conditions in modern industrialized production without natural contact with the mother hen and the nest environment. The mechanism of CE is still not fully understood, but the protection has been shown to be primarily physical and provided by bacteria that adhere to the GI epithelial cells.

The only consistently effective CE preparations till today have been undefined, mixed cultures of faecal or cecal origin. Attempts to develop effective pure-culture preparations for protective purposes have generally been disappointing. Although a few commercial

CE products are available and used since long in some countries and their use is tolerated in the European Union, undefined CE products are still prohibited in many countries, including the United States. Therefore, there is an urgent need to create a special category for the licensing of undefined CE products as suggested by WHO.

11 Where to look for further information

Papers dealing with this subject which are published in peer reviewed journals can be found in PubMed on the internet using key words competitive exclusion, chicken, poultry, *Salmonella*, *Campylobacter* and faecal microbiota transplantation for the use in humans. Additional information not published in peer reviewed journals can be found e.g. in proceedings from scientific conferences and symposia and in text books dealing with poultry sciences and prevention and control of infectious diseases in the poultry production, in particular *Salmonella* and *Campylobacter.*

12 References

Aho, M. and Hirn, J. (1988), 'Prevalence of campylobacteria in the Finnish broiler chicken chain from the producer to the consumer', *Acta Veterinaria Scandinavica* 29, 451–62.

Aho, M., Nuotio, L., Nurmi, E. and Kiiskinen, T. (1992), 'Competitive exclusion of campylobacters from poultry with K-bacteria and Broilact', *International Journal of Food Microbiology*, 15, 265–75.

Al-Zenki, S. F., Al-Nasser, A. Y., Al-Saffer, A. E., Abdullah, F. K., Al-Bahouh, M. E., Al-Haddad, A. S., Alomirah, H. and Mashaly, M. (2009), 'Effects of using a chicken-origin competitive exclusion culture and probiotic cultures on reducing *Salmonella* in broilers', *Journal of Applied Poultry Research*, 18, 23–9.

Apajalahti, J., Kettunen, A. and Graham, H. (2004), 'Characteristics of the gastrointestinal microbial communities, with special reference to chicken', *World's Poultry Science*, 60, 223–31.

Bailey, J. S. (1987), 'Factors affecting microbial competitive exclusion in poultry', *Food Technology*, July, 88–92.

Bailey, J. S., Blankenship, L. C., Stern, N. J., Cox, N. A. and McHan, F. (1988), 'Effect of anticoccidial and antimicrobial feed additives on prevention of salmonella colonization of chicks treated with anaerobic cultures of chicken feces', *Avian Diseases*, 32, 324–9.

Bailey, J. S., Cason, J. A. and Cox, N. A. (1998), 'Effect of *Salmonella* in young chicks on competitive exclusion', *Poultry Science*, 77, 394–9.

Bailey, J. S and Line, E. (2001), '*In ovo* gentamicin and Mucosal Starter Culture to control *Salmonella* in broiler production', *Journal of Applied Poultry Research*, 10, 376–9.

Bamba, H., Toyoshima, K. and Kamiya, M. (1997), 'Protective properties of the treatments of cecal contents on *Salmonella* typhimurium colonization in the bowel of growing quail chicks', *Research Bulletin of the Aichi-ken Agricultural Research Center*, 29, 355–8.

Barnes, E. M. (1972), 'The avian intestinal flora with particular reference to the possible ecological significance of the cecal anaerobic bacteria', *American Journal of Clinical Nutrition*, 25, 1477–9.

Barnes, E. M. (1977), 'Ecological concepts of the anaerobic flora in the avian intestine', *American Journal of Clinical Nutrition*, 30, 1793–8.

Barnes, E. M. and Impey, C. S. (1970), 'The isolation and properties of the predominant anaerobic bacteria in the caeca of chickens and turkeys', *British Poultry Science*, 11, 467–81.

Barnes, E. M. and Impey, C. S. (1971), 'The isolation of anaerobic bacteria from chicken caeca with particular reference to members of the family *Bacteroidaceae*', in Shapton, D.A. and Board, R.G. (eds), *Isolation of Anaerobes*. London UK: Academic Press, pp. 115–123.

Barnes, E. M., Impey, C. S. and Cooper, D. M. (1980a), 'Competitive exclusion of salmonellas from the newly hatched chick', *The Veterinary Record,* 106, 61.

Barnes, E. M., Impey, C. S. and Cooper, D. M. (1980b), 'Manipulation of the crop and intestinal flora of the newly hatched chick', *American Journal of Clinical Nutrition,* 33, 2426–33.

Barnes, E. M., Impey, C. S. and Stevens, J. H. (1979), 'Factors affecting the incidence and anti-salmonella activity of the anaerobic caecal flora of the young chick', *Journal of Hygiene, Cambridge,* 82, 263–83.

Barnes, E. M., Mead, G. C., Barnum, D. A. and Harry, E. G. (1972), 'The intestinal flora of the chicken in the period 2 to 6 weeks of age, with particular reference to the anaerobic bacteria', *British Poultry Science* 13, 311–26.

Barrow, P. A., Smith, H. W. and Tucker, J. F. (1984), 'The effect of feeding diets containing avoparcin on the excretion of salmonellas by chickens experimentally infected with natural sources of salmonella organisms', *Journal of Hygiene, Cambridge*, 93, 439–44.

Bielke, L. R., Elwood, A. L., Donoghue, D. J., Donoghue, A. M., Newberry, L. A., Neighbor, N. K. and Hargis, B. M. (2003), 'Approach for selection of individual enteric bacteria for competitive exclusion in turkey poults', *Poultry Science*, 82, 1378–82.

Blankenship, L. C., Bailey, J. S., Cox, N. A., Stern, N. J., Brewer, R. and Williams, O. (1993), 'Two-step mucosal competitive exclusion flora to diminish salmonellae in commercial broiler chickens', *Poultry Science*, 72, 1667–72.

Bolder N. M. and Palmu, L. (1995), 'Effect of antibiotic treatment on competitive exclusion against *Salmonella enteritidis* PT4 in broilers', *The Veterinary Record*, 137, 350–1.

Bolder, N. M., van Lith, L. A. J. T., Putirulan, F. F., Jacobs-Reitsma, W. F. and Mulder, R. W. A. W. (1992), 'Prevention of colonization by *Salmonella enteritidis* PT4 in broiler chickens', *International Journal of Food Microbiology*, 15, 313–17.

Bolder, N. M., Vereijken, P. F. G., Putirulan, F. F. and Mulder, R. W. A. W. (1995), 'The effect of competitive exclusion on the *Salmonella* contamination of broilers (a field study)', in Briz, R. C. (ed.), *Proceedings of the 2nd Annual Meeting of EC COST Working Group No. 2*. Zaregosa, Spain: Graficas Imprinter.

Burns, R. (1995), 'Commercial competitive exclusion bacterial mix ready soon', *Feedstuffs*, 13 March, 27.

Cameron, D. M. and Carter, J. N. (1992), 'Evaluation of the efficacy of BROILACT in preventing infection of broiler chicks with *S. enteritidis* PT4', *International Journal of Food Microbiology*, 15, 319–26.

Cameron, D. M., Carter, J. N., Mansell, P. and Redgrave, V. A. (1997), 'Floor-pen efficacy study with Aviguard against *Salmonella typhimurium* DT 104 colonization in turkeys', *Proceedings of the Salmonella and Salmonellosis Symposium*, Ploufragan, France, pp. 481–5.

Chen, M., Stern, N. J., Bailey, J. S. and Cox, N. A. (1998), 'Administering mucosal competitive exclusion flora for control of Salmonellae', *The Journal of Applied Poultry Research*, 7, 384–91.

Coloe, P. J., Bagust, T. J. and Ireland, L. (1984), 'Development of the normal gastrointestinal microflora of specific pathogen-free 0chickens', *Journal of Hygiene, Cambridge*, 92, 79–87.

Corrier, D. E., Byrd, J. A., Hargis, B. M., Hume, M. E., Bailey, R. H. and Stanker, L. H. (1999), 'Presence of *Salmonella* in the crop and ceca of broiler chickens before and after preslaughter feed withdrawal', *Poultry Science*, 78, 45–9.

Corrier, D. E., Hinton Jr., A., Hargis, B. and DeLoach, J. R. (1992), 'Effect of used litter from floor pens of adult broilers on salmonella colonization of broiler chicks', *Avian Diseases*, 36, 897–902.

Corrier, D. E., Hinton Jr., A., Kubena, L. F., Ziprin, R. L. and DeLoach, J. R. (1991), 'Decreased *Salmonella* colonization in turkey poults inoculated with anaerobic cecal microflora and provided dietary lactose', *Poultry Science,* 70, 1345–50.

Corrier, D. E., Nisbet, D. J., Scanlan, C. M., Hollister, A. G., Caldwell, D. J., Thomas, L. A., Hargis, B. M., Tomkins, T. and Deloach, J. R. (1995a), 'Treatment of commercial broiler chickens with a characterized culture of cecal bacteria to reduce salmonellae colonization', *Poultry Science*, 74, 1093–101.

Corrier, D. E., Nisbet, D. J., Scanlan, C. M., Hollister, A. G. and Deloach, J. R. (1995b), 'Control of *Salmonella typhimurium* colonization in broiler chicks with continuous-flow characterized mixed culture of cecal bacteria', *Poultry Science*, 74, 916–24.

Cox, N. A. and Bailey, J. S. (1993), 'Introduction of bacteria in ovo', *United States Patent 5,206,015*.

Cox, N. A., Bailey, J. S. and Berrang, M. E. (1997), 'The presence of *Listeria monocytogenes* in the integrated poultry industry', *Journal of Applied Poultry Research*, 6(1), 116–19.

Cox, N. A., Bailey, J. S., Mauldin, J. M. and Blankenship, L. C. (1990), 'Presence and impact of salmonellae contamination in commercial broiler hatcheries', *Poultry Science*, 69, 1606–9.

Cox, N. A., Bailey, J. S., Mauldin, J. M. and Blankenship, L. C. (1991), 'Extent of salmonellae contamination in breeder hatcheries', *Poultry Science*, 70, 416–18.

Cox, N. A., Bailey, J. S., Mauldin, J. M. and Blankenship, L.C. (1992), '*In ovo* administration of a competitive exclusion treatment to broiler embryos', *Poultry Science*, 71, 1781–4.

Cox, N. A., Bailey, J. S. and Stern, N. J. (2001), 'Effectiveness on an undefined mucosal competitive exclusion treatment to control *Salmonella* in turkeys during brooding', *The Journal of Applied Poultry Research*, 10, 319–22.

Cox, N. A., Craven, S. E., Musgrove, M. T., Berrang, M. E. and Stern, N. J. (2003), 'Effect of sub-therapeutic levels of antimicrobials in feed on the intestinal carriage of *Campylobacter* and *Salmonella* in turkeys', *Journal of Applied Poultry Research*, 12, 32–6.

Craven, S. E, Stern, N. J., Cox, N. A., Bailey, J. S. and Berrang, M. (1999), 'Cecal carriage of Clostridium perfringens in broiler chickens given Mucosal Starter Culture', *Avian Diseases*, 43, 484–90.

Criado, M. T., Saintz, V., del Rio, M. C., Ferreiros, C. M., Criado, J., Carballo J. and Souto, M. J. (1989), 'Failure of adherence to buccal cells and surface hydrophobicity as virulence markers in *Neisseria meningitides*', *Medical Microbiology & Immunology*, 178, 53–9.

Danzeisen, J. L., Kim, H. B., Isaacson, R. E., Tu, Z. J. and Johnson, T. J. (2011), 'Modulations of the chicken cecal microbione and metagenome in response to anticoccidial and growth promoter treatment', *PloS ONE*, 6(11), 1–14.

Deruyttere, L., Klaasen, J., Froyman, R. and Day, C. A. (1997), 'Field study to demonstrate the efficacy of Aviguard against intestinal *Salmonella* colonization in broilers', *Proceedings of the Salmonella and Salmonellosis Symposium*, Ploufragan, France, pp. 523–5.

De Oliveira, G. H., Berchieri Jr., A. and Barrow, P. A. (2000), 'Prevention of *Salmonella* infection by contact using intestinal flora of adult birds and/or a mixture of organic acids', *Brazilian Journal of Microbiology*, 31, 116–20.

EFSA (European Food Safety Authority) (2008), Scientific Opinion of the 'Panel on Biological Hazards on a request from the Health and Consumer Protection', Directorate General, European Commission on Microbiological Risk Assessment in feedingstuffs for foodproducing animals, *The EFSA Journal*, 720.

EFSA (European Food Safety Authority) (2011), Panel of Biological Hazards (BIOHAZ), 'Scientific Opinion on *Campylobacter* in broiler meat production: control options and performance objectives and/or targets at different stages of the food chain', *The EFSA Journal*, 9, 2105.

EFSA and ECDC (European Food Safety Authority and European Centre for Disease Prevention and Control) (2015), The European Union Summary Report on 'Trends and Sources of Zoonoses, Zoonotic Agents and Food-borne Outbreaks in 2013', *The EFSA Journal*, 13, 3991.

Elwinger, K., Schneitz, C., Berndtson, E., Fossum, O., Teglöf, B. and Engström, B. (1992), 'Factors affecting the incidence of necrotic enteritis, caecal carriage of *Clostridium perfringens* and bird performance in Broiler chicks', *Acta Veterinaria Scandinavica*, 33, 369–78.

Ferreira, A. J. P, Ferreira, C. S. A., Knobl, T., Moreno, A. M., Bacarro, M. R., Chen, M., Robach, M. and Mead, G. C. (2003), 'Comparison of three commercial competitive-exclusion products for controlling *Salmonella* colonization of broilers in Brazil', *Journal of Food Protection*, 66, 490–2.

Ficken, M. D. and Wages, D. P. (1997), 'Necrotic enteritis', in Calnek, B. W. (ed.), *Diseases of Poultry*. Ames: Mosby-Wolfe, pp. 261–264.

Filho, R. L. A., Sampaio, H. M., Barros, M. R., Gratâo, P. R. and Cataneo, A. (2003), 'Use of cecal microflora cultured under aerobic or anaerobic conditions in the control of experimental infection of chicks with *Salmonella* Enteritidis', *Veterinary Microbiology*, 92, 237–44.

Fons, M., Gomez, A, and Karjalainen, T. (2000), 'Mechanisms of colonisation and colonisation resistance of the digestive tract', *Microbial Ecology in States of Health and Disease*, 2, 240–6.

Fuller, R. and Turvey, A. (1971), 'Bacteria associated with the intestinal wall of the fowl (*Gallus domesticus*)', *Journal of Applied Microbiology*, 34, 617–22.

Ghazikhanian, G. Y., Bland, M. C., Hofacre, C. L. and Froyman, R. (1997), 'Floor-pen study to determine the effect of Aviguard application to day-old turkey poults in reduction of clinical disease and intestinal colonization caused by *S. kedougou* infection', *Proceedings of the Salmonella and Salmonellosis Symposium*, Ploufragan, France, pp. 531–3.

Ginns, C. A., Browning, G. F., Benham, M. L., Anderson, G. A. and Whithear, K. G. (1996), 'Antimicrobial resistance and epidemiology of *Escherichia coli* in broiler chickens', *Avian Pathology* 25, 591–605.

Gleeson, T. M., Stavric, S. and Blanchfield, B. (1989), 'Protection of chicks against *Salmonella* infection with a mixture of pure cultures of intestinal bacteria', *Avian Diseases*, 33, 636–42.

Goren, E., de Jong, W. A., Doornenbal, P., Bolder, N. M., Mulder, R. W. A. W. and Jansen, A. (1988), 'Reduction of salmonella infection of broilers by spray application of intestinal microflora: a longitudinal study', *The Veterinary Quarterly*, 10, 249–55.

Goren, E., de Jong, W. A., Doornenbal, P., Koopman, J. P. and Kennis, H. M. (1984a), 'Protection of chicks against *Salmonella infantis* infection induced by strict anaerobically cultured intestinal microflora', *The Veterinary Quarterly*, 6, 22–6.

Goren, E., de Jong, W. A., Doornenbal, P., Koopman, J. P. and Kennis, H. M. (1984b), 'Protection of chicks against salmonella infection induced by spray application of intestinal microflora in the hatcher', *The Veterinary Quarterly*, 6, 73–9.

Greenberg, B. (1969), '*Salmonella* suppression by known populations of bacteria in flies', *J. Bacteriology*, 99, 629–35.

Gustafson, R. H. and Kobland, J. D. (1984), 'Factors influencing salmonella shedding in broiler chicken', *Journal of Hygiene, Cambridge*, 92, 385–94.

Hakkinen, M. and Schneitz, C. (1996), 'Efficacy of a commercial competitive exclusion product against chicken pathogenic *Escherichia coli* and *E. Coli* O157:H7', *The Veterinary Record*, 139, 139–41.

Hakkinen, M. and Schneitz, C. (1999), 'Efficacy of a commercial competitive exclusion product against Campylobacter jejuni', *British Poultry Science*, 40, 619–21.

Higgins, J. P., Higgins, S. E, Vicente, J. L., Wolfenden, A. D., Tellez, S. G. and Hargis, B. M. (2007), 'Temporal effects of lactic acid bacteria probiotic culture on *Salmonella* in neonatal chicks', *Poultry Science*, 86, 1662–6.

Hinton Jr., A., Corrier, D.E., Ziprin, R.L., Spates, G.E. and DeLoach, J.R. (1991a), 'Comparison of the efficacy of cultures of cecal anaerobes as inocula to reduce *Salmonella typhimurium* colonization in chicks with or without dietary lactose', *Poultry Science*, 70, 67–73.

Hinton, M. and Mead, G. C. (1991), '*Salmonella* control in poultry: the need for the satisfactory evaluation of probiotics for this purpose', *Letters in Applied Microbiology*, 13, 49–50.

Hinton, M., Mead, G. C. and Impey, C. S. (1991b), 'Protection of chicks against environmental challenge with *Salmonella enteritidis* by "competitive exclusion" and acid-treated feed', *Letters in Applied Microbiology*, 12, 69–71.

Hirn, J., Nurmi, E., Johansson, T. and Nuotio, L. (1992), 'Long-term experience with competitive exclusion and salmonellas in Finland', *International Journal of Food Microbiology*, 15, 281–5.

Hofacre, C. L., Beacorn, T., Collett, S. and Mathis, G. (2003), 'Using competitice exclusion, mannan-oligosaccharide and other intestinal products to control necrotic enteritis', *Journal of Applied Poultry Research*, 12, 60–4.

Hofacre, C. L., Froyman, R., George, B., Goodwin, M. A. and Brown, J. (1998), 'Use of Aviguard, Virginiamycin, or Bacitracin MD against Clostridium perfringens-associated necrotizing enteritis', *Journal of Applied Poultry Research*, 7, 412–18.

Hofacre, C. L., Primm, N. D., Vance, K., Goodwin, M. A. and Brown, J. (2000), 'Comparison of a lyophilized chicken-origin competitive exclusion culture, a lyophilized probiotic, and fresh turkey cecal material against *Salmonella* colonization', *Journal of Applied Poultry Research*, 9, 195–203.

Hofshagen, M. and Kaldhusdal, M. (1992), 'Barley inclusion and avoparcin supplementation in broiler diets. 1. Effect on small intestinal bacterial flora and performance', *Poultry Science*, 71, 959–69.

Holt, P. S., Macri, P. and Porter Jr., R. E. (1995), 'Microbial analysis of the early *Salmonella enteritidis* infection in molted and unmolted hens', *Avian Diseases*, 39, 55–63.

Holt, P. S. and Porter Jr., R. E. (1993), 'Effect of induced molting on the recurrence of a previous *Salmonella enteritidis* infection', *Poultry Science*, 72, 2069–78.

Hughes, R. J. (2008), 'Relationship between digesta transit time and apparent metabolisable energy value of wheat in chickens', *British Poultry Science*, 49, 716–20.

Humbert, F., Lalande, F., L'Hospitalier, R., Salvat, G. and Bennejean, G. (1991), 'Effect of four antibiotic additives on the *Salmonella* contamination of chicks protected by an adult caecal flora', *Avian Pathology*, 20, 577–84.

Hume, M. E., Byrd, J. A., Stanker, L. H. and Ziprin, R. L. (1998a), 'Reduction of caecal Listeria monocytogenes in Leghorn chicks following treatment with a competitive exclusion culture (Preempt™)', *Letters in Applied Microbiology*, 26, 432–6.

Hume, M. E., Corrier, D. E., Nisbet, D. J. and Deloach, J. R. (1998b), 'Early salmonella challenge time and reduction in chick cecal colonization following treatment with a characterized competitive exclusion culture', *Journal of Food Protection*, 61, 673–6.

Husu, J. R., Beery, J. T., Nurmi, E. and Doyle, M. P. (1990), 'Fate of Listeria monocytogenes in orally dosed chicks', *International Journal of Food Microbiology*, 11, 259–69.

Impey, C. S. and Mead, G. C. (1989), 'Fate of salmonellas in the alimentary tract of chicks pretreated with a mature caecal microflora to increase colonization resistance', *Journal of Applied Microbiology*, 66, 469–75.

Impey, C. S., Mead, G. C. and George, S. M. (1982), 'Competitive exclusion of salmonellas from the chick caecum using a defined mixture of bacterial isolates from the caecal microflora of an adult bird', *Journal of Hygiene, Cambridge*, 89, 479–90.

Impey, C. S., Mead, G. C. and George, S. M. (1984), 'Evaluation of treatment with defined and undefined mixtures of gut microorganisms for preventing *Salmonella* colonization in chicks and turkey poults', *Food Microbiology*, 1, 143–7.

Johnson, C.T. (1992), The use of an antimicrobial and competitive exclusion combination in *Salmonella*-infected pullet flocks', *International Journal of Food Microbiology*, 15, 293–8.

Kaldhusdal, M. and Hofshagen, M. (1992), 'Barley inclusion and avoparcin supplementation in broiler diets. 2. Clinical, pathological, and bacteriological findings in a mild form of necrotic enteritis', *Poultry Science*, 71, 1145–53.

Kaldhusdal, M., Schneitz, C., Hofshagen, M. and Skjerve, E. (2001), 'Reduced incidence of Clostridium perfringens-associated lesions and improved performance in broiler chickens treated with normal intestinal bacteria from adult fowl', *Avian Diseases*, 45, 149–56.

Khan, A. A., Nawaz, M. S., Summage West, C., Khan, S. A. and Lin, J. (2005), 'Isolation and molecular characterization of fluoroquinolone-resistant Escherichia coli from poultry litter', *Poultry Science*, 84, 61–6.

Klasing, K. C. (1998), 'Nutritional modulation of resistance to infectious diseases', *Poultry Science*, 77, 1119–25.

Kubena, L. F., Byrd, J. A., Moore, R. W., Ricke, S. C. and Nisbet, D. J. (2005), 'Effects of drinking water treatment on susceptibility of laying hens to *Salmonella* enteritidis during forced molt', *Poultry Science*, 84, 204–11.

Lafont, J. P., Brée, A., Naciri, M., Yvoré, P., Guillot, J. F. and Chaslus-Dancla, E. (1983), 'Experimental study of some factors limiting "competitive exclusion" of salmonella in chickens', *Research in Veterinary Science*, 34, 16–20.

La Ragione, R. M., Narbad, A., Gasson, M. J. and Woodward, M. J. (2004), '*In vivo* characterization of Lactobacillus johnsonii F19785 for use as a defined competitive exclusion agent against bacterial pathogens in poultry', *Letters in Applied Microbiology*, 38, 197–205.

Linton, A. H., Al-Chalaby, Z. A. M. and Hinton, M. H. (1985), 'Natural subclinical salmonella infection in chickens: A potential model for testing the effects of various procedures on salmonella shedding', *The Veterinary Record*, 116, 361–4.

Lloyd, A. B., Cumming, R. B. and Kent, R. D. (1974), 'Competitive exclusion as exemplified by *Salmonella typhimurium*', *Proceedings of the Australian Poultry Science Convention*, Hobart, Tasmania, pp. 185–6.

Lloyd, A. B., Cumming, R. B. and Kent, R. D. (1977), 'Prevention of *Salmonella typhimurium* infection in poultry by pretreatment of chickens and poults with intestinal extracts', *Australian Veterinary Journal*, 53, 82–7.

Lumpkins, B. S., Batal, A. B. and Lee, M. D. (2010), 'Evaluation of the bacterial community and intestinal development of different genetic lines of chickens', *Poultry Science*, 89, 1614–21.

Macri, N. P., Porter Jr., R. E. and Holt, P.S. (1997), 'The effects of induced molting on the severity of acute intestinal inflammation caused by *Salmonella enteritidis*', *Avian Diseases*, 41, 117–24.

Maijala, R., Ranta, J., Seuna, E. and Peltola, J. (2005), 'The efficiency of the Finnish *Salmonella* control programme', *Food Control*, 16, 669–75.

Majowicz, S. E., Musto, J., Scallan, E., Angulo, F. J., O'Brein, S. J. and Jones, T. F. (2010), 'The global burden of nontyphoidal *Salmonella* gastroenteritis', *Clinical Infectious Diseases*, 50, 882–9.

Martin, C., Dunlap, E., Caldwell, S., Barnhart, E., Keith, N. and Deloach, J. R. (2000), 'Drinking water delivery of a defined competitive exclusion culture (Preempt) in 1-day-old broiler chicks', *The Journal of Applied Poultry Research*, 9, 88–91.

McReynolds, J. L., Caldwell, D. Y., Barnhart, E. T., Deloach, J. R., McElroy, A. P., Moor, R. W., Hargis, B. M. and Caldwell, D. J. (2000), 'The effect of *in ovo* or day-of-hatch subcutaneous antibiotic administration on competitive exclusion culture (Preempt) estblishment in neonatal chickens', *Poultry Science*, 79, 1524–30.

Mead, G. C. (2000), 'Prospects for "competitive exclusion" treatment to control salmonellas and other foodborne pathogens in poultry', *The Veterinary Journal*, 159, 111–23.

Mead, G. C. (2002), 'Factors affecting intestinal colonisation of poultry by campylobacter and role of microflora control', *World's Poultry Science Journal*, 58, 169–78.

Mead, G. C. and Adams, B. W. (1975), 'Some observations on the caecal microflora of the chick during the first two weeks of life', *British Poultry Science*, 16, 169–76.

Mead, G. C. and Barrow P. A. (1990), '*Salmonella* control in poultry by 'competitive exclusion' or immunization', *Letters in Applied Microbiology*, 10, 221–7.

Mead, G. C., Barrow, P. A., Hinton, M. H., Humbert, F., Impey, C. S., Lahellec, C., Mulder, R. W. A. W., Stavric, S. and Stern, N. J. (1989a), 'Recommended assay for treatment of chicks to prevent *Salmonella* colonization by competitive exclusion', *Journal of Food Protection*, 52, 500–2.

Mead, G. C. and Impey, C. S. (1986), 'Current progress in reducing salmonella colonization of poultry by competitive exclusion', *Journal of Applied Microbiology*, Symposium Supplement 61, 65S–75S.

Mead, G. C., Schneitz, C. E., Nuotio, L. O. and Nurmi, E. V. (1989b), 'Treatment of chicks using competitive exclusion to prevent transmission of *Salmonella enteritidis* in delivery boxes', *IXth International Congress of the World Veterinary Poultry Association* (Abstract), Brighton, UK, p. 115.

Mead, G. C., Scott, M. J., Humphrey, T. J. and McAlpine, K. (1996), 'Observations on the control of *Campylobacter jejuni* infection of poultry by "competitive exclusion"', *Avian Pathology*, 25, 69–79.

Meijerhof, R. and Hulet, R. M. (1997), '*In ovo* injection of competitive exclusion culture in broiler hatching eggs', *The Journal of Applied Poultry Research*, 6, 260–6.

Menconi, A., Wolfenden, A. D., Shivaramaiah, S., Terraes, J. C., Urbano, T., Kuttel, J., Cremer, C., Hargis, B. M. and Tellez, G. (2011), 'Effect of lactic acid bacteria probiotic culture for the treatment of *Salmonella enterica* serovar Heidelberg in neonatal broiler chickens and turkey poults', *Poultry Science*, 90, 561–5.

Meynell, G. G. (1963), 'Antibacterial mechanisms of the mouse gut. II. The role of Eh and volatile fatty acids in the normal gut', *British Journal of Experimental Pathology*, 44, 209–19.

Meynell, G. G. and Subbaiah, T. V. (1963), 'Antibacterial mechanisms of the mouse gut I: Kinetics of infection by *Salmonella* typhi-murium in normal and streptomycin-treated mice studied with abortive transductans', *British Journal of Experimental Pathology*, 44, 197–208.

Milner, K. C. and Shaffer, M. F. (1952), 'Bacteriologic studies of experimental salmonella infections in chick', *The Journal of Infectious Diseases*, 90, 81.

Newman, K. E. and Spring, P. (1996), 'Effect of a commercial competitive exclusion culture (AviFree) on *Salmonella typhimurium* concentration in broiler chicks', *12th Annual symposium on Biotechnology in the Feed Industry* (Poster), Enclosure Code AVI 2.1.

Nisbet, D. J., Corrier, D. E. and DeLoach, J. R. (1995), 'Probiotic for control of *Salmonella*', *United States Patent 5,478,557*.

Nisbet, D. J., Corrier, D. E., Ricke, S. C., Hume, M. E., Byrd II, J. A. and DeLoach, J. R. (1996a), 'Maintenance of the biological efficacy in chicks of a cecal competitive-exclusion culture against *Salmonella* by continuous-flow fermentation', *Journal of Food Protection*, 59, 1279–83.

Nisbet, D. J., Corrier, D. E., Ricke, S. C., Hume, M. E., Byrd II, J. A. and DeLoach, J. R. (1996b), 'Cecal propionic acid as a biological indicator of the early establishment of a microbial ecosystem inhibitory to *Salmonella* in chicks', *Anaerobe*, 2, 345–50.

Nisbet, D. J., Corrier, D. E., Scanlan, C. M., Hollister, A. G., Beier, R. C. and Deloach, J. R. (1993), 'Effect of a defined continuous-flow derived bacterial culture and dietary lactose on *Salmonella* typhimurium colonization in broiler chickens', *Avian Diseases* 37, 1017–25.

Nisbet, D. J., Tellez, G. I., Lowry, V. K., Anderson, R. C., Garcia, G., Nava, G., Kogut, M. H., Corrier, D. E. and Stanker, L. H. (1998), 'Effect of a commercial competitive exclusion culture (Preempt) on mortality and horizontal transmission of *Salmonella gallinarum* in broiler chickens', *Avian Diseases*, 42(4), 651–6.

Nuotio, L., Schneitz, C., Halonen, U. and Nurmi, E. (1992), 'Use of competitive exclusion to protect newly-hatched chicks against intestinal colonisation and invasion by *Salmonella enteritidis* PT4', *British Poultry Science*, 33, 775–9.

Nuotio, L., Schneitz, C. and Nilsson, O. (2013), 'Effect of competitive exclusion in reducing the occurrence of *Escherichia coli* producing extended-spectrum β-lactamases in the ceca of broiler chicks', *Poultry Science*, 92, 250–4.

Nurmi, E. and Nuotio, L. (1994), 'Safety requirements for commercial competitive exclusion products', *Proceedings of The 9th European Poultry Conference*, The World's Poultry Science Association, Vol. II, Glasgow, UK, pp. 90–3.

Nurmi, E., Nuotio, L. and Schneitz, C. (1992), 'The competitive exclusion concept: development and future', *International Journal of Food Microbiology*, 15, 237–40.

Nurmi, E. V. and Rantala, M. (1973), 'New aspects of *Salmonella* infection in broiler production', *Nature*, 241, 210.

Nurmi, E. V., Schneitz, C. E. and Mäkelä, P. H. (1987), 'Process for the production of a bacterial preparation for the prophylaxis of intestinal disturbances in poultry', *United States patent 4,689, 226*.

Palmu, L. and Camelin, I. (1997), 'The use of competitive exclusion in broilers to reduce the level of *Salmonella* contamination on the farm and at the processing plant', *Poultry Science*, 76, 1501–5.

Pan, D. and Yu, Z. (2014), 'Intestinal microbiome of poultry and its interaction with host and diet', *Gut Microbes*, 5(1), 108–19.

Pedroso, A. A., Menten, J. F. M., Lambais, M. R., Racanicci, A. M. C., Longo, F. A. and Sorbara, J. O. B. (2006), 'Intestinal community and growth performance of chickens fed diets containing antibiotics', *Poultry Science*, 85, 747–52.

Pivnick, H., Barnum, D., Stavric, S., Gleeson, T. and Blanchfield, B. (1985), 'Investigations on the use of competitive exclusion to control *Salmonella* in poultry', in Snoeyenbos, G. H. (ed.), *Proceedings of the International Symposium on Salmonella*. USA: American Association of Avian Pathologists, pp. 80–7.

Pivnick, H. and Nurmi, E. (1982), 'The Nurmi Concept and its role in the control of *Salmonella* in poultry', in Davies, R. (ed.), *Developments in Food Microbiology – 1*, Essex, UK: Applied Science Publishers Ltd.

Raevuori, M., Seuna, E. and Nurmi, E. (1978), 'An epidemic of *Salmonella infantis* infection in Finnish broiler chickens in 1975', *Acta Veterinaria Scandinavica*, 19, 317–30.

Rantala, M. and Nurmi, E. (1973), 'Prevention of the growth of *Salmonella infantis* in chicks by the flora of the alimentary tract of chickens', *British Poultry Science*, 14, 627–30.

Rantala, M. (1974), 'Cultivation of a bacterial flora able to prevent the colonization of *Salmonella* infantis in the intestines of broiler chickens, and its use', *Acta Path. Micro. Scand. Sect B*, 82, 75–80.

Raminez, G. A., Sarlin, L. L., Caldwell, D. J., Yezak, C. R., Hume, D. E., Corrier, D. E., DeLoach, J. R. and Hargis, B. M. (1997), 'Effect of feed withdrawal on the incidence of salmonella in the crops and ceca of market age broiler-chickens', *Poultry Science*, 76, 654–6.

Rao, K. and Safdar, N. (2016), 'Fecal microbiota transplantation for the treatment of Clostridium difficile infection', *Journal of Hospital Medicine*, 11(1), 56–61.

Revolledo, L., Ferreira, A. J. P. and Mead, G. C. (2006), 'Prospects in *Salmonella* control: Competitive exclusion, probiotics, and enhancement of avian intestinal immunity', *Journal of Applied Poultry Research*, 15, 341–51.

Revolledo, L., Ferreira, C. S. A. and Ferreira, A. J. P. (2009), 'Prevention of *Salmonella* Typhimurium colonization and organ invasion by combination treatment in broiler chicks', *Poultry Science*, 88, 734–43.

Reynolds, D. J., Davies, R. H., Richards, M. and Wray, C. (1997), 'Evaluation of combined antibiotic and competitive exclusion treatment in broiler breeder flocks infected with *Salmonella enterica* serovar Enteritidis', *Avian Pathology*, 26, 83–95.

Riddle, C. and Kong, X. (1992), 'The influence of diet on necrotic enteritis in broiler chicks', *Avian Diseases*, 36, 499–503.

Rigby, C. E. and Pettit, J. R. (1980), 'Observations on competitive exclusion for preventing *Salmonella typhimurium* infection of broiler chickens', *Avian Diseases*, 24, 604–15.

Robyn, J., Rasschaert, G., Hermans, D., Pasmans, F. and Heyndrickx, M. (2013), '*In vivo* broiler experiments to assess anti-*Campylobacter jejuni* activity of a live *Enterococcus faecalis* strain', *Poultry Science*, 92, 265–71.

Rolfe, R. D. (1991), 'Population dynamics of the intestinal tract', in Blankenship, L. C. (ed.), *Colonization Control of Human Bacterial Enteropathogens in Poultry*, San Diergo, USA: Academic Press, Inc., pp. 59–75.

Savage, D. C. (1983), 'Mechanisms by which indigenous microorganisms colonize gastrointestinal epithelial surfaces', *Progress in Food and Nutrition Science*, 7, 65–74.

Schleifer, J. H. (1985), 'A review of the efficacy and mechanism of competitive exclusion for the control of salmonella in poultry', *World's Poultry Science Journal*, 41, 72–83.

Schneitz, C. (1992), 'Automated droplet application of a competitive exclusion preparation', *Poultry Science*, 71, 2125–8.

Schneitz, C. and Hakkinen, M. (1998a), 'Comparison of two different types of competitive exclusion products', Letters *in* Applied Microbiology, 26, 338–41.

Schneitz, C., Hakkinen, M., Nuotio, L., Nurmi, E. and Mead, G. (1990), 'Droplet application for protecting chicks against salmonella colonisation by competitive exclusion', *The Veterinary Record*,126, 510.

Schneitz, C. and Hakkinen, M. (2016a), 'The efficacy of a commercial competitive exclusion product on *Campylobacter* colonization in broiler chickens in a 5-week pilot-scale study', *Poultry Science*, 1–4.

Schneitz, C., Kiiskinen, T., Toivonen, V. and Näsi, M. (1998b), 'Effect of Broilact on the physico-chemical conditions and nutrient digestibility in the gastrointestinal tract of broilers', *Poultry Science*, 77, 426–32.

Schneitz, C., Koivunen, E., Tuunainen, P. and Valaja, J. (2016b), 'The effects of a competitive exclusion product and two probiotics on *Salmonella* colonization and nutrient digestibility in broiler chickens', accepted for publication in *Journal of Applied Poultry Research*, 1–11.

Schneitz, C. and Mead, G.C. (2010), 'Competitive exclusion treatment in poultry management', in Guerrero-Legarreta, I. and Hui, Y. H. (eds), *Handbook of Poultry Science and Technology*, USA: John Wiley & Sons, Inc, pp. 19–14.

Schneitz, C., Nuotio, L. and Lounatmaa, K. (1993), 'Adhesion of *Lactobacillus acidophilus* to avian intestinal epithelial cells mediated by the crystalline bacterial cell surface layer (S-layer)', *Journal of Applied Microbiology*, 74, 290–4.

Schneitz, C. and Nuotio, L. (1992), 'Efficacy of different microbial preparations for controlling salmonella colonisation in chicks and turkey poults by competitive exclusion', *British Poultry Science*, 33, 207–11.

Schneitz, C., Nuotio, L., Kiiskinen, T. and Nurmi, E. (1991), 'Pilot-scale testing of the competitive exclusion method in chickens', *British Poultry Science*, 32, 877–80.

Schneitz, C. and Renney, D. J. (2003), 'Effect of a commercial competitive exclusion product on the colonization of *Salmonella* infantis in day-old pheasant chicks', *Avian Diseases*, 47, 1448–51.

Schoeni, J. J. and Wong, C. L. (1994), 'Inhibition of *Campylobacter jejuni* colonization in chicks by defined competitive exclusion bacteria', *Applied and Environmental Microbiology*, 1191–7.

Seuna, E. (1979), 'Sensitivity of young chickens to *Salmonella typhimurium var. copenhagen* and *S. infantis* infection and the preventive effect of cultured intestinal microflora', *Avian Diseases*, 23, 392–400.

Seuna, E., Raevuori, M. and Nurmi, E. (1978), 'An epizootic of *Salmonella typhimurium* var. *copenhagen* in broilers and the use of cultured chicken intestinal flora for its control', *British Poultry Science*, 19, 309–14.

Shin, J. W., Kang, J. K., Jang, K.-I. and Kim, K. Y. (2002), 'Intestinal colonization characteristics of Lactobacillus spp. isolated from chicken cecum and competitive inhibition against *Salmonella* typhimurium', *Journal of Microbiology and Biotechnology*, 12, 576–82.

Simjee, S., McDermot, P. F., White, D. G., Hofacre, C., Berghaus, R. D., Carter, P. J., Stewart, L., Liu, T. and Maier, M. and Maurer, J. J. (2007), 'Antibacterial suspectibility and distribution of antimicrobial-resistance genes among Enterococcus and coagulase-negative Staphylococcus isolates recovered from poultry litter', *Avian Diseases*, 51, 884–92.

Smith, H. and Green, S. I. (1980), 'Effect of feed additives on the incidence of naturally acquired salmonella in turkeys', *The Veterinary Record* 107, 289.

Smith, H. W. and Tucker, J. F. (1978),'The effect of antimicrobial feed additives on the colonization of the alimentary tract of chickens by *Salmonella typhimurium*', *Journal of Hygiene, Cambridge*, 80, 217–31.

Smith, H. W. and Tucker, J. F. (1980), 'Further observations on the effect of feeding diets containing avoparcin, bacitracin and sodium arsenilate on the colonization of the alimentary tract of poultry by *Salmonella* organisms', *Journal of Hygiene, Cambridge*, 84, 137–50.

Snoeyenbos, G. H., Weinack, O. M. and Smyser, C. F. (1978), 'Protecting chicks and poults from salmonellae by oral administration of "normal" gut microflora', *Avian Diseases*, 22, 273–86.

Snoeyenbos, G. H., Weinack, O. M. and Smyser, C. F. (1979), 'Further studies on competitive exclusion for controlling salmonellas in chickens', *Avian Diseases*, 24, 904–14.

Snoeyenbos, G. H., Weinack, O. M. and Soerjadi, A. S. (1983), 'Our current understanding of the role of native microflora in limiting some bacterial pathogens of chickens and turkeys', *Australian Veterinary Poultry Association and International Union of Immunological Societies, Proceedings No. 66, Disease Prevention and Control in Poultry Production*, Sydney, Australia, pp. 45–51.

Snoeyenbos, G. H., Weinack, O. M., Soerjadi-Liem, A. S., Miller, B. M., Woodward, D. E. and Weston, C. R. (1985), 'Large-scale trials to study competitive exclusion of salmonella in chickens', *Avian Diseases*, 29, 1004–11.

Soerjadi, A.S., Stehman, S.M., Snoeyenbos, G.H., Weinack, O.M. and Smyser, C.F. (1981b), 'The influence of lactobacilli on the competitive exclusion of paratyphoid salmonellae in chickens', *Avian Diseases* 25, 1027–33.

Soerjadi, A. S., Rufner, R., Snoeyenbos, G. H. and Weinack, O. M. (1982b), 'Adherence of salmonellae and native gut microflora to the gastrointestinal mucosa of chicks', *Avian Diseases*, 26(3), 576–84.

Soerjadi, A. S., Stehman, S. M., Snoeyenbos, G. H., Weinack, O. M. and Smyser, C. F. (1981a), 'Some measurements of protection against paratyphoid *Salmonella* and *Escherichia coli* by competitive exclusion in chickens', *Avian Diseases*, 24, 706–12.

Soerjadi, A. S., Snoeyenbos, G. H. and Weinack, O. M. (1982a), 'Intestinal colonization and competitive exclusion of *Campylobacter fetus subsp. jejuni* in young chicks', *Avian Diseases*, 26, 520–4.

Soerjadi-Liem, A. S., Snoeyenbos, G. H. and Weinack, O. M. (1984a), 'Comparative studies on competitive exclusion of three isolates of *Campylobacter fetus subsp. jejuni* in chickens by native gut microflora', *Avian Diseases*, 28, 139–46.

Soerjadi-Liem, A. S., Snoeyenbos, G. H. and Weinack, O. M. (1984b), 'Establishment and competitive exclusion of *Yersinia enterocolitica* in the gut of monoxenic and holoxenic chicks', *Avian Diseases*, 28, 256–60.

Spencer, J. L. and Garcia, M. M. (1995), 'Resistance of chicks and poults fed vermicompost to caecal colonization by *Salmonella*', *Avian Pathology*, 24, 157–70.

Stavric, S. (1987), 'Microbial colonization control of chicken intestine using defined cultures', *Food Technology*, 41, 93–8.

Stavric, S. (1992), 'Defined cultures and prospects', *International Journal of Food Microbiology*, 15, 245–63.

Stavric, S. and D'Aoust, J.-Y. (1993), 'Undefined and defined bacterial preparations for the competitive exclusion of *Salmonella* in poultry – a review', *Journal of Food Protection*, 56, 173–80.

Stavric, S., Buchanan, B. and Gleeson, T. M. (1992), 'Competitive exclusion of *Escherichia coli* O157:H7 from chicks with anaerobic cultures of faecal microflora', *Letters in Applied Microbiology*, 14, 191–3.

Stavric, S., Gleeson, T. M. and Blanchfield, B. (1991), 'Effect of avian intestinal microflora possessing adhering and hydrophobic properties on competitive exclusion of *Salmonella typhimurium* from chicks', *Journal of Applied Microbiology*, 12, 414–21.

Stavric, S., Gleeson, T. M., Blanchfield, B. and Pivnick, H. (1985), 'Competitive exclusion of *Salmonella* from newly hatched chicks by mixtures of pure bacterial cultures isolated from fecal and cecal contents of adult birds', *Journal of Food Protection*, 48, 778–82.

Stavric, S., Gleeson, T. M., Blanchfield, B. and Pivnick, H. (1987), 'Role of adhering microflora in competitive exclusion of *Salmonella* from young chicks', *Journal of Food Protection*, 50 (11), 928–32.

Stern, N. J. (1990), 'Influence of competitive exclusion on chicken cecal colonization by *Campylobacter jejuni*', *Poultry Science*, 69(Suppl.1), 130.

Stern, N. J. (1994), 'Mucosal competitive exclusion to diminish colonization of chickens by Campylobacter jejuni', *Poultry Science*, 73, 402–7.

Stern, N. J., Bailey, J. S., Cox, N. A. and Blankenship, L. C. (1995), 'Mucosal competitive exclusion flora', *United States Patent 5,451,400*.

Sternberg-Lewerin, S., Boquist, B., Engström, P. and Häggblom, P. (2005), 'The effective control of *Salmonella* in Swedish poultry', in Mead, G.C. (ed.), *Food Safety in the Poultry Industry*, Cambridge, UK: Woodhouse Publishing in Food Science and Technology, CRC Press.

SWARM (2011), *Swedish Veterinary Antimicrobial Resistance Monitoring 2010*, in Bengtsson, B., Greko, C., Grönlund-Andersson, U. and Landen, A. (eds), Uppsala Sweden: The National Veterinary Institute (SVA), pp. 29–30.

SWARM (2012), *Swedish Veterinary Antimicrobial Resistance Monitoring 2011*, in Bengtsson, B., Greko, C., Nilsson, O. and Landen, A. (eds), Uppsala Sweden: The National Veterinary Institute (SVA), pp. 43–6.

Telg, B. E. and Caldwell, D. J. (2009), 'Efficacy testing of a defined competitive exclusion product in combination with fructooligosaccharide for protection against *Salmonella* Typhimurium challenge in broiler chicks', *Journal of Applied Poultry Research*, 18, 521–8.

van der Waaij, D., Berghuis-de Vries, J. M. and Lekkerkerk-van der Wees, J. E. C. (1971), 'Colonization resistance of the digestive tract in conventional and antibiotic-treated mice', *Journal of Hygiene*, 69, 405–511.

van der Wielen, P. W. J. J., Biesterveld, S., Notermans, S., Hofstra, H., Urlings, B. A. P. and van Knapen, F. (2000), 'Role of volatile fatty acids in development of the cecal microflora in broiler chickens during growth', *Applied and Environmental Microbiology*, 66 (6), 2536–40.

van der Wielen, P. W. J. J., Keuzenkamp, D. A., Lipman, L. J. A., van der Knapen, F. and Biesterveld, S. (2002), 'Spatial and temporal variation of the intestinal bacterial community in commercially raised broiler chickens during growth', *Microbial Ecology*, 44, 286–93.

Wei, S., Gutek, A., Lilburn, M. and Yu, Z. (2013), 'Abundance of pathogens in the gut and litter of broiler chickens as affected by bacitracin and litter management', *Veterinary Microbiology*, 166, 595–601.

Weinack, O. M., Snoeyenbos, G. H. and Smyser, C. F. (1979), 'A supplemental test system to measure competitive exclusion of salmonellae by native microflora in the chicken gut', *Avian Diseases*, 23, 1019–30.

Weinack, O. M., Snoeyenbos, G. H., Smyser, C. F. and Soerjadi, A. S. (1981), 'Competitive exclusion of intestinal colonization of *Escherichia coli* in chicks', *Avian Diseases*, 25, 696–705.

Weinack, O. M., Snoeyenbos, G. H., Smyser, C. F. and Soerjadi, A. S. (1982), 'Reciprocal competitive exclusion of *Salmonella* and *Escherichia coli* by native intestinal microflora of the chicken and turkey', *Avian Diseases*, 26, 585–95.

Weinack, O. M., Snoeyenbos, G. H., Smyser, C. F. and Soerjadi-Liem, A. S. (1984), 'Influence of *Mycoplasma gallisepticum*, infectious bronchitis, and cyclophosphamide on chickens protected by native intestinal microflora against *Salmonella typhimurium* or *Escherichia coli*', *Avian Diseases*, 28, 416–25.

Weinack, O. M., Snoeyenbos, G. H. and Soerjadi-Liem, A. S. (1985a), 'Further studies on competitive exclusion of *Salmonella typhimurium* by lactobacilli in chicken', *Avian Diseases*, 29, 1273–7.

Weinack, O. M., Snoeyenbos, G. H., Soerjadi-Liem, A. S. and Smyser, C. F. (1985b), 'Influence of temperature, social and dietary stress on development and stability of protective microflora in chickens against *S. typhimurium*', *Avian Diseases*, 29, 1177–83.

WHO (1997), 'The medical impact of the use of antimicrobials in food animals', Report of a WHO Meeting, Berlin, Germany, WHO/EMC/ZOO/97.4.

WHO (2014), 'Antimicrobial resistance: global report on surveillance', Geneva Switzerland, ISBN 978 92 4 156474 8, Available at www.who.int/mediacentre/news/releases/2014/amr-report/en.

WHO (World Health Organization) (1994), 'Report of the WHO-FEDESA-FEP workshop on competitive exclusion, vaccination and antimicrobials in *Salmonella* control in poultry', Geneva, Switzerland, WHO/CDS/VPH/94.134.

WHO (World Health Organization) (1980), 'Report of the WHO/WAVFH Round Table Conference on the Present Status of the *Salmonella* Problem (Prevention and Control)', Bilthoven, the Netherlands, WHO/VPH/81.27.

Wierup, M. (2013), '*Salmonella* in feed', in Barrow, P. A. and U Methner, U. (eds), *Salmonella in Domestic Animals*, Wallingford, UK: CABI International, p. 377.

Wierup, M., Engström, B., Engvall, A. and Wahlström, H. (1995), 'Control of *Salmonella* enteritidis in Sweden', *International Journal of Food Microbiology*, May, 25(3), 219–26.

Wierup, M., Wahlström, H. and Engström, B. (1992), 'Experience of a 10-year use of competitive exclusion treatment as part of the *Salmonella* control programme in Sweden', *International Journal of Food Microbiology*, 5, 287–91.

Wierup, M., Wold-Troell, M., Nurmi, E. and Hakkinen, M. (1988), 'Epidemiological evaluation of the salmonella-controlling effect of a nationwide use of a competitive exclusion culture in poultry', *Poultry Science*, 67, 1026–33.

Wigley, P, (2004), 'Genetic resistance to *Salmonella* infection in domestic animals', *Research in Veterinary Science*, 76, 165–9.

Wolfenden, A. D., Vicente, J. L., Bielke, L. R., Pixley, C. M., Higgins, S. E., Donoghue, D. J., Donoghue, A. M., Hargis, B. M. and Tellez, G. (2007), 'Effect of defined competitive exclusion culture for prophylaxis and reduction of horizontal transmission of *Salmonella enteritidis* in broiler chickens', *International Journal of Poultry Science*, 6 (7), 489–92.

Yeoman, C. J., Chia, N., Jeraldo, P. Sipos, M., Goldenfeld, N. D. and White, B. A. (2012), 'The microbiome of the chicken gastrointestinal tract', *Animal Health Research Reviews*, 13, 89–99.

Zhang, G., Ma. L. and Doyle, M. P. (2007a), 'Potential competitive exclusion Bacteria from poultry inhibitory to Campylobacter jejuni and *Salmonella*', *Journal of Food Protection*, 70, 867–73.

Zhang, G., Ma. L. and Doyle, M. P. (2007b), 'Salmonellae reduction in poultry by competitive exclusion bacteria Lactobacillus salivarius and Streptococcus crispatus', *Journal of Food Protection*, 70, 874–8.

Chapter 8

Leg disorders in poultry: bacterial chondronecrosis with osteomyelitis (BCO)

Robert F. Wideman Jr., University of Arkansas, USA

1 Introduction

Selection for rapid growth has consistently been implicated in the susceptibility of broilers and turkeys to leg weakness (Wise, 1970a; Nestor, 1984; Kestin et al., 1999, 2001; Sorensen, 1992; Nestor and Anderson, 1998). The highest incidences of leg weakness typically occur in flocks with the fastest growth, and lameness incidences usually are reduced by management strategies that decelerate early growth rates (Riddell, 1983; Havenstein et al., 1994; McNamee et al., 1999; Williams et al., 2000; Kestin et al., 2001; Bradshaw et al., 2002; Julian, 2005).

Growth-associated leg weakness encompasses a variety of distinct pathologies including

- tibial dyschondroplasia (Leach and Nesheim, 1965; Riddell, 1976; Sheridan et al., 1978; Walser et al., 1982; Zhang et al., 1995; Kuhlers and McDaniel, 1996),
- spondylolisthesis (Wise, 1970b, 1973; Riddell, 1973, 1976; Khan et al., 1977),
- valgus–varus deformities (Mercer and Hill, 1984; Sorensen, 1992),
- twisted leg or perosis (Somes, 1969; Haye and Simons, 1978; Mercer and Hill, 1984; Sorensen, 1992),
- bacterial osteomyelitis (McNamee et al., 1998; McNamee and Smyth, 2000; Wideman et al., 2012, 2013).

These distinctively different pathologies are linked by the hypothesis that, in susceptible subsets of the population, the skeleton does not mature rapidly enough to support

http://dx.doi.org/10.19103/AS.2016.0011.23

exponential increases in body mass (Wise, 1970a; Kestin et al., 2001). The concept of 'skeletal maturation' broadly includes processes such as chondrocyte proliferation and differentiation within the growth plates, bone mineralization and remodelling, vasculogenesis and angiogenesis, and tendon and ligament development.

This chapter focuses on a specific category of lameness known as bacterial chondronecrosis with osteomyelitis (BCO), widely recognized as one of the most common causes of lameness in broilers (McNamee and Smyth, 2000). The pathogenesis of BCO reflects the fact that bacterial infection and necrosis typically develop in bones that are subjected to repeated mechanical stress, primarily but not exclusively, including the proximal femora, the proximal tibiae and the flexible thoracic vertebrae (Wise, 1971; Nairn, 1973; Julian, 1985; Wyers et al., 1991; McNamee et al., 1998; McNamee and Smyth, 2000; Stalker et al., 2010; Kense and Landman, 2011; Wideman et al., 2012; Wideman and Prisby, 2013). Several topics are addressed. The pathogenesis of BCO is summarized. Experimental models that successfully trigger BCO are reviewed. Sources of bacteria that infect the bones are discussed, with the primary emphasis on bacterial translocation across the gastrointestinal epithelium. Finally, we review the efficacy of probiotics as a prophylactic treatment for BCO.

2 The pathogenesis of BCO

There are a number of stages which lead to BCO. High growth rates in poultry lead to the development of unstable growth plates in bone. When subjected to mechanical stress, these develop microfractures which allow bacteria to infect the growth plate. The resulting damage leads to lameness.

2.1 Osteochondrosis

Growth of the leg bones includes elongation accomplished via growth plates located at both ends of the shaft (diaphysis), as well as marked increases in the overall diameter attributable to highly dynamic remodelling of cortical bone (Wideman and Prisby, 2013). As demonstrated by Applegate and Lilburn (2002), the femora and tibiae increase approximately fourfold in length during the first six weeks posthatch, with mid-shaft diameters increasing three- to fivefold during the same interval. Other investigators have published similar estimates of rapid leg bone growth in broilers (Wise, 1970a; Riddell, 1975; Thorp, 1988c; Bond et al., 1991; Williams et al., 2000; Yalcin et al., 2001; Yair et al., 2012). As shown in Fig. 1E, the growing ends of avian bones are comprised of a cartilagenous epiphysis, a growth plate or physis consisting of columns of chondrocytes (cartilage cells) in sequential maturational layers and a metaphysis consisting of calcifying chondrocytes and the newly formed spicules of trabecular bone that provide a support scaffolding for the growth plate. The pace of bone elongation is determined by the rate of mitosis within the proliferating zone of the growth plate, which, as it thickens, expands the physeal–epiphyseal boundary longitudinally at the end of the shaft (Wideman and Prisby, 2013). Maximal growth rates in poultry are associated with the presence of unusually long, unevenly aligned and mechanically fragile columns of unmineralized chondrocytes extending from the growth plate into the metaphysis. The proximal ends of the leg bones elongate twice as fast and have significantly thicker and more fragile growth plates when compared with the distal ends of the same bones (Hales, 1727; Church and Johnson, 1964; Lutfi, 1970a; Kember and Kirkwood, 1987; Leach and Gay, 1987; Thorp, 1988c; Kirkwood et al., 1989a,b; Kember et al., 1990).

Mechanical stress chronically exerted on thick, unstable growth plates can create osteochondrotic clefts or microfractures between and within the cartilage layers (e.g. epiphyseal–physeal osteochondrosis or osteochondrosis dissecans). The microfractures often transect local blood vessels, thereby causing focal ischaemia and necrosis. Osteochondrosis is routinely observed during histological evaluations of the leg bones and vertebrae of apparently healthy broilers and turkeys, leading to the assumption that osteochondrotic microfractures of the epiphyseal–physeal cartilage are a common noninfective, noninflammatory skeletal disorder (Wise, 1970b, 1973; McCaskey et al., 1982; Riddell et al., 1983; Duff, 1990a,b; Thorp, 1994). However, osteochondrotic clefts and associated regions of focal necrosis clearly constitute wound sites with an exposed collagen matrix that facilitates colonization by opportunistic bacterial pathogens. Accordingly, although clinical lameness has not been directly correlated with osteochondrosis per se, the resulting wound sites in fragile growth plates do provide suitable niches for secondary bacterial infections, thereby enhancing the susceptibility to osteomyelitis (Carnaghan, 1966; Wise, 1971; McNamee et al., 1998; Martin et al., 2011; Wideman and Prisby, 2013; Wideman, 2014).

2.2 Haematogenous bacterial distribution

Bacteria are transmitted to chicks from breeder parents, contaminated egg shells or hatchery sources (Skeeles, 1997; McCullagh et al., 1998; Rodgers et al., 1999; McNamee and Smyth, 2000; Stalker et al. 2010; Kense and Landman, 2011) or enter the chick's circulation via translocation through the integument, respiratory system or gastrointestinal tract (Mutalib et al., 1983a,b; Andreasen et al., 1993; Thorp et al., 1993; McNamee et al., 1999; Martin et al., 2011). They spread haematogenously and can reach the growth plate via terminal capillary loops that have a discontinuous (fenestrated) endothelium with openings large enough to permit cellular elements in the blood to pass into the cartilaginous matrix (Beaumont, 1967; Lutfi, 1970b; Hunt et al., 1979; Howlett, 1980; Howlett et al., 1984; Emslie and Nade, 1983, 1985). Haematogenously distributed bacteria possessing the specific ability to bind to exposed bone collagen appear in some cases to be more virulent in their capacity to trigger osteomyelitis (Kibenge et al., 1983; Smeltzer and Gillaspy, 2000). The bacteria form obstructive emboli in the epiphyseal and metaphyseal vascular plexuses, adhere directly to the exposed cartilage matrix and colonize osteochondrotic clefts and zones of necrosis. Bacterial foci and sequestrae within infected bone tissue are notoriously inaccessible to antibiotics and the immune system (Emslie and Nade, 1983; Emslie et al., 1983, 1984; Kibenge et al., 1983; Speers and Nade, 1985; Alderson et al., 1986a,b; Alderson and Nade, 1987; Thorp, 1988b; Thorp et al., 1993; McNamee et al., 1998, 1999; McNamee and Smyth, 2000; Smeltzer and Gillaspy, 2000; Kense and Landman, 2011; Martin et al., 2011).

2.3 Bacterial osteomyelitis and clinical lameness

Bacterial sequestrae rapidly expand into focal zones of necrosis or large fibrinonecrotic abscesses in the metaphysis of infected bones, destroying the vasculature and eliminating struts of metaphyseal trabecular bone that normally provide structural support to the growth plates (Emslie et al., 1983, 1984; Emslie and Nade, 1983, 1985; Alderson et al.,

1986b; Thorp, 1988a; Daum et al., 1990; Wyers et al., 1991; Thorp et al., 1993; Skeeles, 1997; McNamee et al., 1998; Joiner et al., 2005; Wideman et al., 2012; Wideman and Prisby, 2013). Terminal BCO presents as necrotic degeneration and bacterial infection primarily within the proximal ends of the femora and tibiae, as well as in the growth plates of the flexible thoracic vertebrae (Fig. 1 and 2).

High incidences of femoral, tibial and vertebral BCO lesions have been observed in lame broilers from commercial flocks. The distal ends of the femora and tibiae are affected less frequently (Emslie et al., 1983; Thorp and Waddington, 1997; McNamee et al., 1999). It is not unusual for BCO to affect only one leg while the contralateral leg appears macroscopically normal (McNamee et al., 1998; Dinev, 2009; Wideman, 2014). Femoral, tibial and vertebral BCO lesions usually occur together in commercial flocks, although in

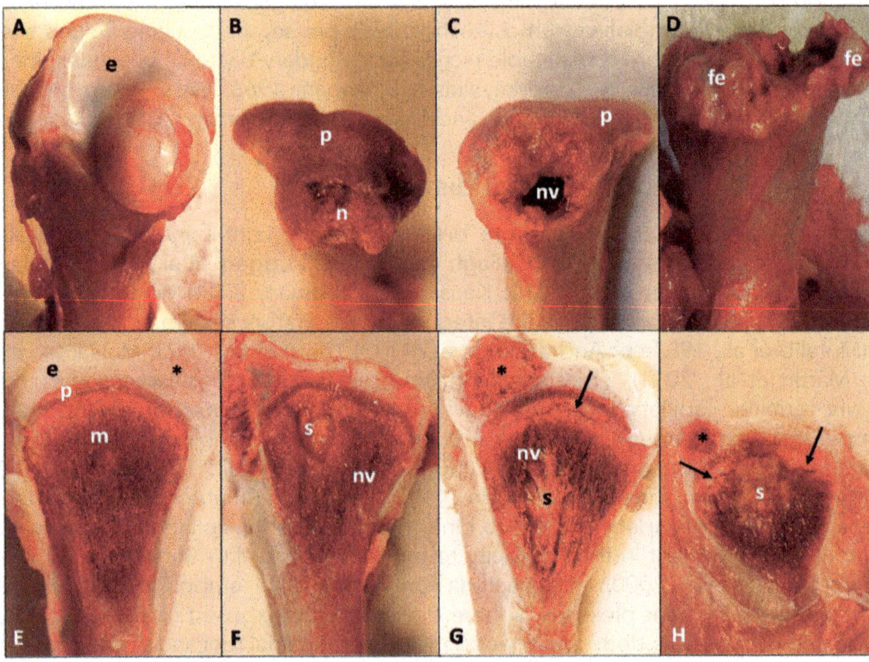

Figure 1 Stages of proximal femoral degeneration (**A–D**) and proximal tibial degeneration (**E–H**) leading progressively to BCO. (**A**) Normal proximal femur with white cap of epiphyseal cartilage (e). (**B**) Femoral head separation (FHS: epiphyseolysis) with the epiphysis remaining in the socket when the femur was disarticulated, revealing the underlying surface of the growth plate or physis (p) and an early region of necrosis (n). (**C**) Fracturing of the growth plate (p) revealing a necrotic void (nv) within the metaphysis. (**D**) Terminal femoral head necrosis in which the femoral epiphysis, physis and most of the metaphysis remained attached to the acetabulum when the diaphysis weakened by widely dispersed necrosis was fractured during disarticulation, revealing copious fibrinonecrotic exudate (fe). (**E**) Normal proximal tibia showing the epiphysis (e) with a secondary centre of ossification (*) and the physis/growth plate (p) fully supported by struts of trabecular bone in the metaphyseal zone (m). (**F–H**) Bacterial infiltration and sequestrae (s), necrotic voids (nv) and microfractures below the growth plate (arrows) provide macroscopic evidence of bone damage associated with osteomyelitis.

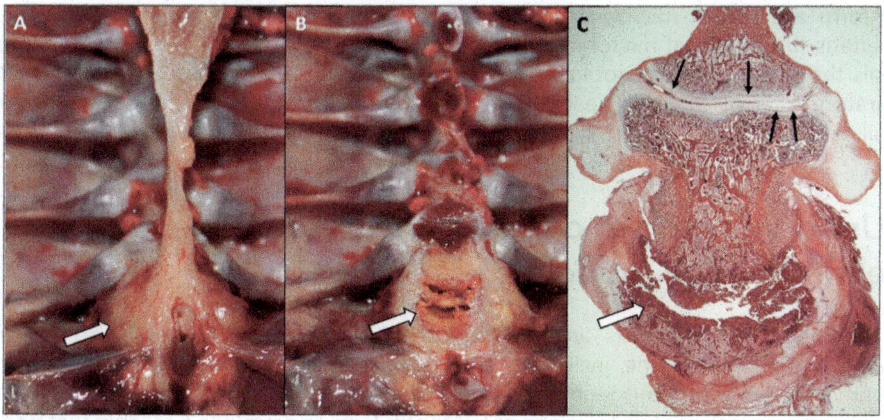

Figure 2 Example of vertebral BCO from a broiler that exhibited spinal cord compression and paraplegic hock-resting posture due to grossly abscessed vertebral bodies (white arrows). (**A**) Externally the flexible thoracic vertebral body exhibited marked nodular swelling with a distinctive yellow discolouration of the translucent remnants of the vertebral bodies. (**B**) Internally the caseous abscessation filled the caudal portion of the vertebral body. (**C**) Histologically the caudal epiphysis and physis are absent and have been replaced by the necrotic abscess; areas of osteochondrosis in the cranial articular cartilage layers are indicated by black arrows.

some outbreaks femoral and tibial lesions may be more common than vertebral lesions, whereas the opposite may be true in other outbreaks (Stalker et al., 2010; Kense and Landman, 2011; Wideman, personal observations).

Multiple opportunistic organisms have been isolated from BCO lesions, including several *Staphylococcus* spp., *Escherichia coli* and *Enterococcus cecorum*, often in mixed cultures with other bacteria, including *Salmonella* spp. (Smith, 1954; Carnaghan, 1966; Wise, 1971; Nairn and Watson, 1972; Griffiths et al., 1984; Andreasen et al., 1993; Tate et al., 1993; Thorp et al., 1993; McNamee et al., 1998; Butterworth, 1999; Smeltzer and Gillaspy, 2000; Joiner et al., 2005; Dinev, 2009; Stalker et al., 2010; Kense and Landman, 2011; Martin et al., 2011; Wijetunge et al., 2012).

Recently, Jiang et al. (2015) used molecular profiling of 16S ribosomal RNA (16S rRNA) gene sequences to comprehensively analyse the structure and diversity of microbial communities in the proximal femora and tibiae from clinically healthy broilers and from lame broilers with obvious BCO lesions. Complex microbial communities were detected in all samples, including bones that appeared to be macroscopically normal. Further analyses demonstrated major differences in the microbial communities of different bones (femur versus tibia) and different lesion categories (macroscopically normal versus gross BCO lesions). The genus *Staphylococcus* was overrepresented in bones with BCO lesions, along with the genera *Enterobacter* and *Serrotia*.

These results indicate that complex microbial communities exist in the growth plates and metaphyses of macroscopically normal proximal femora and tibiae and that BCO lesions are associated with a less diverse community with overrepresentation by potentially pathogenic genera (Jiang et al., 2015). There is the likelihood that numerous bacterial species routinely enter the bloodstream and are sequestered in the growth plates, where they form a complex microbial community. An obvious source of these translocating bacteria is the intestinal microbiome (Jiang et al., 2015), although recent evidence also reveals the presence of microbial communities within the conducting airways and lower respiratory tract of clinically healthy poultry (Nehme et al., 2005; Shabbir et al., 2015; Sohail et al., 2015).

3 Understanding and treating BCO: the role of experimental models

The aetiology, pathogenesis and treatment strategies for BCO have been difficult to investigate because the incidence often is low or nonexistent in research flocks. Experimental models for triggering BCO are needed to confirm key triggering mechanisms and stressors, reveal innate limitations or susceptibilities and provide a reliable test bed for evaluating practical prophylactic and therapeutic strategies.

3.1 Pathogen exposure models

Attempts to initiate BCO via a respiratory challenge have been only marginally successful (Devriese et al., 1972; Nicoll and Jensen, 1986; Jensen et al., 1987; Mutalib et al., 1983b; McNamee et al., 1999; Martin et al., 2011), whereas BCO has repeatedly been reproduced by injecting broilers and turkeys intravenously with appropriate strains of *Staphylococcus* spp. in quantities sufficient to cause sustained bacteraemia. The artificially induced bacteraemia resembles naturally occurring sepsis. Terminal metaphyseal and epiphyseal blood vessels rapidly become occluded by bacterial emboli and phagocytic cells, leading to focal necrosis and bacterial colonization of the ischaemic cartilage. Intravenous pathogen injections typically trigger high incidences of lameness attributable to synovitis and BCO within 24–72 hours postinjection (Smith, 1954; Carnaghan, 1966; Wise, 1971; Nairn, 1973; Emslie and Nade, 1983, 1985; Emslie et al., 1983; Kibenge et al., 1983; Mutalib et al., 1983a; Griffiths et al., 1984; Alderson et al., 1986b; Daum et al., 1990; Martin et al., 2011).

Smith (1954) induced bacteraemia and lameness by injecting *Staphylococcus* spp. intravenously, but none of the strains produced any effect when injected intraperitoneally or subcutaneously. Wise (1971) isolated *Staphylococcus aureus* and *Staphylococcus albus* from vertebral BCO abscesses in a turkey and a broiler, respectively. When these isolates were inoculated intravenously into broilers, only the *S. albus* isolate triggered vertebral BCO lesions (Wise, 1971). Griffiths et al. (1984) isolated *S. aureus* from lame commercial broilers. When inoculated intravenously into healthy broilers, this isolate exhibited a broad tropism for rapidly growing bone and triggered vertebral BCO as well as proximal femoral and tibial BCO (Griffiths et al., 1984). Alderson and Nade (1987) re-created septic arthritis and focal osteomyelitis by injecting the hock joints of four-week-old broilers with a strain of *S. aureus* isolated from lame chickens. Al-Rubaye et al. (2014) isolated *Staphylococcus agnetis* from bone and blood samples of broilers with clinical BCO. Administering this isolate in the drinking water increased the incidence of BCO threefold when compared with that in uninoculated broilers (Al-Rubaye et al., 2014).

During outbreaks of lameness in broiler and broiler breeder flocks in the European Union and North America, *E. cecorum* was implicated as the pathogen responsible for causing arthritis, tenosynovitis and femoral and vertebral BCO (Stalker et al., 2010; Kense and Landman, 2011; Martin et al., 2011; Wijetunge et al., 2012). Necropsies of lame broilers exhibiting a hock-sitting posture revealed nodular swelling, abscessation and osteomyelitis of the T4 vertebral body, accompanied by compression of the spinal cord. Lung tissue was focally adherent to the abscessed vertebrae. *E. coli* and Gram-positive cocci were isolated from the spinal abscesses, and the cocci were identified by 16S rRNA sequencing as *E. cecorum* (Stalker et al., 2010). Martin et al. (2011) used *E. cecorum* isolated from a commercial outbreak of vertebral BCO to inoculate male broiler breeder chicks via the air sac, and the intravenous and the oral routes. Oral transmission of *E. cecorum* was reported by these investigators to be the most productive method for reproducing the characteristic gross and microscopic lesions of vertebral BCO (Martin et al., 2011).

As a caveat to the use of direct pathogen exposure for triggering BCO, pre-existing osteochondrotic microfracturing of the epiphyseal–physeal cartilage may not be an essential permissive condition if naturally occurring sepsis or experimental pathogen inoculations cause bacteraemia. For example, intravenously injecting an effective dose of *Staphylococcus* spp. typically triggers sustained bacteraemia that causes bacterial emboli to form quite rapidly in the terminal epiphyseal and metaphyseal vascular plexuses throughout the skeleton, leading to widely dispersed osteomyelitis and a rapid (within one to three days) onset of lameness. This form of generalized/septic osteomyelitis apparently does not depend on the creation of osteochondrotic clefts (wound sites) as niches for bacterial colonization. Models of pathogen inoculation that very quickly overwhelm the body's defence mechanisms and trigger generalized septicaemia are unlikely to be useful for detecting differences in BCO susceptibility among broiler lines that may primarily depend on differences in innate susceptibility to osteochondrosis. Nor would such models be useful for demonstrating relative efficacies of practical prophylactic treatments that improve barrier integrity and thereby reduce bacterial translocation.

3.2 Mechanical models

These models for triggering BCO were designed to create sustained footing instability and thereby persistently exert excessive mechanical stress on susceptible leg and vertebral joints. Based on our current understanding, amplified footing instability creates conducive wound sites (e.g. osteochondrotic clefts, focal necrosis) in the epiphyseal–physeal cartilage. The mechanical approach avoids the need to purposefully inoculate the birds with known pathogens based on the assumption that opportunistic pathogens already are present but quiescent within the bird's tissues or in microbial communities or in the environment. For example, ramps placed under nipple waterers forced broilers to walk up a 30% slope to drink and, by eight weeks of age, triggered a higher cumulative incidence of BCO (27%) than the spontaneous (6%) incidence in pens without ramps (Wideman, 2014).

Portable obstacles known as 'speed bumps' have been used to consistently trigger BCO in broilers (Fig. 3) (Gilley et al., 2014). Speed bumps are constructed in the shape of an isosceles triangular prism, with sides having slopes ranging from 33 to 66%. They are designed to be installed between feeders and waterers in litter flooring facilities, thereby forcing the birds to climb up and then down the slopes as they move back and forth to eat and drink. Speed bumps trigger threefold higher incidences of BCO when compared with incidences in broilers reared on litter flooring without speed bumps (Gilley et al., 2014). Covering the entire surface of a pen floor with flat wire flooring panels subjects broilers to chronic footing instability as well as to behavioural stress attributable to denying the

Figure 3 Mechanical models for triggering BCO are designed to persistently exert excessive mechanical stresses on susceptible leg and vertebral joints, thereby creating conducive wound sites (e.g. osteochondrotic clefts, focal necrosis) in the epiphyseal–physeal cartilage. Portable obstacles known as 'speed bumps' (above) trigger threefold higher incidences of BCO when compared with those in broilers reared on litter flooring without speed bumps. Speed bumps are constructed in the shape of an isosceles triangular prism and are placed between feeders and waterers to force broilers to climb up and then down the slopes as they move back and forth to eat and drink (Gilley et al., 2014).

birds access to litter (*vide infra*). Incidences of lameness between 20 and 60% are reliably induced by this challenge, depending on the genetic susceptibility and hatchery source of the broilers being evaluated.

All of the mechanical models consistently triggered pathognomonic BCO lesion progression, primarily within the proximal femora and proximal tibiae. Vertebral BCO was rarely observed in these experiments until a recent study in which the chicks and feed were obtained from a commercial source. In that experiment, 80% of the lameness was attributed to proximal femoral and tibial BCO and 20% of the lameness was attributed to typical vertebral BCO lesions (Wideman et al., 2012, 2013, 2014, 2015a,b; Gilley et al., 2014; Prisby et al., 2014; Jiang et al., 2015; Wideman, personal observations).

Most of the lameness triggered by mechanical models develops between three and eight weeks of age, after broilers attain body weights sufficient to produce osteochondrotic microfractures, as has been reported for field outbreaks of BCO (McNamee et al., 1998; Dinev, 2009). Females tend to be marginally less susceptible to lameness than males when both genders are reared together (Wideman et al., 2014). Lameness progresses very rapidly in broilers that appeared to be healthy during the preceding 24–48 h, as previously reported by Joiner et al. (2005). Broilers tend to exhibit relatively mild BCO lesions when they are euthanized at the earliest onset of clinical symptoms (hesitancy to stand, eagerness to sit, slight wing-tip dipping), whereas birds that are permitted to live until they become fully immobilized (unable to move to feed or water) exhibit much more severe lesions.

It is also apparent from necropsying survivors that severe lesions may occasionally be present in very large, apparently robust individuals that exhibit no signs of lameness or leg weakness (Wideman et al., 2012, 2014). We speculate that broilers and turkeys purposefully avoid exhibiting overt symptoms of lameness in order to avoid being victimized by the predatory behaviour of their flock mates. Indeed, gait scoring may not consistently or quantitatively reveal skeletal pathologies or abnormalities that are readily detected during the post-mortem examination of individual birds (Naas et al., 2009; Sandilands et al., 2011; Wideman, unpublished observations). The pathogenesis of BCO cannot be instantaneous and apparently healthy broilers often possess subclinical lesions primarily consisting of the earliest macroscopic BCO lesion pathology. Subclinical lesions are equally likely to develop in males and females in left and right legs, and the status of the proximal femur does not determine the status of the ipsilateral or contralateral proximal tibia and vice versa (Wideman et al., 2012, 2013, 2014, 2015a,b; Gilley et al., 2014). These observations are consistent with the hypothesis that subclinical mechanical damage need not trigger overt lameness until the damaged area becomes infected. The resulting bacterial proliferation, immunological assault by responding phagocytes (macrophages and heterophils) and widespread lysis and necrosis of the metaphyseal trabecular bone and vasculature then culminate in intolerable discomfort and terminal lameness (Howlett, 1980; Duff, 1984b; Thorp et al., 1993; Wideman and Prisby, 2013).

Mechanical models have been used to generate practical strategies for reducing BCO incidences in commercial flocks. In two independent experiments, standard broiler crosses that grow rapidly at an early age developed higher incidences of lameness when compared with high yield crosses that initially tend to grow more slowly, directly demonstrating the association between rapid early growth and susceptibility to BCO (Wideman et al., 2013). Two independent experiments were also conducted to evaluate the influence of commercial sire lines on the susceptibility of broiler crosses to BCO. When grown to heavy yield weights, broilers derived from one sire line were consistently more likely to develop higher incidences of BCO than broilers derived from a second sire line, and these experimental results were confirmed by the experiences of broiler integrators using the same sires and crosses (Wideman et al., 2014).

The wire flooring model was used to demonstrate that enrofloxacin, a potent fluoroquinolone antimicrobial, is therapeutically effective for treating BCO (Wideman et al., 2015a). A proprietary 25-hydroxy vitamin D_3 product was also demonstrated to effectively reduce BCO incidences in broilers reared on wire flooring. This product is currently widely used prophylactically to minimize leg weakness in broilers and turkeys (Wideman et al., 2015b). By consistently triggering moderate to high incidences of BCO in research flocks, mechanical models have provided a reliable experimental protocol for comparing genetic lines and chick sources and for evaluating practical treatment strategies.

3.3 Stress-mediated immunosuppression models

Immunosuppression has been implicated in the aetiology of spontaneous BCO outbreaks in commercial poultry flocks (Mutalib et al., 1983a,b; Andreasen et al., 1993; Butterworth, 1999; McNamee et al., 1998, 1999; Huff et al., 2000; McNamee and Smyth, 2000). Environmental stressors and immunosuppression contribute to the eruption of opportunistic pathogens in the proximal tibial joints of turkeys that develop turkey osteomyelitis complex (TOC). Bacterial infection of the proximal tibiae is a characteristic symptom of TOC, and in an experimental setting, TOC can be triggered by injecting turkey poults with repeated immunosuppressive doses of dexamethasone, a synthetic glucocorticoid (Wyers et al., 1991; Huff et al., 1998, 1999, 2000, 2005, 2006).

Based on the pathogenic similarities between BCO and TOC, we conducted three experiments to determine if dexamethasone injections might trigger BCO in broilers. In all three experiments, dexamethasone-injected broilers developed lameness primarily associated with lesions of the proximal tibiae (Wideman and Pevzner, 2012). Within this context it is highly likely that, in addition to creating footing instability, wire flooring models also elicit stress and immunosuppression. Elevated flooring systems that deprive birds of access to floor litter stimulate chronic stress, including immunosuppression (El-Lethey et al., 2003).

4 Sources and routes of bacterial colonization

4.1 Epidemiology

The current hypothesis for the pathogenesis of BCO proposes that exponential increases in early body weight impose excessive torque and shear stress on structurally immature epiphyseal and physeal cartilage, creating osteochondrotic wound sites that serve as ideal niches for colonization by haematogenously distributed opportunistic bacteria. This hypothesis does not specify the original sources of the bacterial pathogens. Sources and routes of entry of the offending bacterial pathogens may vary widely, as was suggested by a multiyear epidemiological study of sequential *Campylobacter jejuni* outbreaks in a broiler farm in the United Kingdom (Pearson et al., 1996). A contaminated water distribution system was identified as the initial source of a single *Campylobacter* serotype that was broadly distributed throughout all sheds on the farm. Marked overall reductions in *Campylobacter* isolation rates were achieved by cleaning and disinfecting the water lines and reservoirs, but thereafter, a limited number of different *Campylobacter* serotypes began to emerge intermittently. Notably these serotypes developed in some but not in all of the sheds. Specific *C. jejuni* serotypes were associated with the hatchery supplying the one-day-old chicks, as well as with grandparent flocks and broiler breeders. The investigators postulated that low-level vertical transmission from the broiler breeders to relatively few infected chicks would then lead to horizontal transmission of the same serotype throughout a shed (Pearson et al., 1996).

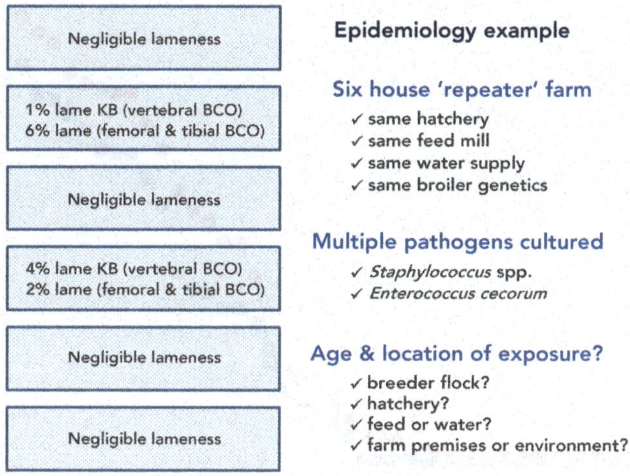

Figure 4 Outbreaks of BCO on so-called 'repeater' farms often exhibit similar patterns. Incidences of femoral, tibial and vertebral BCO tend to vary widely among separate barns on the same farm, with some barns exhibiting little or no lameness and other barns develop high BCO incidences attributable to *Staphylococcus* spp. and *E. cecorum*. Different barns may be affected during sequential flock cycles, strongly implicating the breeder flock or hatchery as potential factors contributing to early bacterial contamination. Flock managers often report that initially a few broilers develop BCO lameness at a focal location within a barn, after which lameness appears to spread horizontally throughout the affected barn but not into adjacent barns that are managed by the same personnel and provided feed from the same feed mill. This epidemiological pattern appears to reflect vertical transmission of pathogenic bacteria to a few chicks, followed by horizontal transmission of the same pathogen throughout a barn (Wideman, personal observations).

Outbreaks of BCO on so-called 'repeater' farms often exhibit similar patterns (Fig. 4). Incidences of femoral, tibial and vertebral BCO tend to vary widely among separate barns on the same farm, with some barns exhibiting little or no lameness and other barns develop high BCO incidences attributable to *Staphylococcus* spp. and *E. cecorum*. Different barns may be affected during sequential flock cycles, strongly implicating the breeder flock, hatchery or posthatch stress as potential factors contributing to early bacterial contamination. Flock managers often report that initially a few broilers develop BCO lameness at a focal location within a barn, after which lameness appears to spread horizontally throughout the affected barn but not into adjacent barns that are managed by the same personnel and provided feed from the same feed mill (Wideman, personal observations). Again, this epidemiological pattern appears to reflect low-level vertical transmission of pathogenic bacteria to a few chicks, followed by horizontal transmission of the same pathogen throughout a barn.

4.2 Vertical versus horizontal transmission

Breeder flocks and hatcheries have repeatedly been suspected of being the source of *Staphylococcus* spp. and *Enterococcus* spp. that subsequently are isolated from lame

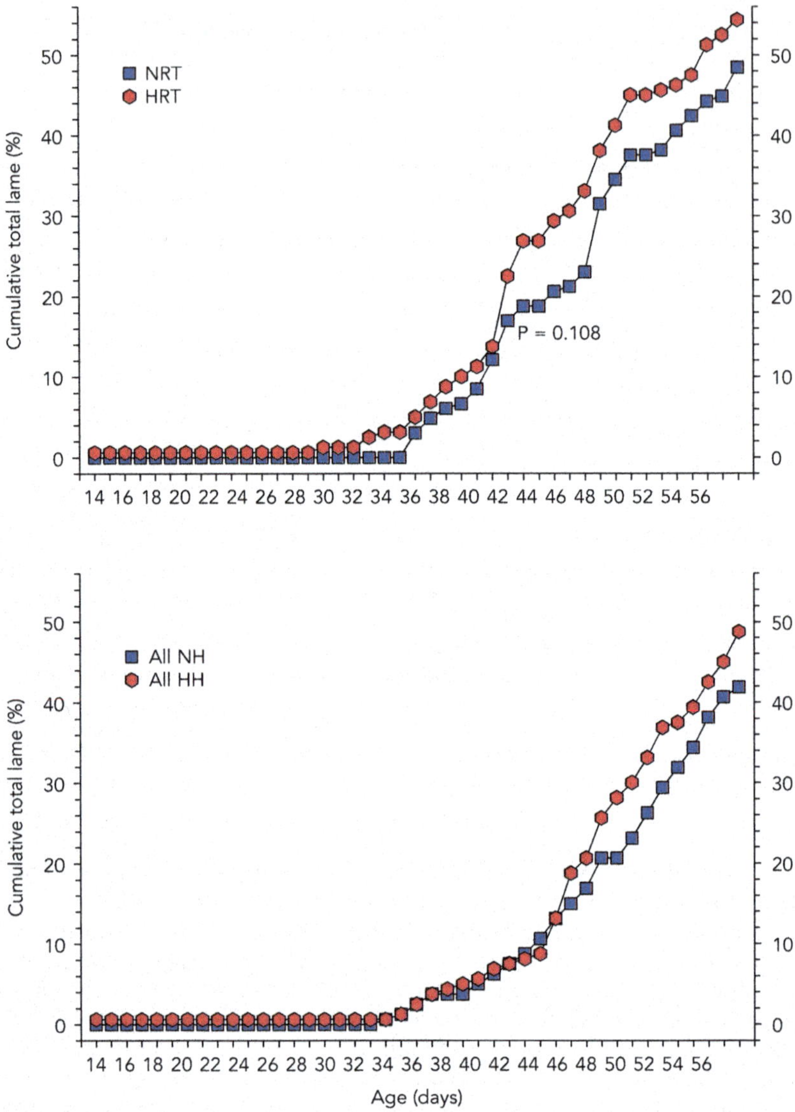

Figure 5 Newly hatched chicks were selected as having normal (<105.0 °F) or high (>105.5 °F) rectal temperatures (upper panel: NRT or HRT, respectively) in a commercial hatchery, or on days 18 through 21 of incubation the hatcher temperature in the University of Arkansas research hatchery was set to a normal (100 °F) or a high (101 °F) temperature (lower panel: NH or HH, respectively). The chicks were wing banded and co-mingled in pens with flat wire flooring from day 1 through day 56 to trigger high incidences of lameness attributable to BCO. Chicks with normal rectal temperatures or that were exposed to normal hatcher temperatures (NRT and NH) tended to have lower incidences of BCO lameness than chicks with high rectal temperatures or that were exposed to high hatcher temperatures (HRT and HH).

broilers (Skeeles, 1997; McCullagh et al., 1998; Rodgers et al., 1999; McNamee and Smyth, 2000; Stalker et al., 2010; Kense and Landman, 2011). Various bacterial species, including *Staphylococcus* and *Enterococcus*, have been isolated from spleens harvested from broiler embryos at day 18 of incubation. Broiler breeder hens are continuously subjected to stresses such as feed restriction, aggressive males, slat (nonlitter) flooring and oviposition. Sustained stress may promote bacterial translocation and bacteraemia sufficient to cause vertical transmission of the resident microbial communities from breeder hens to their progeny via the yolk follicles or reproductive tract (Humphrey and Lanning, 1988; Cox et al., 2002, 2005b; Liljebjelke et al., 2005; Funkhouser and Bordenstein, 2013). The broiler breeder hen's reproductive tract undergoes reverse peristalsis and therefore is susceptible to ascending infection by intestinal bacteria or contaminated semen, facilitating inoculation of the albumen and shell membranes before the shell is deposited (Reiber et al., 1995; Buhr et al., 2002; Cox et al., 2005a). Soiled eggshells can facilitate bacterial penetration through the shell and commensal colonization of the intestinal tracts of apparently healthy, newly hatched chicks (Clark and Bueschkens, 1985).

Opportunities for posthatch horizontal transmission abound. Certain hatcheries, particularly older hatcheries, are notorious for producing chicks that are susceptible to BCO. Our ongoing research using the wire flooring model repeatedly has implicated breeder flocks, hatchery sources or chick quality in the susceptibility of broilers to BCO. Indeed, our experimental chicks are routinely culled on day 14 because necropsies of runts and culls during the first two weeks posthatch often reveal evidence of systemic bacterial infection, including omphalitis, sepsis and osteomyelitis. Placing chicks on wire flooring on the day of hatch followed by heavy culling on day 14 has evolved into a standard experimental protocol for delaying the onset of clinical BCO until days 35–40 (Wideman et al., 2012, 2013, 2014, 2015a,b).

Current industry efforts to reduce the incidence of BCO in problematic (repeater) broiler complexes include improving hatchery sanitation, minimizing heat stress in the hatchery and during brooding, prescribing antibiotic treatments for chicks immediately posthatch and reducing the photoperiod to modestly reduce early growth performance. For these approaches to be effective, they must be addressing triggering mechanisms or early bacterial colonization that confer subclinical susceptibility at hatch or prior to hatching. For example, newly hatched chicks that had higher rectal temperatures or that were exposed to a high temperature in the hatcher on days 18 through 21 of incubation tended to develop higher incidences of BCO lameness on wire flooring when compared with chicks having normal rectal temperatures or that were exposed to a normal temperature in the hatcher (Fig. 5). These observations suggest that bacterial pathogens responsible for causing BCO may be present but quiescent very early pre- or posthatch, waiting for conducive wound sites to develop in larger heavier birds before clinical lameness attributable to BCO emerges. In that case, the epidemiology of BCO outbreaks beginning at 5 weeks of age would be difficult to understand if the primary source of their bacterial inoculation occurred *in ovo* or in the hatchery.

4.3 Bacterial translocation

For bacteria to be distributed haematogenously to osteochondrotic microfractures, they first must penetrate the body's defences and persist within the tissues. The ease with which viable bacteria can be routinely isolated from the blood and tissues of poultry pre- and posthatch readily discredits the common misconceptions that blood and tissues are essentially sterile in birds that appear to be clinically healthy and that the immune system efficiently destroys invading pathogens (Buhr et al., 2002; Cox et al., 2005a, 2006a,b, 2007; Clark et al., 2010; Richardson et al., 2011). In fact, broadly diverse microbial communities are now known to persist in the respiratory tract, unabsorbed yolk, and blood and bones of apparently healthy poultry, and these communities appear to be tolerated by the immune system (Nehme et al., 2005; Cox et al., 2006b,c; Jiang et al., 2015; Shabbir et al., 2015; Sohail et al., 2015).

Potential routes of bacterial penetration include vertical transmission from the breeder hen or horizontal transmission from the environment (e.g. the hatchery, growout facilities, accumulated litter, feed, water, bioaerosols) followed by translocation through the integument, respiratory epithelium or intestinal epithelium. Many poultry health professionals consider osteomyelitis to be a secondary infection that develops several weeks after a severe respiratory challenge (e.g. mycoplasma infection; poor hatchery sanitation; vigorous reaction to respiratory vaccine), an enteric challenge (e.g. coccidiosis; enteritis), an immunosuppressive challenge (e.g. stress; chicken anaemia virus; infectious bursal disease virus) or infection of the integument (e.g. wet navels; foot pad or hock lesions). Dermal lesions and omphalitis can lead to generalized sepsis and thus secondarily to broadly distributed, generalized osteomyelitis. For example, trimming the toes of recently hatched turkey poults provided a portal of entry for *S. aureus* to initiate bilateral osteomyelitis and synovitis in their femora and tibiae (Alfonso and Barnes, 2006). Also, injecting *S. aureus* directly into a broiler's hock joint triggered septic arthritis and bacterial colonization of the local epiphyseal and physeal vasculature (Alderson and Nade, 1987).

However, although compromised intestinal, respiratory and integument barriers may clearly precede and contribute to osteomyelitis outbreaks in commercial flocks, it is not necessary to induce mucosal barrier dysfunction as a precondition for triggering BCO in broilers with mechanical models or oral administration of pathogenic bacteria. Experiments have repeatedly demonstrated the ease with which orally administered or indigenous enteric bacteria can be translocated across the gastrointestinal epithelium to be carried by the circulatory system, perhaps adherent to or engulfed by phagocytes, to organs and tissues throughout the body (Lissner et al., 1983; Hand et al., 1984; Desmidt et al., 1998; Cossart and Sansonetti, 2004; Bailey et al., 2005; Cox et al., 2006b; Martin et al., 2011; Al-Rubaye et al., 2014).

5 The role of probiotics

Translocation of orally administered or indigenous enteric bacteria and haematogenous distribution by the circulatory system (likely engulfed by phagocytes) to organs and tissues throughout the body has been demonstrated repeatedly (Carter and Collins, 1974; Berg, 1992; Ando et al., 2000; Kremer et al., 2011; Richardson et al., 2011). Key components of the intestinal barrier to bacterial translocation are illustrated in Fig. 6 (upper panel). Goblet cells secrete protective mucus that lubricates the passage of digesta through the lumen and can hinder attachment by pathogenic bacteria to the glycocalyx of the brush border microvilli (Byrne et al., 2007). The inner mucin layer also serves as a matrix in which antimicrobial proteins (defensins) and secretory immunoglobulins (IgA) are sequestered to create an immune exclusion barrier protecting the luminal surfaces of the

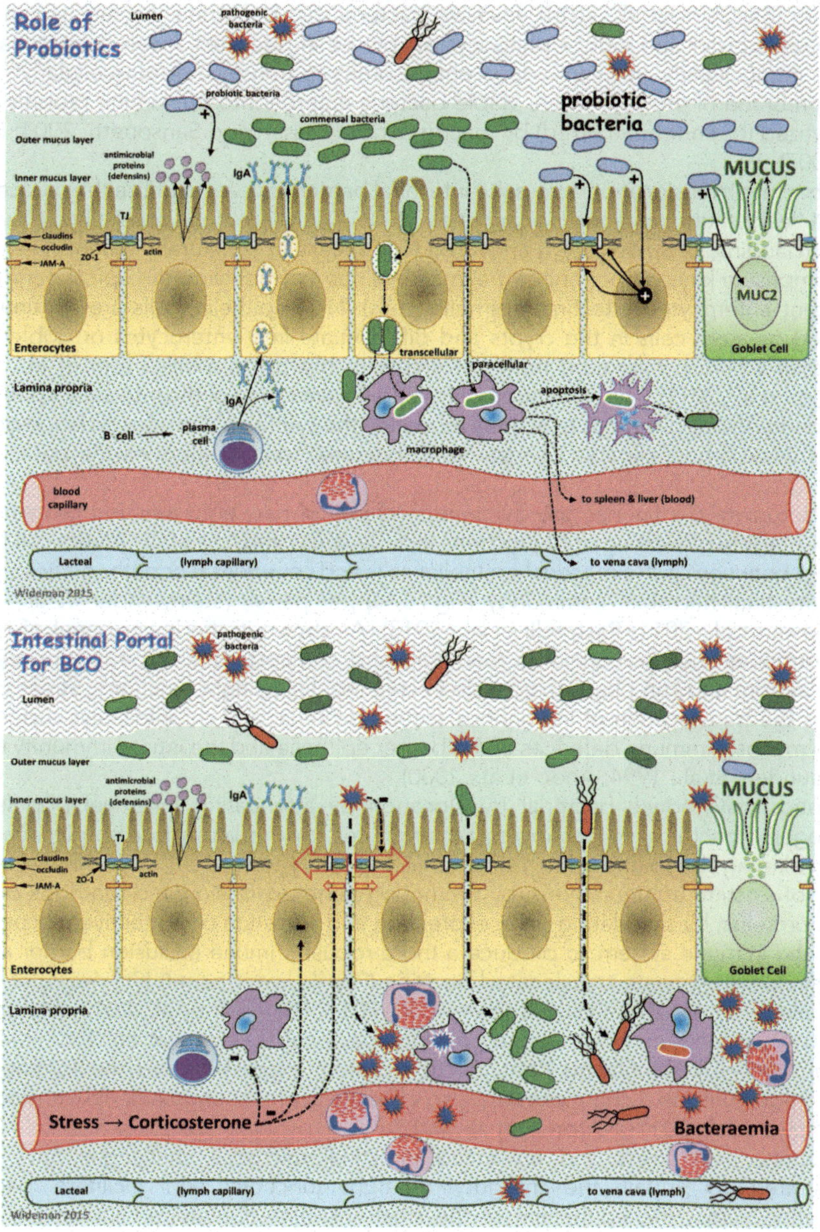

Figure 6 Role of probiotics and the immune system in preventing bacterial translocation across the healthy gastrointestinal tract (upper panel) and routes of bacterial translocation during dysbiosis (lower panel). Diagrams based on Steinwender at al. (2001), Ulluwishewa et al. (2011), Quinteiro-Filho et al. (2012a,b), Pastorelli et al. (2013) and Murugesan et al. (2014).

epithelium (Bar-Shira et al., 2014). Invasive bacteria that penetrate the mucus can bind to the brush border and induce phagocytosis by the enterocytes. Invasive bacteria may replicate intracellularly and can induce phagosomes to translocate transcellularly to the basal membrane of the cell, where viable bacteria emerge into the lamina propria to be phagocytized (but not neutralized) by macrophages (Cossart and Sansonetti, 2004; Byrne et al., 2007).

Most bacterial translocation across the intestinal barrier appears to utilize the paracellular route via 'leaky' tight junctions (Soriani et al., 2006). Tight junctions consisting of over 40 proteins (claudins, occludin, ZO-1) anchored to intracellular structural elements (including actin) comprise a key component of the intestinal barrier by sealing the apical perimeters of adjacent enterocytes (intestinal epithelial cells). New epithelial cells are continuously generated by stem cells in the crypts and differentiate into enterocytes or goblet cells during their 72–96 h migration from the crypts to the tip of a villus (Uni et al., 1995; Bohorquez et al., 2011). This continuous, life-long regeneration of the intestinal epithelium creates ongoing opportunities to acutely modulate the synthesis and consolidation of tight junctions into barriers capable of minimizing bacterial translocation via the paracellular route.

Factors known to influence the dynamic synthesis of new tight junction proteins and alter the integrity of existing tight junctions include stress and 'crosstalk' (direct cell to cell signalling) between gastrointestinal epithelial cells and commensal or pathogenic bacteria of the intestinal microbial community (Fig. 6, upper panel) (Steinwender et al., 2001; Ulluwishewa et al., 2011; Pastorelli et al., 2013). As shown in the lower panel of Fig. 6, stress hormones such as cortisol in mammals and corticosterone in birds suppress the immune system and can play a major role in creating 'leaky' tight junctions. Stress clearly facilitates the translocation of indigenous enteric bacteria, and chronic stress suppresses both the mucosal immune defences of the barrier epithelia and the innate immunity (Berg, 1992; Saunders et al., 1994; Ando et al., 2000).

Probiotics may reduce bacterial translocation by attenuating intestinal populations of pathogenic bacteria, by occupying binding sites on the brush border glycocalyx to prevent binding by pathogenic bacteria (competitive exclusion), by enhancing the integrity of existing tight junctions, by promoting gene expression and synthesis of tight junction proteins, by stimulating gene expression and secretion of protective mucus or by priming the immune system to produce a more robust immune exclusion barrier and to more effectively eliminate translocated bacteria (Fig. 6, lower panel) (Steinwender at al., 2001; Smirnov et al., 2005; Ulluwishewa et al., 2011; Pastorelli et al., 2013; Murugesan et al., 2014). Indeed, commensal and probiotic bacterial species that enhance intestinal barrier integrity by stimulating tight junction protein expression and the formation of occlusive tight junctional complexes are also effective in preventing bacterial translocation (Ulluwishewa et al., 2011; Pastorelli et al., 2013).

In broilers, heat stress and enhanced intestinal microbial challenges have been demonstrated to impair the integrity of tight junctions and facilitate bacterial translocation across the epithelium of the small intestine (Quinteiro-Filho et al., 2010, 2012a,b; Murugesan et al., 2014). It has also been demonstrated that probiotics alone or in combination with prebiotics can attenuate intestinal barrier dysfunction in broilers challenged by heat stress or pathogenic bacteria (Sohail et al., 2010, 2012; Murugesan et al., 2014; Song et al., 2014). For example, as shown in Fig. 7, the intestinal epithelium from broiler chicks reared on reused/dirty litter to simulate a typical commercial enteric challenge had a lower permeability to macromolecules (e.g. bacterial lipopolysaccharide),

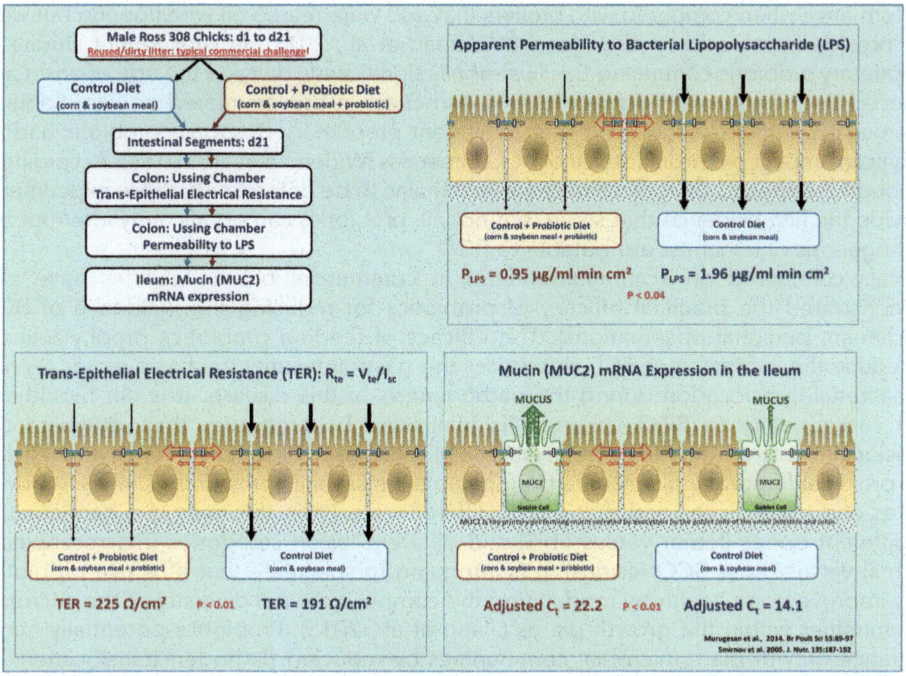

Figure 7 Broiler chicks reared on reused/dirty litter to simulate a typical commercial enteric challenge were fed a control diet or the same diet containing a probiotic. Intestinal segments were harvested on day 21, and Ussing chambers were used to assess intestinal permeability. Probiotic inclusion in the diet reduced the epithelial permeability to macromolecules (e.g. bacterial lipopolysaccharide) and increased the transepithelial resistance (reflecting improved tight junction barrier function). The probiotic also increased ileal mucin gene expression levels when compared with ileal samples from chicks fed the control diet (Murugesan et al., 2014).

a higher transepithelial resistance and a higher mucin gene expression level when the feed contained a probiotic (Murugesan et al., 2014).

In view of concerns regarding the development of antibiotic resistance in bacteria commonly associated with osteomyelitis (McNamee and Smyth, 2000; Waters et al., 2011), it was our hypothesis that probiotics potentially can provide a plausible alternative for reducing the incidence of BCO when the intestinal tract is the site of bacterial translocation. Five independent experiments conducted over the course of two years and using four different broiler lines demonstrated that prophylactically administering probiotics in the feed beginning at one day of age significantly reduced the incidence of lameness attributable to BCO (Wideman et al., 2012).

The first four of these experiments evaluated a proprietary probiotic containing *Enterococcus faecium*, *Bifidobacterium animalis*, *Pediococcus acidilactici* and *Lactobacillus reuteri*. The fifth experiment evaluated a proprietary single-microbe probiotic containing *E. faecium*. Prophylactically providing these probiotics in the feed consistently reduced the incidence of BCO lameness by at least 50% and without attenuating growth performance when compared with broilers that also were reared on wire flooring but were not provided probiotics in their feed (Wideman et al., 2012). In subsequent studies, a proprietary probiotic containing *Bacillus subtilis* significantly delayed the age of onset and reduced the cumulative incidence of BCO lameness in broilers reared on wire flooring, whereas experiments conducted with a different proprietary *B. subtilis* probiotic had no significant impact on the incidence of BCO lameness (Wideman et al., 2015a). Accordingly, although the specific biological mechanism remains to be determined, these experiments provide the first evidence that some, but not all, probiotics can significantly interrupt the pathogenesis of lameness attributable to BCO.

Trials conducted on accumulated litter in commercial broiler facilities have also demonstrated the practical efficacy of probiotics for reducing the incidence of BCO (Wideman, personal observations). The efficacy of feeding probiotics prophylactically to reduce the incidence of BCO implicates the gastrointestinal tract as a major avenue of bacterial translocation during the pathogenesis of this disease. It is our hypothesis that susceptibility to BCO lameness is minimized by probiotics that attenuate the translocation of pathogenic bacteria into the bloodstream. Based on recent evidence that the proximal femora and tibiae harbour complex microbial communities in their growth plates and metaphyses and that major differences exist in the microbial communities in different bones (femur versus tibia) and different lesion categories (macroscopically normal versus gross BCO lesions), it is intriguing to speculate that effective probiotics may improve bone health by modulating the composition and diversity of the microbial communities within the growth plates (Jiang et al., 2015). Probiotics potentially might influence growth plate microbial communities by reducing pathogen translocation, by enhancing the immune response to translocating microbial species or via translocation of the probiotic species to the growth plates followed by direct modulation of local microbial communities *in situ*.

6 Summary and conclusions

The current hypothesis for the pathogenesis of BCO proposes that exponential increases in early body weight impose excessive torque and shear stress on structurally immature epiphyseal and physeal growth plate cartilage, creating osteochondrotic wound sites that serve as ideal niches for colonization by haematogenously distributed opportunistic bacteria. We do not understand how pathogenic bacteria circulating in the blood avoid elimination by the immune system, or how complex microbial communities are allowed to persist in the growth plates of affected bones without being eliminated by an inflammatory response. Lameness is triggered when pathogenic bacteria create large abscesses primarily within the proximal ends of the femora and tibiae as well as in the flexible thoracic vertebrae. Experimental models for triggering BCO are essential for confirming key triggering mechanisms, revealing innate susceptibilities and providing a reliable test bed for evaluating practical prophylactic and therapeutic strategies. Attempts to initiate BCO via a respiratory challenge have been only marginally successful, whereas osteomyelitis has repeatedly been reproduced by oral challenges and by injecting broilers and turkeys intravenously with appropriate strains of pathogenic bacteria. Intravenous injections of bacterial pathogens acutely overwhelm the body's defence mechanisms, initiate sustained septicaemia and tend to trigger widely dispersed osteomyelitis instead of the focal femoral, tibial and vertebral lesions that are pathognomonic for BCO. Intravenous injection models are unlikely to be useful for detecting differences in BCO susceptibility among broiler lines that may primarily depend on differences in innate susceptibility to osteochondrosis, nor are such models useful for demonstrating relative efficacies of practical prophylactic treatments that improve barrier integrity and thereby reduce bacterial translocation. Mechanical models based on wire flooring consistently trigger BCO by creating sustained footing instability that persistently exerts excessive mechanical stress on susceptible leg and vertebral joints. Mechanical models also act as chronic physiological stressors that promote indigenous bacterial translocation and proliferation attributed to stress-mediated immunosuppression. By consistently triggering significant incidences of BCO in research flocks, mechanical models have provided a reliable experimental protocol for comparing genetic lines and chick sources and for evaluating practical treatment strategies. Mechanical models avoid the need to purposefully inoculate birds with known pathogens based on the assumption that the offending pathogens already are present but quiescent within the bird or in the environment, waiting for an effective triggering scenario (e.g. stress, immunosuppression) and attainment of sufficient body mass to create a conducive wound site (e.g. osteochondrotic clefts, focal necrosis). Pathogenic bacteria may be transmitted from breeder parents, contaminated eggshells or hatchery sources or may enter the circulation via translocation through the integument, respiratory system or intestinal tract. Translocation of orally administered or indigenous enteric bacteria and haematogenous distribution by the circulatory system throughout the body has been demonstrated repeatedly. Most bacterial translocation across the intestinal barrier appears to utilize the paracellular route via 'leaky' tight junctions. Probiotics are known to improve tight junction integrity and reduce bacterial translocation. Furthermore, recent experiments provide the first evidence that some, but not all, probiotics can significantly interrupt the pathogenesis of lameness attributable to BCO. The efficacy of feeding probiotics prophylactically to reduce the incidence of BCO implicates the gastrointestinal tract as a major avenue of bacterial translocation during the pathogenesis of this disease. The roles of probiotics in improving the barrier function of the intestinal epithelium remain to be revealed. It is intriguing to speculate that probiotic species may translocate into the bloodstream to modulate internal microbial communities in various organs and tissues, including the growth plates of bones. Further research is necessary to study the role of probiotics in arresting the pathogenesis of BCO.

7 Where to look for further information

- Avian leg bone growth: Wideman, R. F., and R. D. Prisby. 2013. Bone circulatory disturbances in the development of spontaneous bacterial chondronecrosis with osteomyelitis: A translational model for the pathogenesis of femoral head necrosis. *Frontiers in Science (Front. Endocrin.)* 3: 183. doi:10.3389/fendo.2012.00183.
- Osteochondrosis review: Ytrehus, B., C. S. Carlson, and S. Ekman. 2007. Etiology and pathogenesis of osteochondrosis. *Vet. Pathol.* 44: 429–48.
- BCO review: McNamee, P. T., and J. A. Smyth. 2000. Bacterial chondronecrosis with osteomyelitis ('femoral head necrosis') of broiler chickens: A review. *Avian Pathol.* 29: 253–70.

- BCO review: Wideman, R. F. 2015. Bacterial chondronecrosis with osteomyelitis and lameness in broilers: A review. *Poult. Sci.* dx.doi.org/10.3382/ps/pev320.
- Intestinal tight junctions and probiotics: Ulluwishewa, D., R. C. Anderson, W. C. McNabb, P. J. Moughan, J. M. Wells, and N. C. Roy. 2011. Regulation of tight junction permeability by intestinal bacteria and dietary components. *J. Nutr.* 141: 769–76.
- Annual meetings: Symposium on Gut Health in Production of Food Animals (http://www.guthealthsymposium.com).

8 References

Alderson, M., and S. Nade. 1987. Natural history of acute septic arthritis in an avian model. *J. Orthopaed. Res.* 5: 261–74.

Alderson, M., K. Emslie, D. Speers, and S. Nade. 1986a. Transphyseal blood vessels exist in avian species. *J. Anat.* 146: 217–24.

Alderson, M., D. Speers, K. Emslie, and S. Nade. 1986b. Acute haematogenous osteomyelitis and septic arthritis – a single disease. *J. Bone Joint Surg.* 68B: 268–74.

Alfonso, M., and H. J. Barnes. 2006. Neonatal osteomyelitis associated with *Staphylococcus aureus* in turkey poults. *Avian Dis.* 50: 148–51.

Al-Rubaye, A. A., K. Estill, R. F. Wideman, and D. D. Rhoads. 2014. 16S rRNA-based diagnosis and whole-genome sequencing of bacteria cultured from lame broilers with osteomyelitis. *Poult. Sci.* 93 (E-Suppl. 1): 314P.

Ando, T., R. F. Brown, R. D. Berg, and A. J. Dunn. 2000. Bacterial translocation can increase plasma corticosterone and brain catecholamine and indoleamine metabolism. *Am. J. Physiol. Reg. Integr. Comp. Physiol.* 279: R2164–72.

Andreasen, J. R., C. B. Andreasen, M. Anwer, and A. E. Sonn. 1993. Heterophil chemotaxis in chickens with natural staphylococcal infections. *Avian Dis.* 37: 284–9.

Applegate, T. J., and M. S. Lilburn. 2002. Growth of the femur and tibia of a commercial broiler line. *Poult. Sci.* 81: 1289–94.

Bailey, J. S., N. A. Cox, D. E. Cosby, and L. J. Richardson. 2005. Movement and persistence of *Salmonella* in broiler chickens following oral or intracloacal inoculation. *J. Food Protect.* 68: 2698–2701.

Bar-Shira, E., I. Cohen, O. Elad, and A. Friedman. 2014. Role of goblet cells and mucin layer in protecting maternal IgA in precocious birds. *Develop. Comp. Immunol.* 44: 186–94.

Beaumont, G. D. 1967. The intraosseous vasculature of the ulna of *Gallus domesticus*. *J. Anat.* 101: 543–54.

Berg, R. D. 1992. Bacterial translocation from the gastrointestinal tract. *J. Med.* 23: 217–44.

Bohorquez, D. V., N. E. Bohorquez, and P. R. Ferket. 2011. Ultrastructural development of the small intestinal mucosa in the embryo and turkey poult: A light and electron microscopy study. *Poult. Sci.* 90: 842–55.

Bond, P. L., T. W. Sullivan, J. H. Douglas, and L. G. Robeson. 1991. Influence of age, sex, and method of rearing on tibia length and mineral deposition in broilers. *Poult. Sci.* 70: 1936–42.

Bradshaw, R. H., R. D. Kirkden, and D. M. Broom. 2002. A review of the aetiology and pathology of leg weakness in broilers in relation to welfare. *Avian Poult. Biol. Rev.* 13: 45–103.

Buhr, R. J., N. A. Cox, N. J. Stern, M. T. Musgrove, J. L. Wilson, and K. L. Hiett. 2002. Recovery of Campylobacter from segments of the reproductive tract of broiler breeder hens. *Avian Dis.* 46: 919–24.

Butterworth, A. 1999. Infectious components of broiler lameness: A review. *W. Poult. Sci. J.* 55: 327–52.

Byrne, C. M., M. Clyne, and B. Bourke. 2007. *Campylobacter jejuni* adhere to and invade chicken intestinal epithelial cells *in vitro*. *Microbiol.* 153: 561–9.

Carnaghan, R. B. A. 1966. Spinal cord compression due to spondylitis caused by *Staphylococcus pyogenes*. *J. Comp. Pathol.* 76: 9–14.

Carter, P. B., and F. M. Collins. 1974. The route of enteric infection in normal mice. *J. Exp. Med.* 139: 1189–1203.

Church, L. E., and L. C. Johnson. 1964. Growth of long bones in the chicken. Rates of growth in length and diameter of the humerus, tibia, and metatarsus. *Am. J. Anat.* 114: 521–38.

Clark, A. G., and D. H. Bueschkens. 1985. Laboratory infection of chicken eggs with *Campylobacter jejuni* by using temperature and pressure differentials. *Appl. Environ. Microbiol.* 49: 1467–71.

Clark, S., R. Porter, B. McComb, R. Lippert, S. Olsen, S. Nohner, and H. L. Shivaprasad. 2010. Clostridial dermatitis and cellulitis: An emerging disease in turkeys. *Avian Dis.* 54: 788–94.

Cossart, P., and P. J. Sansonetti. 2004. Bacterial invasion: The paradigms of enteroinvasive pathogens. *Science* 304: 242–8.

Cox, N. A., N. J. Stern, K. L. Hiett, and M. Berrang. 2002. Identification of a new source of Campylobacter contamination in poultry: Transmission from breeder hens to broiler chickens. *Avian Dis.* 46: 535–41.

Cox, N. A., C. L. Hofacre, J. S. Bailey, R. J. Buhr, J. L. Wilson, K. L. Hiett, L. J. Richardson, M. T. Musgrove, D. E. Cosby, J. D. Tankson, Y. L. Vizzier, P. F. Cray, L. E. Vaughn, P. S. Holt, and D. V. Bourassa. 2005a. Presence of *Campylobacter jejuni* in various organs one hour, one day, and one week following oral or intracloacal inoculations of broiler chicks. *Avian Dis.* 49: 155–8.

Cox, N. A., J. S. Bailey, L. J. Richardson, R. J. Buhr, D. E. Cosby, J. L. Wilson, K. L. Hiett, G. R. Siragusa, and D. V. Bourassa. 2005b. Presence of naturally occurring *Campylobacter* and *Salmonella* in the mature and immature ovarian follicles of late-life broiler breeder hens. *Avian Dis.* 49: 285–7.

Cox, N. A., L. J. Richardson, R. J. Buhr, P. J. Fedorka-Cray, J. S. Bailey, J. L. Wilson, and K. L. Hiett. 2006a. Natural presence of *Campylobacter* spp. in various internal organs of commercial broiler breeder hens. *Avian Dis.* 50: 450–3.

Cox, N. A., L. J. Richardson, R. J. Buhr, J. K. Northcutt, B. D. Fairchild, and J. M. Mauldin. 2006b. Presence of inoculated *Campylobacter* and *Salmonella* in unabsorbed yolks of male breeders raised as broilers. *Avian Dis.* 50: 430–3.

Cox, N. A., L. J. Richardson, R. J. Buhr, J. K. Northcutt, P. J. Fedorka-Cray, J. S. Bailey, B. D. Fairchild, and J. M. Mauldin. 2006c. Natural occurrence of *Campylobacter* species, *Salmonella* serovars, and other bacteria in unabsorbed yolks of market-age commercial broilers. *J. Appl. Poult. Res.* 15: 551–7.

Cox, N. A., L. J. Richardson, R. J. Buhr, J. K. Northcutt, J. S. Bailey, P. F. Cray, and K. L. Hiett. 2007. Recovery of *Campylobacter* and *Salmonella* serovars from the spleen, liver and gallbladder, and ceca of six- and eight-week-old commercial broilers. *J. Appl. Poult. Res.* 16: 477–80.

Daum, R. S., W. H. Davis, K. B. Farris, R. J. Campeau, D. M. Mulvihill, and S. M. Shane. 1990. A model of staphylococcus aureus bacteremia, septic arthritis, and osteomyelitis in chickens. *J. Orthopaed. Res.* 8: 804–13.

Desmidt, M., R. Ducatelle, and F. Haesebrouck. 1998. Serological and bacteriological observations on experimental infection with *Salmonella hadar* in chickens. *Vet. Microbiol.* 60: 259–69.

Devriese, L. A., A. H. Devos, and J. Beumer. 1972. *Staphylococcus aureus* colonization on poultry after experimental spray inoculations. *Avian Dis.* 16: 656–65.

Dinev, I. 2009. Clinical and morphological investigations on the prevalence of lameness associated with femoral head necrosis in broilers. *Br. Poult. Sci.* 50: 284–90.

Duff, S. R. I. 1984. Capital femoral epiphyseal infarction in skeletally immature broilers. *Res. Vet. Sci.* 37: 303–9.

Duff, S. R. I. 1990a. Do different forms of spondylolisthesis occur in broiler fowls? *Avian Pathol.* 19: 279–94.

Duff, S. R. I. 1990b. Diseases of the musculoskeletal system. In F. T. W. Jordan (ed.), *Poultry Diseases*, 3rd ed. (pp. 254–83). London, UK: Bailliere Tindall.

El-Lethey, H., B. Huber-Eicher, and T. W. Jungi. 2003. Exploration of stress-induced immunosuppression in chickens reveals both stress-resistant and stress-susceptible antigen responses. *Vet. Immunol. Immunopathol.* 95: 91–101.
Emslie, K. R., and S. Nade. 1983. Acute hematogenous staphylococcal osteomyelitis: A description of the natural history in an avian model. *Am. J. Pathol.* 110: 333–45.
Emslie, K. R., and S. Nade. 1985. Acute hematogenous staphylococcal osteomyelitis. *Comp. Pathol. Bulletin* 17: 2–3.
Emslie, K. R., N. R. Ozanne, and S. M. L. Nade. 1983. Acute hematogenous osteomyelitis: An experimental model. *J. Pathol.* 141: 157–67.
Emslie, K. R., L. M. Fenner, and S. M. L. Nade. 1984. Acute haematogenous osteomyelitis: II. The effect of a metaphyseal abscess on the surrounding blood supply. *J. Pathol.* 142: 129–34.
Funkhouser, L. J., and S. R. Bordenstein. 2013. Mom knows best: The universality of maternal microbial transmission. *PLoS Biol.* 11(8): e1001631. doi::10.1371/journal.pbio.1001631.
Gilley, A. D., H. Lester, I. Y. Pevzner, N. B. Anthony, and R. F. Wideman, Jr. 2014. Evaluating portable wire flooring models for inducing bacterial chondronecrosis with osteomyelitis (BCO) in broilers. *Poult. Sci.* 93: 1354–67.
Griffiths, G. L., W. L. Hopkinson, and J. Lloyd. 1984. Staphylococcal necrosis in the head of the femur in broiler chickens. *Austral. Vet. J.* 61: 293.
Hales, S. 1727. *Static Essays, Vol I: Vegetable Staticks.* London: W. Innys & Woodward.
Hand, W. L., N. L. King-Thompson, and J. D. Johnson. 1984. Influence of bacterial-antibiotic interactions on subsequent antimicrobial activity of alveolar macrophages. *J. Infect. Dis.* 149: 271–6.
Havenstein, G. B., P. R. Ferket, S. E. Scheidler, and B. T. Larson. 1994. Growth, livability and feed conversion of 1957 vs. 1991 broilers when fed 'typical' 1957 and 1991 broiler diets. *Poult. Sci.* 73: 1785–94.
Haye, U., and P. C. M. Simons. 1978. Twisted legs in broilers. *Br. Poult. Sci.* 19: 549–57.
Howlett, C. R. 1980. The fine structure of the proximal growth plate and metaphysis of the avian tibia: Endochondral osteogenesis. *J. Anat.* 130: 745–68.
Howlett, C. R., M. Dickson, and A. K. Sheridan. 1984. The fine structure of the proximal growth plate of the avian tibia: Vascular supply. *J. Anat.* 139: 115–32.
Huff, G. R., W. E. Huff, J. M. Balog, and N. C. Rath. 1998. Effects of dexamethasone immunosuppression on turkey osteomyelitis complex in an experimental *Escherichia coli* respiratory infection. *Poult. Sci.* 77: 654–61.
Huff, G. R., W. E. Huff, J. M. Balog, and N. C. Rath. 1999. Sex differences in the resistance of turkeys to *Escherichia coli* challenge after immunosuppression with dexamethasone. *Poult. Sci.* 78: 38–44.
Huff, G. R., W. E. Huff, N. C. Rath, and J. M. Balog. 2000. Turkey osteomyelitis complex. *Poult. Sci.* 79: 1050–6.
Huff, G. R., W. E. Huff, J. M. Balog, N. C. Rath, N. B. Anthony, and K. E. Nestor. 2005. Stress response differences and disease susceptibility reflected by heterophil to lymphocyte ratio in turkeys selected for increased body weight. *Poult. Sci.* 84: 709–17.
Huff, G. R., W. E. Huff, N. C. Rath, and J. M. Balog, N. B. Anthony, and K. E. Nestor. 2006. Stress-induced colibacillosis and turkey osteomyelitis complex in turkeys selected for increased body weight. *Poult. Sci.* 85: 266–72.
Humphrey, T. J., and D. G. Lanning. 1988. The vertical transmission of salmonellas and formic acid treatment of chicken feed. *Epidem. Inf.* 100: 43–9.
Hunt, C. D., D. A. Ollerich, and F. H. Nielsen. 1979. Morphology of the perforating cartilage canals in the proximal tibial growth plate of the chick. *Anat. Rec.* 194: 143–57.
Jensen, M. M., W. C. Downs, J. D. Morrey, T. R. Nicoll, S. D. Lefevre, and C. M. Meyers. 1987. Staphylococcosis of turkeys. 1. Portal of entry and tissue colonisation. *Avian Dis.* 31: 64–9.
Jiang, T., R. K. Mandal, R. F. Wideman Jr., A. Khatiwara, I. Pevzner, and Y. M. Kwon. 2015. Molecular survey of bacterial communities associated with bacterial chondronecrosis with osteomyelitis in broilers. *PLoS ONE,* doi:10.1371/journal.pone.0124403, 16 April 2015. http://www.ncbi.nlm.nih.gov/pubmed/25881241.

Joiner, K. S., F. J. Hoerr, E. van Santen, and S. J. Ewald. 2005. The avian major histocompatibility complex influences bacterial skeletal disease in broiler breeder chickens. *Vet. Pathol.* 42: 275–81.

Julian, R. J. 1985. Osteochondrosis, dyschondroplasia, and osteomyelitis causing femoral head necrosis in turkeys. *Avian Dis.* 29: 854–66.

Julian, R. J. 2005. Production and growth related disorders and other metabolic diseases of poultry – A review. *Vet. J.* 169: 350–69.

Khan, M. A., N. O. Olson, and D. O. Overman. 1977. Spontaneous spondylolisthesis in embryonic and adult chick. *Poult. Sci.* 56: 689–97.

Kember, N. F., and J. K. Kirkwood. 1987. Cell kinetics and longitudinal bone growth in birds. *Cell Tissue Kinet.* 20: 625–9.

Kember, N. F., J. K. Kirkwood, P. J. Duignan, D. Godfrey, and D. J. Spratt. 1990. Comparative cell kinetics of avian growth plates. *Res. Vet. Sci.* 49: 283–8.

Kense, M. J., and W. J. M. Landman. 2011. *Enterococcus cecorum* infections in broiler breeders and their offspring: Molecular epidemiology. *Avian Pathol.* 40: 603–12.

Kestin, S. C., G. Su, and P. Sorensen. 1999. Different commercial broiler crosses have different susceptibilities to leg weakness. *Poult. Sci.* 78: 1085–90.

Kestin, S. C., S. Gordon, G. Su, and P. Sorensen. 2001. Relationship in broiler chickens between lameness, live weight, growth rate and age. *Vet. Rec.* 148: 195–7.

Kibenge, F. S. B., G. E. Wilcox, and D. A. Pass. 1983. Pathogenicity of four strains of *Staphylococcus aureus* isolated from chickens with clinical tenosynovitis. *Avian Pathol.* 12: 213–20.

Kirkwood, J. K., P. J. Duignan, N. F. Kember, P. M. Bennett, and D. J. Price. 1989a. The growth of the tarsometatarsus bone in birds. *J. Zool., Lond.* 217: 403–16.

Kirkwood, J. K., D. M. J. Spratt, and P. J. Duignan. 1989b. Patterns of cell proliferation and growth rate in limb bones of the domestic fowl (*Gallus domesticus*). *Res. Vet. Sci.* 47: 139–47.

Kremer, C. J., K. M. O'Mera, S. L. Layton, B. M. Hargis, and K. Cole. 2011. Evaluation of recombinant *Salmonella* expressing the flagellar protein flic for persistence and enhanced antibody response in commercial turkeys. *Poult. Sci.* 90: 752–8.

Kuhlers, D. L., and G. R. McDaniel. 1996. Estimates of heritabilities and genetic correlations between tibial dyschondroplasia expression and body weight at two ages in broilers. *Poult. Sci.* 75: 959–61.

Leach, R. M., and M. C. Nesheim. 1965. Nutritional, genetic and morphological studies of an abnormal cartilage formation in young chicks. *J. Nutr.* 86: 236–44.

Leach, R. M. Jr., and C. V. Gay. 1987. Role of epiphyseal cartilage in endochondral bone formation. *J. Nutr.* 117: 784–90.

Liljebjelke, K. A., C. L. Hofacre, T. Liu, D. G. White, S. Ayers, S. Young, and J. J. Maurer. 2005. Vertical and horizontal transmission of Salmonella within integrated broiler production system. *Foodborne Pathog. Dis.* 2: 90–102.

Lissner, C. R., R. N. Swanson, and A. D. O'Brien. 1983. Genetic control of the innate resistance of mice to *Salmonella typhimurium*: Expression of the Ity gene in peritoneal and splenic macrophages isolated *in vitro*. *J. Immunol.* 131: 3003–13.

Lutfi, A. M. 1970a. Study of cell multiplication in the cartilagenous upper end of the tibia of the domestic fowl by tritiated thymidine autoradiography. *Acta Anat.* 76: 454–63.

Lutfi, A. M. 1970b. The mode of growth, fate and function of cartilage canals. *J. Anat.* 106: 135–45.

Martin, L. T., M. P. Martin, and H. J. Barnes. 2011. Experimental reproduction of enterococcal spondylitis in male broiler breeder chickens. *Avian Dis.* 55: 273–8.

McCaskey, P. C., G. N. Rowland, R. K. Page, and L. R. Minear. 1982. Focal failures of endochondral ossification in the broiler. *Avian Dis.* 26: 701–17.

McCullagh, J. J., P. T. McNamee, J. A. Smyth, and H. J. Ball. 1998. The use of pulsed-field gel electrophoresis to investigate the epidemiology of *Staphylococcus aureus* infection in commercial broiler flocks. *Vet. Microbiol.* 63: 275–81.

McNamee, P. T., and J. A. Smyth. 2000. Bacterial chondronecrosis with osteomyelitis ('femoral head necrosis') of broiler chickens: A review. *Avian Pathol.* 29: 253–70.

McNamee, P. T., J. J. McCullagh, B. H. Thorp, H. J. Ball, D. Graham, S. J. McCullough, D. McConaghy, and J. A. Smyth. 1998. Study of leg weakness in two commercial broiler flocks. *Vet. Rec.* 143: 131–5.

McNamee, P. T., J. J. McCullagh, J. D. Rodgers, B. H. Thorp, H. J. Ball, T. J. Connor, D. McConaghy, and J. A. Smyth. 1999. Development of an experimental model of bacterial chondronecrosis with osteomyelitis in broilers following exposure to Staphlococcus aureus by aerosol, and inoculation with chicken anemia and infectious bursal disease viruses. *Avian Pathol.* 28: 26–35.

Mercer, J. T., and W. G. Hill. 1984. Estimation of genetic parameters for skeletal defects in broiler chickens. *Heredity* 53: 193–203.

Murugesan, G. R., N. K. Gabler, and M. E. Persia. 2014. Effects of direct-fed microbial supplementation on broiler performance, intestinal nutrient transport and integrity under experimental conditions with increased microbial challenge. *Br. Poult. Sci.* 55: 89–97.

Mutalib, A., C. Riddell, and A. D. Osborne. 1983a. Studies on the pathogenesis of staphylococcal osteomyelitis in chickens. I. Effect of stress on experimentally induced osteomyelitis. *Avian Dis.* 27: 141–56.

Mutalib, A., C. Riddell, and A. D. Osborne. 1983b. Studies on the pathogenesis of staphylococcal osteomyelitis in chickens. II. Role of the respiratory tract as a route of infection. *Avian Dis.* 27: 157–60.

Naas, I. A., I. C. L. A. Paz, M. S. Baracho, A. G. Menezes, L. G. F. Bueno, I. C. L. Almeida, and D. J. Moura. 2009. Impact of lameness on broiler well-being. *J. Appl. Poult. Res.* 18: 432–9.

Nairn, M. E. (1973). Bacterial osteomyelitis and synovitis in the turkey. *Avian Dis.* 17: 504–17.

Nairn, M. E., and A. R. A. Watson. 1972. Leg weakness of poultry – a clinical and pathological characterisation. Aust. Vet. J. 48: 645–56.

Nehme, P. A., E. K. Barbour, V. K. Sagherian, S. K. Hamadeh, and R. K. Zurayk. 2005. Baseline data on enumerated tracheal bacterial flora and drug susceptibility in chickens reared under different systems. *Inter. J. Appl. Res. Vet. Med.* 3: 372–7.

Nestor, K. E. 1984. Genetics of growth and reproduction in the turkey. 9. Long-term selection for increased 16-week body weight. *Poult. Sci.* 63: 2114–22.

Nestor, K. E., and J. W. Anderson. 1998. Effect of crossing a line selected for increased shank width with two commercial sire lines on performance and walking ability of turkeys. *Poult. Sci.* 77: 1601–7.

Nicoll, T. R., and M. M. Jensen. 1986. Preliminary studies on bacterial interference of staphylococcosis of chickens. *Avian Dis.* 31: 140–4.

Pastorelli, L., C. De Salvo, J. R. Mercado, M. Vecchi, and T. T. Pizarro. 2013. Central role of the gut epithelial barrier in the pathogenesis of chronic intestinal inflammation: Lessons learned from animal models and human genetics. *Frontiers in Science (Front. Immunol.)* 4: 280. doi:10.3389/fimmu.2013.002080.

Pearson, A. D., M. H. Greenwood, R. K. A. Feltham, T. D. Healing, J. Donaldson, D. M. Jones, and R. R. Colwell. 1996. Microbial ecology of *Campylobacter jejuni* in a United Kingdom supply chain: Intermittent common source, vertical transmission, and amplification by flock propagation. *Appl. Environ. Microbiol.* 62: 4614–20.

Prisby, R., T. Menezes, J. Campbell, T. Benson, E. Samraj, I. Pevzner, and R. F. Wideman, Jr. 2014. Kinetic examination of femoral bone modeling in broilers. *Poult. Sci.* 93: 1122–9.

Quinteiro-Filho, W. M., A. Ribeiro, V. Ferraz-de-Paula, M. L. Pinheiro, M. Sakai, L. R. M. Sa, A. J. P. Ferreira, and J. Palermo-Neto. 2010. Heat stress impairs performance parameters, induces intestinal injury, and decreases macrophage activity in broiler chickens. *Poult. Sci.* 89: 1905–14.

Quinteiro-Filho, W. M., M. V. Rodrigues, A. Ribeiro, V. Ferraz-de-Paula, M. L. Pinheiro, L. R. M. Sa, A. J. P. Ferreira, and J. Palermo-Neto. 2012a. Acute heat stress impairs performance parameters and induces mild intestinal enteritis in broiler chickens: Role of acute hypothalamic-pituitary-adrenal axis of activation. *J. Anim. Sci.* 90: 1986–94.

Quinteiro-Filho, W. M., A. V. S. Gomes, M. L. Pinheiro, A. Ribeiro, V. Ferraz-de-Paula, C. S. Astolfi-Ferreira, A. J. P. Ferreira, and J. Palermo-Neto. 2012b. Heat stress impairs performance and induces intestinal inflammation in broiler chickens infected with *Salmonella* Enteritidis. *Avian Pathol.* 41: 421–7.

Reiber, M. A., J. A. McInroy, and D. E. Conner. 1995. Enumeration and identification of bacteria in chicken semen. *Poult. Sci.* 74: 795–9.

Richardson, L. J., N. A. Cox, R. J. Buhr, and M. A. Harrison. 2011. Isolation of *Campylobacter* from circulating blood of commercial broilers. *Avian Dis.* 55: 375–8.

Riddell, C. 1973. Studies on spondylolisthesis (kinky-back) in broiler chickens. *Avian Pathol.* 2: 295–304.

Riddell, C. 1975. Studies on the pathogenesis of tibial dyschondroplasia in chickens. II. Growth rate of long bones. *Avian Dis.* 19: 490–6.

Riddell, C. 1976. Selection of broiler chickens for high and low incidence of tibial dyschondroplasia with observations on spondylolisthesis and twisted legs (perosis). *Poult. Sci.* 55: 145–51.

Riddell, C. 1983. Pathology of the skeleton and tendons of broiler chickens reared to roaster weights. I. crippled chickens. *Avian Dis.* 27: 950–62.

Riddell, C., M. W. King, and K. R. Gunasekera. 1983. Pathology of the skeleton and tendons of broiler chickens reared to roaster weights. II. Normal chickens. *Avian Dis.* 27: 980–91.

Rodgers, J. D., J. J. McCullagh, P. T. McNamee, J. A. Smyth, and H. J. Ball. 1999. Comparison of *Staphylococcus aureus* recovered from personnel in a poultry hatchery and in broiler parent farms with those isolated from skeletal disease in broilers. *Vet. Microbiol.* 69: 189–98.

Sandilands, V., S. Brocklehurst, N. Sparks, L. Baker, R. McGovern, B. Thorp, and D. Pearson. 2011. Assessing leg health in chickens using a force plate and gait scoring: How many birds is enough? *Vet. Rec.* 168: 77–83.

Saunders, P. R., U. Kosecka, D. M. McKay, and M. H. Perdue. 1994. Acute stressors stimulate ion secretion and increase epithelial permeability in rat intestine. *Am. J. Physiol.* 267: G794–9.

Shabbir, M. Z., T. Malys, Y. V. Ivanov, J. Park, M. A. B. Shabbir, M. Rabbani, T. Yaqub, and E. T. Harvill. 2015. Microbial communities present in the lower respiratory tract of clinically healthy birds in Pakistan. *Poult. Sci.* 94: 612–20.

Sheridan, A. K., C. R. Howlett, and R. W. Burton. 1978. The inheritance of tibial dyschondroplasia in broilers. *Br. Poult. Sci.* 19: 491–9.

Skeeles, K. J. 1997. Staphylococcosis. In B. W. Calnek, H. J. Barnes, C. W. Beard, L. R. McDougald, and Y. M. Saif (ed.), *Diseases of Poultry 10th Edition*, ch. 11 (pp. 247–53). Ames, IA, USA: Iowa State University Press.

Smeltzer, M. S., and A. F. Gillaspy. 2000. Molecular pathogenesis of staphylococcal osteomyelitis. *Poult. Sci.* 79: 1042–9.

Smirnov, A., R. Perez, E. Amit-Romach, D. Sklan, and Z. Uni. 2005. Mucin dynamics and microbial populations in chicken small intestine are changed by dietary probiotic and antibiotic growth promoter supplementation. *J. Nutr.* 135: 187–92.

Smith, H. W. 1954. Experimental staphylococcal infection in chickens. *J. Pathol. Bacteriol.* 67: 81–7.

Sohail, M. U., A. Ijaz, M. S. Yousaf, K. Ashraf, H. Zaneb, M. Aleem, and H. Rehman. 2010. Alleviation of cyclic heat stress in broilers by dietary supplementation of mannan-oligosaccharide and *Lactobacillus*-based probiotic: Dynamics of cortisol, thyroid hormones, cholesterol, C-reactive protein, and humoral immunity. *Poult. Sci.* 89: 1934–8.

Sohail, M. U., M. E. Hume, J. A. Byrd, D. J. Nisbet, A. Ijaz, A. Sohail, M. Z. Shabbir, and H. Rehman. 2012. Effect of supplementation of prebiotic mannan-oligosaccharides and probiotic mixture on growth performance of broilers subjected to chronic heat stress. *Poult. Sci.* 91: 2235–40.

Sohail, M. U., M. E. Hume, J. A. Byrd, D. J. Nisbet, M. Z. Shabbir, A. Ijaz, and H. Rehman. 2015. Molecular analysis of the caecal and tracheal microbiome of heat stressed broilers supplemented with prebiotic and probiotic. *Avian Pathol.* 44: 67–74.

Somes, R. G., Jr. 1969. Genetic perosis in the domestic fowl. *J. Hered.* 60: 163–6.

Song, J., K. Xiao, Y. L. Ke, L. F. Jiao, C. H. Hu, Q. Y. Diao, B. Shi, and X. T. Zou. 2014. Effect of a probiotic mixture on intestinal microflora, morphology, and barrier integrity of broilers subjected to heat stress. *Poult. Sci.* 93: 581–8.

Sorensen, P. 1992. The genetics of leg disorders. In C. C. Whitehead (ed.), *Bone Biology and Skeletal Disorders in Poultry* (pp. 213–29). Abingdon, United Kingdom: Carfax Publishing Company.

Soriani, M., I. Santi, A. Taddei, R. Rappuoli, G. Grandi, and J. L. Tedford. 2006. Group B *Streptococcus* crosses human epithelial cells by a paracellular route. *J. Infect. Dis.* 193: 241–50.

Speers, D. J., and S. M. L. Nade. 1985. Ultrastructural studies of adherence of Staphylococcus aureus in experimental acute hematogenous osteomyelitis. *Infection and Immunity* 49: 443–6.

Stalker, M. J., M. L. Brash, A. Weisz, R. M. Ouckama, and D. Slavic. 2010. Arthritis and osteomyelitis associated with *Enterococcus cecorum* infection in broiler and broiler breeder chickens in Ontario, Canada. *J. Vet. Diagn. Invest.* 22: 643–5.

Steinwender, G., G. Schimpl, B. Sixl, and H. H. Wenzl. 2001. Gut-derived bone infection in neonatal rat. *Pediatr. Res.* 50: 767–71.

Tate, C. R., W. C. Mitchell, and R. G. Miller. 1993. *Staphylococcus hyicus* associated with turkey stifle joint osteomyelitis. *Avian Dis.* 37: 905–7.

Thorp, B. H. 1988a. Vascular pattern of the developing knee joint in the domestic fowl. *Res. Vet. Sci.* 44: 89–99.

Thorp, B. H. 1988b. Vascular pattern of the developing intertarsal joint in the domestic fowl. *Res. Vet. Sci.* 44: 100–11.

Thorp, B. H. 1988c. Relationship between the rate of longitudinal bone growth and physeal thickness in the growing fowl. *Res. Vet. Sci.* 45: 83–5.

Thorp, B. H. 1994. Skeletal disorders in the fowl: A review. *Avian Pathol.* 23: 203–36.

Thorp, B. H., and D. Waddington. 1997. Relationships between the bone pathologies, ash and mineral content of long bones in 35-day-old broiler chickens. *Res. Vet. Sci.* 62: 67–73.

Thorp, B. H., C. C. Whitehead, L. Dick, J. M. Bradbury, R. C. Jones, and A. Wood. 1993. Proximal femoral degeneration in growing broiler fowl. *Avian Pathol.* 22: 325–42.

Ulluwishewa, D., R. C. Anderson, W. C. McNabb, P. J. Moughan, J. M. Wells, and N. C. Roy. 2011. Regulation of tight junction permeability by intestinal bacteria and dietary components. *J. Nutr.* 141: 769–76.

Uni, Z., Y. Noy, and D. Sklan. 1995. Post hatch changes in morphology and function of the small intestines in heavy- and light-strain chicks. *Poult. Sci.* 74: 1622–9.

Walser, M. M., F. L. Cherms, and H. E. Dziuk. 1982. Osseous development and tibial dyschondroplasia in five lines of turkeys. *Avian Dis.* 26: 265–70.

Waters, A. E., T. Contente-Cuomo, J. Buchhagen, C. M. Liu, L. Watson, K. Pierce, J. T. Foster, J. Bowers, E. M. Driebe, D. M. Engelthaler, P. S. Keim, and L. B. Price. 2011. Multi-drug resistant *Staphylococcus aureus* in US meat and poultry. *Clin. Infect. Dis.* 52: 1227–30.

Wideman, R. F. Jr. 2014. Bacterial chondronecrosis with osteomyelitis and lameness in broilers: Pathogenesis, experimental models, and preventative treatments. In W. H. A. Abdelrahman (ed.), *Probiotics in Poultry Production, Concept and Applications* (pp. 91–128). Sheffield, UK: 5m Publishing Ltd.

Wideman, R. F., and I. Pevzner. 2012. Dexamethasone triggers lameness associated with necrosis of the proximal tibial head and proximal femoral head in broilers. *Poult. Sci.* 91: 2464–74.

Wideman, R. F., and R. D. Prisby. 2013. Bone circulatory disturbances in the development of spontaneous bacterial chondronecrosis with osteomyelitis: a translational model for the pathogenesis of femoral head necrosis. *Frontiers in Science (Front. Endocrin.)* 3: 183. doi:10.3389/fendo.2012.00183.

Wideman, R. F., K. R. Hamal, J. M. Stark, J. Blankenship, H. Lester, K. N. Mitchell, G. Lorenzoni, and I. Pevzner. 2012. A wire flooring model for inducing lameness in broilers: Evaluation of probiotics as a prophylactic treatment. *Poult Sci.* 91: 870–83.

Wideman, R. F. Jr., A. Al-Rubaye, A. Gilley, D. Reynolds, H. Lester, D. Yoho, J. D. Hughes Jr., and I. Y. Pevzner. 2013. Susceptibility of four commercial broiler crosses to lameness attributable to bacterial chondronecrosis with osteomyelitis. *Poult. Sci.* 92: 2311–25.

Wideman, R. F. Jr., A. Al-Rubaye, D. Reynolds, D. Yoho, H. Lester, C. Spencer, J. M. Hughes, and I. Y. Pevzner. 2014. Bacterial chondronecrosis with osteomyelitis (BCO) in broilers: Influence of sires and straight-run vs. sex-separate rearing. *Poult. Sci.* 93: 1675–87.

Wideman, R. F. Jr., A. Al-Rubaye, Y. M. Kwon, J. Blankenship, H. Lester, K. N. Mitchell, I. Y. Pevzner, T. Lohrmann, and J. Schleifer. 2015a. Prophylactic administration of a combined prebiotic and probiotic, or therapeutic administration of enrofloxacin, to reduce the incidence of bacterial chondronecrosis with osteomyelitis in broilers. *Poult. Sci.* 94: 25–36.

Wideman, R. F. Jr., J. Blankenship, I. Y. Pevzner, and B. J. Turner. 2015b. Efficacy of 25-OH vitamin D_3 prophylactic administration for reducing lameness in broilers grown on wire flooring. *Poult. Sci.* 94: 1821–7.

Wijetunge, D. S., P. Dunn, E. Waller-Pendleton, V. Lintner, H. Lu, and S. Kariyawasam. 2012. Fingerprinting of poultry isolates of *Enterococcus cecorum* using three molecular typing methods. *J. Vet. Diag. Invest.* 24: 1166–71.

Williams, B., S. Solomon, D. Waddington, B. Thorp, and C. Farquharson. 2000. Skeletal development in meat-type chickens. *Br. Poult. Sci.* 41: 141–9.

Wise, D. R. 1970a. Carcass conformation comparisons of growing broilers and laying strain chickens. *Br. Poult. Sci.* 11: 325–32.

Wise, D. R. 1970b. Spondylolisthesis ('kinky back') in broiler chickens. *Res. Vet. Sci.* 11: 447–51.

Wise, D. R. 1971. Staphylococcal osteomyelitis of the avian vertebral column. *Res. Vet. Sci.* 12: 169–71.

Wise, D. R. 1973. The incidence and aetiology of avian spondylolisthesis (kinky-back). *Res. Vet. Sci.* 14: 1–10.

Wyers, M., Y. Cherel, and G. Plassiart. 1991. Late clinical expression of lameness related to associated osteomyelitis and tibial dyschondroplasia in male breeding turkeys. *Avian Dis.* 35: 408–14.

Yair, R., Z. Uni, and R. Shahar. 2012. Bone characteristics of late-term embryonic and hatchling broilers: Bone development under extreme growth rate. *Poult. Sci.* 91: 2614–20.

Yalcin, S., S. Ozkan, E. Coskuner, G. Bilgen, Y. Delen, Y. Kurtulmus, and T. Tanyalcin. 2001. Effects of strain, maternal age and sex on morphological characteristics and composition of tibial bone in broilers. *Br. Poult. Sci.* 42: 184–90.

Zhang, X., G. R. McDaniel, Z. S. Yalcin, and D. L. Kuhlers. 1995. Genetic correlations of tibial dyschondroplasia incidence with carcass traits in broilers. *Poult. Sci.* 74: 910–15.

Part 2

Animal welfare

Chapter 9

Understanding poultry behaviour

M. M. Makagon and R. A. Blatchford, University of California-Davis, USA

1 Introduction

The study of animal behaviour – called ethology – investigates the evolutionary, mechanistic, developmental and adaptive significance of the actions and reactions of animals. Within the broad scope of this scientific discipline, 'applied ethology' focuses specifically on animals that regularly interact with or are managed by humans. Applied ethology applies behavioural principles and research methods to problems such as pest and predator control, management of economically important species, wildlife conservation and animal welfare, among others (Mench, 1992). As it pertains to poultry production, a context in which humans have a high degree of control over the animals' reproduction and housing environments, research questions focus mainly on the mechanistic and developmental causes of behaviour. Although given less consideration in the scientific literature, research on the evolutionary underpinning of poultry behaviour can and has informed breeding programmes (e.g. breeding for tameness was an important part of the domestication process), and put behaviours observed in captivity into a broader context. An example of the latter is the hen's preference for enclosed nest boxes (Appleby and McRae, 1986), which has likely evolved in response to predator avoidance pressures.

Behaviour is a complex phenomenon that occurs at the individual level, as well as at the group level (e.g. distribution patterns, coordinated movements and social networks). It can be elicited by a wide variety of neurobiological and hormonal mechanisms, as well as

http://dx.doi.org/10.19103/AS.2016.0011.24

by environmental stimuli (Mench, 1992). Experimental, observational and epidemiological approaches are used to evaluate how various internal and external factors influence the onset, frequency and structure of a behaviour. Common research techniques include the quantification of behavioural time budgets, as well as preference and motivation tests. Beyond providing information about the underpinnings of observed behaviours, behavioural approaches and methodologies are often used to evaluate behavioural outputs as part of research projects across a variety of disciplines, including physiology, nutrition, genetics and animal welfare. In some instances, behaviour has been used as a non-invasive proxy for health (e.g. walking ability and sickness behaviour), or emotional states (e.g. preferences and abnormal behaviour). In the context of poultry production, behavioural studies have been particularly useful for optimizing management strategies and understanding the welfare of commercially raised poultry. Such research has led to a better understanding of the way in which human-dictated management decisions such as stocking density, resource and space allowances, social groupings, early experiences, lighting and feeding schedules, and transportation events are perceived and experienced by poultry.

This chapter aims to provide an overview of the contributions that the study of animal behaviour has and can make to poultry science, and to provide general guidance on the design, execution and interpretation of behavioural research. We begin with an overview of commonly used research methods and methodological challenges that should be considered before behavioural data is collected. Next, we provide specific examples of how behavioural approaches have been used to explore the importance and consequences (positive and negative) of perch access. We offer some thoughts on current and future trends in applied ethology and conclude with a list of resources on the topics covered here.

2 Key methodological challenges: defining experimental unit, sample size and behaviour of interest

For individuals new to the field of applied ethology the most common challenges are those related to disciplinary-based differences in experimental design. Defining the appropriate experimental unit and sample size for behavioural studies is not always intuitive. Determining which behavioural assessment tools are most appropriate for addressing a particular research question and ensuring that their interpretations are firmly grounded in behavioural principles can also be of issue. Each of these challenges has direct implications for the validity of the results, highlighting the importance of addressing them prior to data collection. For example, inappropriate assignment of experimental units or replicates can lead to problems such as pseudoreplication (see Davies and Gray, 2015 for a more in-depth discussion on the limitations of pseudoreplication), and calls into question the power of statistical analysis and the external validity of a study as described in Martin and Bateson (2007). In the next two sections, we expand and provide guidance on several common pitfalls associated with animal behaviour research that we have seen in manuscripts submitted for publications, and too often in published research. Recognizing that the broad scope of this topic prevents us from addressing it fully, we have included a list of additional resources later in this chapter.

2.1 Defining the experimental unit and sample size

The importance of selecting the appropriate 'experimental unit' and 'sample size' is ubiquitous across scientific disciplines. However, what constitutes an experimental unit or appropriate sample size can vary according to research field, even when the research is conducted within a single experimental framework. To illustrate this point, let us first consider a study that aims to evaluate the effects of providing broilers the opportunity to self-select their diets on the birds' performance. For authenticity, our general experimental set up will follow the one used by Syafwan et al. (2012) in their well-designed research into free-choice feeding. Broilers are randomly allocated across 12 identical pens, and are housed in groups of 6. The pens are randomly assigned to one of two dietary treatments. Each group of birds receives either a pre-balanced feed ration distributed across 3 feeders or a choice of diets (1 per feeder) that include a high-energy, high-protein, or a pre-balanced option. The variables of interest, all of which are measured on a per week per pen basis, include feed intake, protein and energy intake, and weight gain. In this scenario, both the treatments and the outcome variables are attributed to the pen. Therefore the pen serves as the appropriate experimental unit, and replications of pen measures over time are typically handled through a statistical analysis that accounts for repeated measures.

Next, let us examine several of the common challenges that can arise when a behaviour, such as the amount of time that broilers spend at each feeder, is mapped onto a predesigned study like the feeding trial described above. An erroneous assumption that is often made is that behaviour is intrinsic to the individual, and therefore that birds within pens serve as the experimental units for behavioural observations. While it is true that behaviour is most often observed at the individual level (though it can also be observed at the level of a dyad or a larger group), it is important to bear in mind that the experimental unit should, by definition, correspond to the grouping to which treatments were applied. Although in our example dietary treatments were assigned to pens, it is not necessarily the case that the pen is the appropriate experimental unit, either. An important consideration is whether or not the measures taken on different individuals within a pen or individuals housed in adjacent pens can be assumed independent of one another. Generally speaking, behavioural choices can be modified through social learning. The specific question that should be asked in the context of our hypothetical study is whether the amount of time that broilers housed in one pen spend at each of their feeders is independent of the feeder choices made by birds housed in adjacent pens. If not, precautions should be taken to reduce this influence, for example by ensuring that birds in adjacent pens cannot see each other's choices. Environmental cues, such as location effects (i.e. preferences for particular feeder locations) can further confound behavioural results. Location effects can be accounted for, for example, by rotating the locations of the feed type across feeders over the course of the experiment. Importantly, when pen serves as the experimental unit and when more than one bird per pen is sampled, which is typically the case, social influences of pen-mates should also be considered. One approach is to average the results obtained on individual birds within a pen; another is to use a nested or other appropriate statistical approach for the analysis.

In ethology, decisions regarding sampling size include identifying the number of birds per pen and/or pen replicates to observe, the number of times each group should be examined over the study period, and the duration of the observations made at each of these time points. When determining sample size, it is important to acknowledge that behaviour can vary widely between individuals and groups, as well as within an individual across time. Large sample sizes are often needed to maintain external validity (Still, 1982). It is equally important to note that behavioural variation may be different than the amount of individual variation observed in other measures (e.g. growth rate). Therefore, when possible, power analyses should be used to help determine the number of units needed for the behavioural component of a study.

Behaviour is not necessarily measured on the same timescale as other variables. In our hypothetical feed choice study production measures such as growth rate can, for example, be examined by weighing birds on the first and last day within a week. Compared to body weight, which typically deviates slightly within and across days, behaviour can be highly dynamic. Therefore, observing feeding behaviour at only one moment per week (e.g. just before the animals are weighed) would more than likely produce skewed results, as would observing behaviour daily but within a very small time frame (e.g. several minutes per day). Generally speaking, determinations of how many times each animal or group should be sampled across the experimental timeline should give consideration to the properties of the behaviour of interest. This includes whether the behaviour is a state or event, whether it occurs frequently or rarely, and how its performance changes with time of day, season, bird age, or previous experience. Chapter 3 in Martin and Bateson's (2007) book offers additional tips pertinent to determining sample size, including how to cope with individual differences in behaviour, how much information to collect and when to observe.

2.2 Defining the behaviour of interest

Another general experimental consideration has to do with properly 'identifying and defining behaviours of interest'. Firstly, the behaviours chosen for observation should be directly related to the question being asked. For instance, if a researcher wants to know how enrichment affects aggressive pecking in laying hens, observing comfort behaviour such as preening is likely not very informative. What constitutes an 'aggressive peck' should instead be defined in a way that allows for multiple observers to recognize the behaviour with high within and between rater reliability (Martin and Bateson, 2007). When writing behavioural definitions, it is helpful to remember that people often have their own predetermined ideas of what a particular behaviour should look like. Using the example of time spent at the feeder, while one researcher may consider a bird to be 'at the feeder' when its head is over the feeder's edge, another may define it as time actively pecking at and swallowing food, and yet a third as time when the beak touches the feed. Characteristics of a good behavioural definition include descriptive, objective language (e.g. 'the hen is lethargic, moves slowly and reluctantly, and has milky eyes' vs. 'the hen looks sick'). Optimally, definitions should also take into account and allow for comparability across studies.

It is useful to refer to the behavioural definitions when interpreting results to avoid over extrapolation of data. This is especially true when the observed behaviour is used as a proxy for another measure, or when another measure is used as a proxy for behaviour. An example of the former is the use of gait score, a measure of how easily a bird is able to walk, to estimate the presence and severity of lameness. Although gait score may be correlated to lameness, the underlying causes of lameness are multifactorial and may not always be reflected in the gait score. Similar issues can arise when proxy measure, such as the loss of feathers on the head and vent areas, is used to estimate a behaviour, such as the incidence of aggressive behaviours occurring in a flock. Feather loss can be caused by several factors, of which aggression is only one. While the use of proxy measures can be beneficial, it should be done with care, and ideally be validated using behavioural observations or additional measures.

3 Key methodological challenges: selecting the most appropriate metric and sampling strategy

Once the experimental unit, sample size and behaviours of interest are defined, consideration should be given to the study approach and sampling strategy. Behavioural time budget assessments, and preference and/or motivation tests are some of the most popular methods of studying behaviour. Each approach comes with its own set of strengths and limitations. Traditionally, these research methods have relied on behavioural observations, which have the potential to be extremely time consuming. As noted in our discussion of future directions, there has been a steady increase in the availability of affordable technologies that allow for automated data collection.

'Behavioural time budget analyses' allow for an understanding of how an animal partitions its behaviour over a period, offering insights into how the behaviour is affected by changes in the animal's environment. Researchers can make inferences about the animal's welfare state based on these changes in behaviour. Time budgets have been used to investigate the effects of lighting and enrichment on broiler chickens (Alvino et al., 2009; Schwean-Lardner et al., 2012; Bailie and O'Connell, 2015), as well as the presence of rebound behaviour, indicating a potentially negative welfare state (reviewed in Olsson and Keeling (2005) for dustbathing).

Common pitfalls associated with conducting behavioural time budget analyses have to do with the lack of an appropriate baseline measure and the selection of the behavioural sampling strategy. We refer you to Altmann's (1974), Lehner's (1992), and Martin and Bateson's (2007) texts for thorough descriptions of the array of sampling strategies employed in behaviour research. Here, we limit our discussion to the two most commonly used sampling schemes: focal continuous (all-occurrence) and interval scan sampling. As the name suggests, focal continuous sampling requires the observer to record the behaviour of the bird or group of interest continuously for the predetermined observation period (see discussion of sample size). The benefits of this approach are that it provides the most accurate estimate of behavioural time budgets, and allows the observer to record the start, end and duration of behaviour, as well as its frequency. However, the time required to conduct this type of evaluation can be a limiting factor. For example, continuously observing the feeding behaviour of a single focal bird in each of the 12 experimental pens described in our hypothetical free-choice feeding experiment described above would yield 192 observation hours (assuming observations are only made during 16 hr of light). The192 observation hours is, however, an unrealistically low estimate since observations of just one bird over just one day could yield skewed results. The example serves to illustrate how quickly the actual time required to collect data grows. Observing the same bird daily (daylight hours only) for 3 days per week throughout the 6 week grow-out period would, for example, yield 3456 observation hours.

Instantaneous scan sampling, on the other hand, can be much less time consuming and is often used for this reason (e.g. Syafwan et al., 2012). With this sampling strategy, an observer evaluates the behaviour of all individuals (or a focal individual) within the group of interest at pre-specified time intervals throughout the duration of the sampling period. Arguably, the single most common error made during interval scan sampling is the use of intervals that are inappropriate for or irrelevant to the behaviour of interest. The frequency and duration of the behaviour should be considered in the selection of the sampling intervals, as inappropriate intervals can severely limit information on the duration of behaviours and often the frequency at which they are performed. When the behaviour of interest is short in duration and occurs infrequently, allowing for a large amount of time to pass between instantaneous observations (long interval) is likely to yield poor results. This point was recently highlighted by Daigle and Siegford (2014), who used different sampling schemes and interval lengths to show the impact of sampling scheme on the accuracy of the behavioural results. Importantly, because different behaviours can vary in frequency and duration, sampling frequency should be considered independently for each behaviour of interest. To ensure that the selected sampling scheme is yielding accurate results, the best practice is to compare results obtained using instantaneous sampling to a subsample of data obtained using the continuous sampling method.

'Preference tests' can give a good idea as to how different environments compare from the bird's point of view, and can be an invaluable tool in assessing welfare. It is important to note that the results indicate relative choices among alternatives (e.g. that environment A is preferred over B), but do not address the value of either alternative (how they are perceived on a 'good' to 'poor' spectrum), the importance of the choice or whether the choice will lead to improved health or welfare. Closely related to preference tests are 'motivation tests', which measure an animal's tendency or desire to perform certain behaviours, or access specific resources. Motivation tests often require that the birds perform a task to access a resource (e.g. environment A or B). The biological relevance of the operant response, or task, should align with the outcomes (e.g. pecking is associated with receiving a food reward), as learning the association between the task and outcome may be difficult for poultry if biological relevance is lacking. Savory and Duncan's (1982) research paradigm serves as a good example of this problem. They tested to see how important light or dark was to broiler chickens. The broilers had to peck a light switch to obtain either light or darkness. When in a background of light, the broilers could peck the switch to gain access to 3 minutes of darkness. That broilers were only in darkness for 1% of the time was interpreted to mean that darkness was not important to the birds (they were not willing to work for it). However, it is possible that the lack of response was indicative of a lack of motivation to perform a food-related behaviour to access darkness (which is related to rest, not feeding), and not necessarily a lack of motivation to access dark. Related to preference and motivation tests are aversion tests, which can be used to measure if a preference exists, as well as the strength of motivation or preference. Whereas preference and motivation studies are generally designed to measure positive emotions, aversion tests measure negative emotions such as distress (which underlies most negative states). However, which negative state (i.e. fear, pain, frustration, etc.) is associated with the avoidance behaviour warrants further study.

Kirkden and Pajor (2006) review these three types of tests, pointing out that they allow us to ask four major questions:

1 Is an animal motivated to access or avoid a resource?
2 Does the animal have a preference amongst alternative resources?
3 How strong is the motivation or preference?
4 Is the presence or strength of the preference/motivation influenced by the environment?

The review includes a thorough discussion of benefits as well as limits of each approach. For instance, some preferences and motivation are context dependent. A hen is only motivated to nest when she has an egg to lay, and may only have a preference for nesting material in that context. Results obtained outside of that context may be meaningless, or interpreted as the hen having no preference or little motivation to obtain nesting material.

4 Case study: provision and design of perches

Applied ethology has played an important role in helping researchers understand and address behaviour-based management problems, determine what types of resources should be provided to poultry and realize the impacts of resource design on poultry management and welfare. Research into the consequences of perch provision and perch design is an excellent example of the latter and one that continues to be extensively evaluated by ethologists. In this section we highlight how the specific types of behavioural assessments outlined above have contributed to this research. We include examples from studies of laying hens as well as those of broilers to highlight the importance of considering behavioural differences among chicken strains.

Much of research on perch design and use has sought to assess the needs of laying hens. Time budget analyses have shown that hens spend the majority of their nights roosting on perches, when perch access is available. Perching during the daylight hours varies among individuals, and is influenced by time of day, early experience, as well as perch placement, shape, length and height, among other factors. A number of preference tests have been conducted to evaluate the hens' preferences for specific aspects of perch design. Their results are summarized by Sandilands et al. (2009) and Hester (2014). Changes in behaviour after perch access is removed, and evaluations of the willingness of hens to push through weighted door for perch access have served to explore perching motivation (Olsson and Keeling, 2000 and 2002). The removal of perches increased restlessness among hens, as evidenced by an increase in the amount of time that hens spent moving around and decreased sitting durations during the first hour after lights were turned off (observations were limited to this period; Olsson and Keeling, 2000). Additionally, hens that had previously been raised with perch access were willing to push through weighted doors for access to perches, but not for access to obstructed perches that provided an elevated area without allowing the birds to perch (Olsson and Keeling, 2002). Taken together, these data have been interpreted to indicate that hens that have previously had access to perches, are motivated to perch.

Beyond providing an important behavioural opportunity for laying hens, perch access has been linked to improved overall bone strength and mineralization without affecting production measures (see Hester, 2014 for review). However, behavioural observations, coupled with physiological data, have highlighted the important role that perch designs play in mitigating the overall benefits of perch access. For example, Wechsler and

Huber-Eicher (1998) observed that hens provided with perches that were 45 cm off the ground were at a higher risk of being feather pecked than those that perched at 70 cm, while Stratmann and colleagues (2015) reported increased incidences of keel damage among birds housed with more rigid perches. While the association between keel bone damage and perch design may highlight perches as a contributing factor, the possibility that changing perch design alters perch use, the use of space or other resources thereby influencing keel bone damage cannot be discounted without behavioural observations.

As shown for laying hens, the use of perches by broilers is influenced by a number of factors, including perch design, time of day and bird age (Pettit-Riley and Estevez, 2001; Ventura et al., 2012; Norring et al., 2016). Compared to laying hens, overall perch use by broilers is low (Pettit-Riley and Estevez, 2001; Rodriguez-Aurrekoetxea et al., 2015), peaking at 4–5 weeks of age followed by a decline (Ventura et al., 2012; Bailie and O'Connell, 2015). As discussed earlier in this chapter, the low frequency of perch use should not be interpreted to be indicative of the importance of perch access for broilers. Rather than implying low perching motivation, the overall low incidence perch use and its decline after 5 weeks of age could indicate that the perch designs which have been shown to meet the needs of layers are inadequate. This interpretation gains some support from the observation that broilers do perch when the perches are relatively easy to access (e.g. Bizeray et al., 2002; Pettit-Riley and Estevez, 2001; Ventura et al., 2012). That broilers are significantly more inclined to use elevated platforms with ramps than perches of similar heights (Kaukonnen et al., 2016; Norring et al., 2016) provides additional support for this interpretation. Interestingly, the use of the platforms is higher during the day than at night, which seems to contradict the laying hen literature. However, if we compare the behaviour of laying hens and broilers at comparable ages, we see that this may not be a contradiction at all. Laying hens have been reported to start daytime perching earlier than night-time perching. At 6 weeks of age they can be observed resting under heat lamps at night, not perching (Heikkilä et al., 2006). It is not currently possible to conclude on the relative importance of access to elevated platforms and low, accessible perches for broilers as the necessary motivation tests have not yet been conducted. Additional research is also needed to help optimize perch and platform design. As was the case for laying hens, attention should be given to the influences that the resource has on the entire behavioural repertoire, as well as physiological and health parameters. For example, wooden barrier perches have been found to reduce aggression and disturbances between birds (Ventura et al., 2012), but have no clear benefit in terms of reducing lameness, fear or hock lesions (Ventura et al., 2010).

5 Summary: notes on the contributions of applied ethology to enhanced and sustainable production

The study of poultry behaviour contributes to enhanced and sustainable production by providing tools for assessing how housing, husbandry, and breeding decisions affect the welfare of poultry. As discussed in the next chapter of this volume, poultry welfare includes the consideration of how birds function from three non-discrete perspectives: physiological/health, behavioural and psychological (Fraser et al., 1997; Fraser, 2003). All three can be informed through the study of animal behaviour. For example, there is a close relationship between physiological processes and behavioural phenotypes. It is well known, that the hormone prolactin plays an important role in stimulating brooding behaviour of hens, or that decreased feeding and activity are common early signs of illness or stress. On the flip side, behaviour can also directly impact physiology. For example, panting helps birds cool during heat stress events, decreases carbon dioxide levels in the blood and increases blood pH and results in decreased eggshell quality (Yahav et al., 1997).

Understanding the foundational principles underlying animal behaviour is critical for delineating what constitutes a 'normal' (also referred to as 'natural') species-specific behaviour under commercial housing conditions. Although specific behaviours cannot be added or erased from an animal's behavioural repertoire, the expression of a behaviour can be influenced by environmental factors and genetic selection (e.g. reduction of brooding behaviour). Therefore, determining what constitutes appropriate poultry behaviour cannot be done by simply comparing the behavioural repertoires of commercially housed birds with those of their free-ranging or wild-type counterparts. It is often argued that animals raised in captivity should be allowed to freely perform all behaviours which lead to improved health and those which are important to the animal (Dawkins, 2004). The effects of behaviour on health can be evaluated using epidemiological and experimental approaches. What constitutes important behaviours can be addressed through a combination of preference and motivation tests, as described above, and observation of immediate and long-term changes in behaviours or behavioural time budgets when a behavioural opportunity is provided after being thwarted. Most recently, cognitive approaches are being used to additionally understand the animal's experience and perception of their environments, including their behavioural opportunities.

The majority of research into psychological well-being has focused on identifying situations that result in negative experiences. These often use behavioural fear tests (e.g. tonic immobility, novel object test) or behavioural indices of stress, discomfort or pain. Positive psychological well-being has typically been evaluated in terms of the amount of time dedicated to comfort behaviours, which tend to be associated with a state of relaxation (Zimmerman et al., 2011). Advances in the understanding of positive psychological states of hens are being made with stronger incorporation of cognitive and personality assessments to poultry behaviour research.

6 Future trends in research

Applied poultry behaviour research has traditionally sought to understand how birds perceive and respond to their housing environments, and to define what constitutes 'normal' or 'natural' behaviour. A common goal has been to minimize negative experiences by the birds while optimizing productivity. To do this, behavioural approaches have typically been coupled with environmental, production and physiological metrics. A higher level of integration with other disciplines is a clear current and ongoing research trend. Collaborative efforts between ethologists, engineers and computer scientists have significantly advanced the development and availability of technologies that allow for automated behavioural data collection, and the simultaneous and continuous monitoring of flocks. Many of the emerging research directions, such as investigations into personality and positive affect, have benefited from a deeper integration of principles from behavioural ecology and psychology into applied poultry behaviour science.

The last decade has seen rapid growth in the development and use of a variety of technologies for monitoring the behaviour of individual birds and whole flocks (Siegford and colleagues, 2016, provided an overview of some of the currently available technologies used to track the activities of poultry). Advances in automated tracking of animal behaviour and movement have been done using bird-mounted sensors (e.g. Burel et al., 2002; Zaninelli et al., 2015; Sales et al., 2015), computer-based video and sound analysis tools (e.g. Leroy et al., 2006; Dawkins et al., 2009, 2012; Aydin et al., 2010; Lee et al., 2015), or combinations of both (e.g. Nakarmi et al., 2014). Most bird-mounted sensor technologies, such as radio frequency identification systems (RFIDs) and accelerometers, aim to delineate the behaviour, position or movement of individual birds. Recent applications include monitoring the movement of hens through entry ways leading to indoor versus outdoor enclosures (e.g. Icken et al., 2008; Richards et al., 2011; Gebhardt-Henrich et al., 2014), and into and out of next boxes (e.g. Burel et al., 2002), as well as quantification of physical forces experienced by birds as they move through their environments (Banerjee et al., 2014; Širovnik et al., 2014; Baker et al., in press). Unlike bird-worn sensors, the automated frame-by-frame video image analysis approaches have been most successful for evaluating and differentiating movement patterns of flocks (e.g. Dawkins et al., 2009; Dawkins et al., 2012; Aydin et al., 2010), though efforts are being made to improve the ability of software to detect individual bird behaviours (Leroy et al., 2006; Dawkins et al., 2013; Nakarmi et al., 2014). Importantly, the availability of increasingly more advanced automated data collection systems has led to the development of fully automated flock monitoring systems, a monitoring approach referred to as precision livestock farming (Berckmans, 2006; Wathes et al., 2008). Though still in its infancy, one proposed application of these systems is early detection of avian influenza (Santoshkumar et al., 2013; Okada et al., 2014). The continued advancement of automated data collection and monitoring technologies is dependent on a strong understanding of the biological underpinnings of the monitored behaviours, as well as on close collaborations across ethology, engineering, computer science and other disciplines.

Paralleling the trend towards automation of behavioural data collection, there has been an increase in the incorporation of methodologies from behavioural ecology, human psychology and cognition to the study of domestic poultry behaviour. For example, though commonly used in behavioural ecology, the utility of global information systems (GIS) analysis for tracking the spatial distribution of chicken behaviours and movements in an indoor housing systems has only recently been explored (Daigle et al., 2014). Similarly, social network analysis tools popularly used to analyse social interactions among wild animals and captive primates (Sih et al., 2009) have only recently been used to explore questions related to social preferences of poultry (Makagon et al., 2013; Ringgenberg et al., 2014). The application of cognitive and personality assessment tools has been rapidly growing in other areas of applied animal behaviour research, including research on cattle and pigs, though it has so far been used to only a limited extent in poultry research (e.g. Wichman et al., 2012; Deakin et al., 2016). Use of these metrics to gain a better understanding of the way in which individual traits such as motivation, personality and affect impact how birds respond to their environments is in its infancy (e.g. Favati et al., 2014), and will certainly be an important area of future research.

Applied behaviour research has made important contributions to our understanding of poultry welfare and poultry management. But many aspects of poultry behaviour warrant further research, particularly for strains of poultry that are raised for meat.

Gaining a better understanding of the relationship between walking ability and lameness and determining what types of behaviours broilers are motivated to perform and how best to design resources to meet these needs are a few of the examples mentioned in this chapter. Understanding the social and spatial needs of broilers are two others. However, arguably the greatest area of need and opportunity are behavioural studies of broiler breeders. Perhaps the newly available technologies will help address some of these questions.

7 Where to look for further information

Poultry behaviour, general texts: *The Behavioural Biology of Chickens* (Nicol, 2015); *Poultry Behaviour and Welfare* (Appleby, Mench and Hughes, 2004); *Social Behaviour of Domestic Birds* (Mench and Keeling, 2001).

Applied ethology, general texts: *The Ethology of Domestic Animals: An Introductory Text* (Jensen, 2009); *Principles and Applications of Domestic Animal Behavior* (Price, 2008), *Genetics and the Behavior of Domestic Animals* (Grandin and Deesing, 2014).

Behavioural methods: *Measuring behaviour: An introductory guide* (Martin and Bateson, 2007); Observational study of behavior: Sampling methods (Altmann, 1974); Sampling methods in behavior research (Lehner, 1992); Design and execution of animal behavior research: An overview (Lehner, 1987); When continuous observations just won't do: Developing accurate and efficient sampling strategies for the laying hen (Daigle and Siegford, 2014); Measuring preferences, and the strength of preferences (Duncan, 1992); From an animal's point of view: Motivation, fitness, and animal welfare (Dawkins, 1990).

Key professional societies and related conferences: The International Society for Applied Ethology (www.applied-ethology.org) hosts an annual international conference and biannual North American research meeting; Animal Behavior Society (www.animalbehaviorsociety.org) hosts an annual meeting that is broadly focused on topics related to animal behaviour (basic and applied research). The Poultry Science Association (www.poultryscience.org) holds an annual meeting that includes poultry behaviour research presentations.

Key journals: *Applied Animal Behaviour Science; Poultry Science.*

8 References

Altmann, J. 1974. Observational study of behavior: Sampling methods. *Behavior.* 49: 227–67.

Alvino, G. M., Archer, G. S. and Mench, J. A. 2009. Behavioural time budgets of broiler chickens reared in varying light intensities. *Appl. Anim. Behav. Sci.* 118: 54–61.

Appleby, M. C. and McRae, H. E. 1986. The individual nest box as a super-stimulus for domestic hens. *Appl. Anim. Behav. Sci.* 15: 169–76.

Appleby, M. C., Mench, J. A. and Hughes, B. O. 2004. *Poultry Behaviour and Welfare.* CABI, Oxfordshire, UK.

Aydin, A., Cangar, O., Ozcan, S. E., Bahr, C. and Berckmans, D. 2010. Application of a fully automatic analysis tool to assess the activity of broiler chickens with different gait scores. *Comput. Electr. Agric.* 73: 194–9.

Bailie, C. L. and N. E. O'Connell. 2015. The influence of providing perches and string on activity levels, fearfulness and leg health in commercial broiler chickens. *Animal* 9: 660–8.

Baker, S. L., Robison, C. I., Karcher, D. M., Toscano, M. J., and Makagon, M. M. 2016. Behaviors associated with high impacts experienced at the keel by hens housed in enriched colony systems. In C. Dwayer, M. Haskell and V. Sandilands (eds), *Standing on the shoulders of giants, Proceedings of the 50th Congress of the International Society for Applied Ethology*, p. 387. Wageningen Academic Publishers, The Neatherlands.

Banerjee, D., Daigle, C. L., Dong, B., Wurtz, K., Newberry, R. C., Siegford, J. M. and Biswas, S. 2014. Detection of jumping and landing force in laying hens using wireless wearable sensors. *Poult. Sci.* 93: 2724–33.

Bizeray, D., Estevez, I., Leterrrier, C. and Faure, J. M. 2002. Effects of increasing environmental complexity on the physical activity of broiler chickens. *Appl. Anim. Behav. Sci.* 79: 27–41.

Berckmans, D. 2006. Automatic on-line monitoring of animals by precision livestock farming. In *Livestock Production and Society*. R. Geers and F. Madec (eds), Wageningen Academic Publishers, Wageningen, pp. 287–95.

Burel, C., Ciszuk, P., Wiklund, B. S., Brännäs, E. and Kiessling, A. (2002). Note on a method for individual recording of laying performance in groups of hens. *Appl. Anim. Behav. Sci.* 77: 167–71.

Daigle, C. L., Banerjee, D., Montgomery, R. A., Biswas, S. and Siegford, J. M. (2014). Moving GIS research indoors: Spatiotemporal analysis of agricultural animals. *PLoS ONE* 9: e104002.

Daigle, C. L. and Siegford, J. M. 2014. When continuous observations just won't do: Developing accurate and efficient sampling strategies for the laying hen. *Behav. Proc.* 103: 58–66.

Dawkins, M. S. 1990. From an animal's point of view: Motivation, fitness, and animal welfare. *Behav. Brain Sci.* 13: 1–9.

Dawkins, M. S., Cain, R. and Roberts, S. J. 2012. Optical flow, flock behaviour and chicken welfare. *Anim. Behav.* 84: 219–23.

Dawkins, M. S., Cain, R., Merelie, K. and Roberts, S. J. 2013. In search of the behavioural correlates of optical flow patterns in the automated assessment of broiler chicken welfare. *Appl. Anim. Behav. Sci.* 145: 44–50.

Dawkins, M. S., Lee H., Waitt C. D. and Roberts S. J. 2009. Optical flow patterns in broiler chicken flocks as automated measures of behvaiour and gait. *Appl. Anim. Behav. Sci.* 119: 203–9.

Davies, G. M. and Gray, A. 2015. Don't let spurious accusations of pseudoreplication limit our ability to learn from natural experiments (and other messy kinds of ecological monitoring). *Ecol. Evol.* 5: 5295–304.

Dawkins, M. S. 2004. Using behaviour to assess animal welfare. *Anim. Welf.* 13: S3–7.

Deakin, A., Browne, W. J., Hodge, J. J., Paul, E. S. and Mendl, M. 2016. A screen-peck task for investigating cognitive bias in laying hens. *PLoS ONE*, 11: e0158222.

Duncan, I. J. H. 1992. Measuring preferences and the strength of preferences. *Poult. Sci.* 71: 658–63.

Fraser, D., Weary, D. M., Pajor, E. A. and Milligan, B. N. 1997. A scientific conception of animal welfare that reflects ethical concerns. *Anim. Welf.* 6: 187–205.

Fraser, D. 2003. Assessing animal welfare at the farm and group level: The interplay of science and values. *Anim. Welf.* 12: 433–43.

Favati A., Leimar O. and Løvlie H. 2014. Personality predicts social dominance in male domestic fowl. *PLoS ONE* 9: e103535.

Gebhardt-Henrich, S. G., Toscano, M. J. and Fröhlich, E. K. 2014. Use of outdoor ranges by laying hens in different sized flocks. *Appl. Anim. Behav. Sci.* 155: 74–81.

Grandin, T. and Deesing, M. J. 2014. *Genetics and the Behavior of Domestic Animals*. 2nd edn. Academic Press, Elsevier Inc., London, UK.

Heikkilä, M., Wichman, A., Gunnarsson, S. and Valros, A. 2006. Development of perching behavior in chicks reared in enriched environments. *Appl. Anim. Behav. Sci.* 99: 145–56.

Hester, P. Y. 2014. The effect of perches installed in cages on laying hens. *World's Poult. Sci. J.* 70: 247–64.

Icken, W., Cavero, D., Schmutz, M., Thurner, S., Wendl, G. and Preisinger, R. 2008. Analysis of the free range behaviour of laying hens and the genetic and phenotypic relationships with laying performance. *Br. Poult. Sci.* 49: 533–41.

Jensen, P. 2009. *The Ethology of Domestic Animals: An Introductory Text*. 2nd edn. CABI, Oxfordshire, UK.
Kaukonen, E., Norring, M. and Valos, A. 2016. Perches and elevated platforms in commercial broiler farms: use and effect on walking ability, incidence of tibial dyschondroplasia and bone mineral content. *Animal*, 1.
Kirkden, R. D. and Pajor, E. A. 2006. Using preference, motivation and aversion tests to ask scientific questions about animals' feelings. *Appl. Anim. Behav. Sci.* 100: 29–47.
Lehner, P. N. 1987. Design and execution of animal behavior research: An overview. *J. Anim. Sci.* 65: 1213–19.
Lehner, P. N. 1992. Sampling methods in behavior research. *Poult. Sci.* 71: 643–9.
Lee, J., Noh, B., Jang, S., Park, D., Chung, Y. and Chang, H. H. 2015. Stress detection and classification of laying hens by sound analysis. Asian – Austral. *J. Anim. Sci.* 28: 592.
Leroy, T., Vranken, E., Van Brecht, A., Struelens, E., Sonck, B. and Berckmans, D. 2006. A computer vision method for on-line behavioral quantification of individually caged poultry. *Trans. ASABE*. 49: 795–802.
Martin, P. and Bateson, P. 2007. *Measuring Behavior: An Introductory Guide*. 3rd edn. Cambridge University Press, Cambridge, England.
Makagon, M. M., Robles, S. S. and Mench, J. A. 2013. Pekin duck social behavior: time budgets, synchrony and preferential associations. In Proceedings of the 47th International Congress of the International Society for Applied Ethology (ISAE), p. 49.
Mench, J. A. 1992. Welfare of poultry in modern production systems. *J. Poult. Sci. Rev.* 4: 107–28.
Mench, J. A. and Keeling, L. J. 2001. The social behaviour of domestic birds. In *Social Behaviour of Farm Animals*. L. J. Keeling (ed.), CABI, Oxon, UK, pp. 177–211.
Nakarmi, A. D., Tang, L. and Xin, H. 2014. Automated tracking and behavior quantification of laying hens using 3D computer vision and radio frequency identification technologies. *Trans. ASABE*. 57: 1455–72.
Nicol, C. J. 2015. *The Behavioural Biology of Chickens*. CABI, Oxfordshire, UK.
Norring, M., Kaukonen, E. and Valros, A. 2016. Theuse of perches and platforms by chickens. *Appl. Anim. Behav. Sci.* 184: 91–6.
Okada, H., Suzuki, K., Kenji, T. and Itoh, T. 2014. Applicability of wireless activity sensor network to avian influenza monitoring system in poultry farms. *J. Sensor Tech.* 4: 18–23.
Olsson, I. A. S. and Keeling, L. J. 2000. Night-time roosting in laying hens and the effect of thwarting access to perches. *Appl. Anim. Behav. Sci.* 68: 243–56.
Olsson, I. A. S. and Keeling, L. J. 2002. The push-door for measuring motivation in hens: laying hens are motivated to perch at night. *Anim. Welf.* 11: 11–19.
Olsson, I. A. S. and Keeling, L. J. 2005. Why in earth? Dustbathing behavior in jungle and domestic fowl reviewed from a Tinbergian and animal welfare perspective. *Appl. Anim. Behav. Sci.* 93: 259–82.
Pettit-Riley, R. and Estevez, I. 2001. Effects of density on perching behavior of broiler chickens. *Appl. Anim. Behav. Sci.* 71: 127–40.
Price, E. O. 2008. *Principles and Applications of Domestic Animal Behavior*. CABI, Oxfordshire, UK.
Richards, G. J., Wilkins, L. J., Knowles, T. G., Booth, F., Toscano, M. J., Nicol, C. J. and Brown, S. N. 2011. Continuous monitoring of pop hole usage by commercially housed free-range hens throughout the production cycle. *Vet. Rec.* 169: 338.
Ringgenberg, N., Froehlich, K. F., Harlander-Matauschek, A., Toscano, M. J., Wurbel, H. and Roth, B. A. 2014. Preferential associations between laying hens in nests. In *Applied Ethology 2014: Moving on*. I. Estevez, X. Manteca, R. H. Marin and X. Averos (eds), Wageningen Press, Wageningen, p. 105.
Rodriguez-Aurrekoetxea, A., Leone, A. H. and Estevez, I. 2015. Effects of panels and perches on the behavior of commercial slow-growing free-range meat chickens. *Appl. Anim. Behav. Sci.* 165: 103–11.
Sales, G. T., Green, A. R., Gates, R. S., Brown-Brandl, T. M. and Eigenberg, R. A. 2015. Quantifying detection performance of a passive low-frequency RFID system in an environmental preference chamber for laying hens. *Comp. Electr. Agric.* 114: 261–8.

Sandilands, V., Moinard, C. and Sparks, N. H. C. 2009. Providing laying hens with perches: fulfilling behavioural needs but causing injury? *Br. Poult. Sci.* 50: 395–406.
Santoshkumar, C., K. and Chavhan, S. 2013. Development of wireless sensor node to monitor poultry farm. In *Mobile Communication and Power Engineering* (Proceedings of the Second International Joint Conference, AIM/CCPE). V. V. Das and Y. Chaba (eds), Springer, Berlin, pp. 27–32.
Savory, C. J. and Duncan, I. J. H. 1982. Voluntary regulation of lighting by domestic fowls in skinner boxes. *Appl. Anim. Ethol.* 9: 73–81.
Schwean-Lardner, K., Fancher, B. I. and Classen, H. L. 2012. Impact of daylength on behavioural output in commercial broilers. *Appl. Anim. Behav. Sci.* 137: 43–52.
Siegford, J. M., Berezowski, J., Biswas, S. K., Daigle, C. L., Gebhardt-Henrich, S. G., Hernandez, C. E., Thurner, S. and Toscano, M. J. 2016. Assessing activity and location of individual laying hens in large groups using modern technology. *Animals* 6, 10.
Sih, A., Hanser, S. F. and McHugh, K. A. 2009. Social network theory: new insights and issues for behavioral ecologists. *Behav. Ecol. Sociobiol.* 63: 975–88.
Širovnik, J., Booth, F., Tarlton, J. F. and Toscano M. J. 2014. Use of a bird-mounted tri-axial accelerometer to quantify experienced physical forces during descent in laying hens. 48th Congress of the International Society for Applied Ethology, Vitoria-Gasteiz, Spain.
Still, A. W. 1982. On the number of subjects used in animal behavior experiments. *Anim. Behav.* 30: 873–80.
Stratmann, A., Fröhlich, E. K., Harlander-Matauschek, A., Schrader, L., Toscano, M. J., Würbel, H. and Gebhardt-Henrich, S. G. (2015). Soft perches in an aviary system reduce incidence of keel bone damage in laying hens. *PLoS ONE.* 10: e0122568.
Syafwan, S., Wermink, G. J. D., Kwakkel, R. P. and Verstegen, M. W. A. 2012. Dietary self-selection by broilers at normal and high temperature changes feed intake behavior, nutrient intake and performance. *Poult. Sci.* 91: 537–49.
Ventura, B. A., Siewerdt, F. and Estevez, I. 2010. Effects of barrier perches and density on broiler leg health, fear, and performance. *Poult. Sci.* 89: 1574–83.
Ventura, B. A., Siewerdt, F. and Estevez, I. 2012. Access to barrier perches improves behavior repertoire in broilers. *PLoS ONE*, 7(1): e29826. Wathes, C. M., Kristensen, H. H., Aerts, J. M. and Berckmans, D. 2008. Is precision livestock farming an engineer's daydream or nightmare, an animal's friend or foe, and a farmer's panacea or pitfall? *Comput. Electr. Agric.* 64: 2–10.
Wechsler, B. and Huber-Eicher, B. 1998. The effect of foraging material and perch height on feather pecking and feather damage in laying hens. *Appl. Anim. Behav. Sci.* 58: 131–41.
Wichman, A., Keeling, L. J. and Forkman, B. 2012. Cognitive bias and anticipatory behavior of laying hens housed in basic and enriched pens. *Appl. Anim. Behav. Sci.* 140: 62–9.
Yahav, S. 1997. Blood system response of chickens to changes in environmental temperature. *Poult. Sci.* 76: 627–33.
Zaninelli M., Costa A., Tangorra F. M., Rossi L., Agazzi A. and Savoini G. 2015 Preliminary evaluation of a nest usage sensor to detect double nest occupations of laying hens. *Sensors.* 15: 2680–93.
Zimmerman, P. H., Buijs, S. A. F., Bolhuis, J. E. and Keeling, L. J. 2011. Behaviour of domestic fowl in anticipation of positive and negative stimuli. *Anim. Behav.* 81: 569–77.

Chapter 10

Ensuring the welfare of broilers: an overview

T. B. Rodenburg, Wageningen University, The Netherlands

1 Introduction

Broilers have been intensively selected for high feed efficiency and fast growth over the past 60 to 70 years. This selection has been extremely successful, increasing broiler growth rate between 1957 and 2005 by over 400%, with a concurrent 50% reduction in feed conversion ratio (Zuidhof et al., 2014). However, this selection for highly efficient, fast-growing broilers had negative effects on broiler welfare. Activity levels of fast-growing broilers are low, especially during the second half of their life. Leg health can be problematic, with issues such as lameness, and high incidences of footpad dermatitis and hock burns playing a major role. In the past, also tibial dyschondroplasia caused many health and welfare problems, but the incidence of this metabolic disease has been strongly reduced by selective breeding. Fast-growing broilers are also more sensitive to heart and circulation problems compared with slower-growing genotypes, especially when placed in a suboptimal environment. When subjected to a mild cold stress, the incidence of mortality due to heart and circulation problems in fast-growing broilers was 15% compared to 2% in slower-growing genotypes (van Horne et al., 2004). The majority of mortality cases was caused by ascites, a disease causing circulation problems resulting in fluid accumulation in the peritoneal cavity. Regarding reduction of ascites incidence in commercial broilers, substantial progress has been made in recent years (Thiruvenkadan et al., 2011).

http://dx.doi.org/10.19103/AS.2016.0011.25

In 2000, the EU Scientific Commission on Animal Health and Welfare (SCAHAW) published an authoritative report on the welfare of broilers, which was revised in 2012 (de Jong et al., 2012). The main recommendation of the 2000 version of the report was that, '*Breeders should give a considerably higher priority to health variables in the breeding index, if necessary at the expense of the selection pressure for growth and feed conversion. Breeding which causes very poor welfare should not be permitted and breeders should be responsible for demonstrating that the standards of welfare in the chickens produced by them are acceptable*' (SCAHAW, 2000: 110). In the European Union this has led to the development of systems where slightly slower-growing broiler hybrids are used, which show less welfare problems. In The Netherlands, for instance, this has resulted in the development of a new production system with a slower-growing genotype, more space per bird and access to a covered veranda; products from this system are sold at a premium price under a welfare label supported by the Dutch organisation for animal protection. Apart from genetics, also the relatively high stocking densities in broiler production (over 15 birds/m^2) have received a lot of attention in welfare research. Especially in systems with poor management or housing conditions, high stocking densities can lead to poor litter quality and welfare problems, such as poor leg health and increased mortality rates. The relationship between high stocking density and poor welfare was the reason for the EU to limit the stocking density in broilers to 33 kg/m^2. Only if the producer could show that the company could manage broilers at a higher stocking density without high mortality for seven subsequent flocks (below 3.5% for a slaughter age of 42 days), the producer could be allowed to keep broilers at a higher stocking density to a maximum of 39 kg/m^2. This focus on animal-based measures, such as leg health and mortality, is also supported by newly developed methods to assess broiler welfare, such as the European Welfare Quality method (Welfare Quality Assessment Protocol for Poultry, 2009). Furthermore, practical methods to assess broiler welfare in commercial flocks, such as the transect method (Marchewka et al., 2013), have further increased the opportunities for welfare assessment. Together, increasing attention for broiler health and welfare in breeding and regulation of stocking density in the European Union have led and are expected to lead to an improvement in broiler welfare, while still allowing efficient broiler meat production.

2 Genetic selection

As indicated in the introduction, genetic selection for high feed efficiency and fast growth has been extremely successful in broilers. However, this selection for highly efficient, fast-growing broilers had negative effects on broiler welfare (Rauw et al., 1998; Julian, 1998; Sandoe et al., 1999). In 2000, this led to the SCAHAW report concluding that breeders should put more priority on broiler health, if necessary at the expense of growth and feed conversion and that breeding that causes poor welfare should not be permitted. So, what happened since that recommendation in 2000? One approach has been for breeders to focus more on reducing health and welfare issues. The main focus here has been on improving leg health and reducing lameness. Kapell et al. (2012b) describe that despite low heritabilities for traits such as tibial dyschondroplasia and hock burns and negative correlations with efficiency, substantial progress has been made over the past 25 years in improving leg health through balanced breeding programmes. Similarly, Rekaya et al. (2013) concluded that it is feasible to improve leg health without large negative effects on growth performance. Also regarding footpad and hock dermatitis, reduction through genetic selection seems feasible (Ask, 2010). Ask indicates that contact dermatitis would likely increase when genetic selection for improved growth rate would continue, without attention for leg health in the breeding goal. To be able to improve leg health, it is of course important to record leg health traits not only in the breeding stock, but also in commercial broiler flocks (Hocking, 2014). Kapell et al. (2012a) compared genetic selection for improved footpad health in a pedigree breeding environment and in a sib-testing environment, more similar to a commercial broiler operation. They found a strong genetic correlation of around 0.80 between footpad dermatitis measured in both environments. This indicates that it should be feasible to improve footpad health in commercial flocks through selective breeding in the pedigree birds, although they are kept in a more optimal environment than commercial flocks. Similar approaches as for improving leg health have been taken for reducing ascites in broilers. Pakdel et al. (2005) showed that by including reduction of ascites into the breeding goal, the improvement in body weight gain was reduced from 130 to 111 g, but it was possible to reduce the incidence of ascites in response to cold stress using this second breeding scheme. These studies show that there is potential to improve broiler health and welfare through genetic selection, without strong negative effects on growth performance. At the same time, the clear relationship between genetic selection for fast growth and broiler health and welfare has led to societal pressure to move to broiler production systems using slower-growing genotypes (Dawkins and Layton, 2012), especially in the European Union. For instance, the study by Knowles et al. (2008) indicated that in a comprehensive survey of broiler flocks in the United Kingdom almost 28% of the birds showed poor locomotion at 40 days of age and 3% was almost unable to walk. Bird genotype was identified as one of the main risk factors for poor locomotion. Slower-growing genotypes, such as those used in organic production, are known to have much lower incidences of health and welfare problems (Bessei, 2006). For organic broiler meat production, EU regulations stipulate that slower-growing genotypes should be used with a growing period of 12 weeks.

This makes organic broiler meat production relatively costly due to the high feed costs. However, the lines used to breed the slower-growing strain can also be used to breed a hybrid that shows intermediate growth (6 to 8 weeks to slaughter weight) and still shows a much lower incidence of health and welfare issues compared with fast-growing hybrids (van Horne et al., 2004). This intermediate type is now being used in European broiler production that aims at a market between conventional and organic. This type of broiler meat is usually sold within a specific welfare assurance scheme, such as the Freedom Foods Standards in the United Kingdom, or the 'Beter Leven' standards in The Netherlands. Within these schemes, systems receive a specific number of stars depending on their attention to welfare issues. Within the Dutch middle-market system with one star, producers are required to use the intermediate growing broilers. Recently, Gocsik et al. (2016) compared Dutch conventional, middle-market and organic systems and found that the middle-market system received the highest Welfare Quality score, indicating the best welfare, compared with the other systems (Welfare Quality Assessment Protocol for Poultry, 2009). Regarding cost-efficiency, these middle-market systems outperformed free-range and organic farms, so these systems may be an attractive alternative for efficient broiler meat production with good health and welfare (Gocsik et al., 2016).

3 Stocking density

As indicated in the introduction, a second major topic regarding broiler welfare is the stocking density at which the birds are kept, usually expressed in kilograms of live weight per metre squared. Broilers are kept at relatively high stocking densities of over 15 birds/m^2 or 30 kg/m^2. Towards the end of the growing period, when birds approach slaughter weight and activity levels are low, these high stocking densities can lead to welfare problems. Especially in systems with poor management or housing conditions, high stocking densities can lead to poor litter quality and welfare problems, such as poor leg health and increased mortality rates. However, an influential study by Dawkins et al. (2004) showed that broiler welfare is in fact influenced more by housing conditions and farm management than by stocking density per se. They studied broiler welfare at ten major broiler-producing companies in the United Kingdom. For the study, each producer managed flocks at five different stocking densities, ranging from 30 to 46 kg/m^2. Although there were clear differences in welfare indicators across flocks and producers, these could not be linked to stocking density. Factors such as litter moisture and air ammonia had strong effects on the welfare indicators measured, but some producers also managed to maintain a good litter quality and broiler welfare at the higher stocking densities. A critique on the study by Dawkins et al. (2004) has been that the low stocking density of 30 kg/m^2 was still too high to see any major effects, as earlier studies including stocking densities lower than 30 kg/m^2 did find clear effects on behaviour and welfare (Bessei, 2006). Having said that, the study by Dawkins et al. (2004) does underline the importance of good management: in a modern, well-ventilated and well-heated house with good management it is possible to keep broilers at a higher stocking density than in more basic housing systems. In the European Union, this has led to a rather unique Council Directive (2007/43/EC), where for the first time in history animal-based welfare measurements are included in the legislation. In that Directive, stocking density in broilers is limited to 33 kg/m^2. Only if the producer can show that the company can manage broilers at a higher stocking density without high mortality for seven subsequent flocks (mortality below 3.5% for a slaughter age of 42 days), the producer can be allowed to keep broilers at a higher stocking density of maximum 39 kg/m^2. The legislation also includes specifications about temperature, ammonia and humidity levels inside the house and includes a welfare inspection at slaughter, focusing on signs of footpad dermatitis, parasitism and systemic illness at the farm of origin. This seems a promising direction for legislation on animal welfare: by focusing on animal-based parameters at the farm or at slaughter, more precise information regarding the actual welfare status of the animals can be obtained, compared with just focusing on stocking density or other more descriptive parameters.

4 Case study: hatching environment and early feeding

Currently, major changes are taking place regarding hatching environment and early feeding. Traditionally, broiler chicks hatched at the hatchery remained in the hatcher without access to feed and water until all chicks had hatched. Then they would be processed and transported to the broiler farm, where they would be placed and would receive feed and water. The relatively long period without feed and water (up to three days for early hatching chicks) could have negative effects on growth performance, but is also criticised by animal protection organisations as being inhumane. In recent years, this has led to the development of systems that do not have this long fasting period between hatching and placement, such as the Patio system from Vencomatic (Eersel, The Netherlands) and the HatchBrood system from HatchTech (Veenendaal, The Netherlands). HatchBrood is a system where broiler chicks receive access to feed and water in a lighted environment directly after hatching. Lamot et al. (2014) showed that early access to feed led to a higher feed intake and body weight gain during the first week of life, but the effect disappeared thereafter. In the Patio system, eggs that have been incubated for 18 days are placed on trays at 50 cm above the litter floor. When chicks hatch, they drop down into the litter and once they dry up, they have immediate access to feed, water and conspecifics (Fig. 1, left panel). They do not have to be transported as young chicks and can remain in the Patio system until processing, or they can be moved to a finishing house during a less sensitive phase of the growing period. First results indicated a similar hatchability as in standard hatchers and low mortality rates (between 1% and 2%) (van de Ven et al., 2009). Van de Ven et al. (2013) showed positive effects of early access to feed on chick development and physiology. An added advantage of this type of systems seems to be the reduced need for antibiotic treatment, thereby limiting the risk of increased antibiotic resistance. The causes for this reduced need of antibiotics are still being investigated, but could be related to early access to feed and earlier establishment of gut microbiota and to the reduced stress after hatching. As the Patio system requires a major investment in housing equipment, alternatives have been developed that also allow broiler chicks to hatch at the farm in standard broiler houses, such as the X-Treck system (Vencomatic, Eersel, The Netherlands). In the X-Treck system, a frame is suspended above the litter and the trays with hatching eggs are placed on this frame. After hatching, the frame can be winched up and the system can be managed like a standard broiler house (Fig. 1, right panel). This type of systems is developing rapidly and is increasingly used in commercial practice. Performance, health and welfare of flocks in these systems is currently being monitored and compared to flocks from standard hatcheries. If these systems can indeed lead to better health and welfare and lower antibiotic use, this will mean a major step forward in broiler welfare. Of course, hatching at the broiler farm also has disadvantages, such as the inability to spray vaccinate upon hatching and the impossibility to segregate males and females. These disadvantages should be taken into account when further developing

Figure 1 Broilers hatching in the Patio system (left panel) and the X-Treck system (right panel). In both systems, the eggs trays are suspended over the litter (photos courtesy of Vencomatic Group, Eersel, The Netherlands).

these systems. Furthermore, this type of system requires better management skills of the broiler farmer, especially in the period around hatching. De Jong et al. (2015) recently hatched broiler chicks from the same parent stock flock at different hatcheries and showed that these chicks had a very different performance depending on the hatchery at which the chicks hatched. The same is probably true, and likely even more pronounced, for commercial broiler farms.

5 Case study: alternative and free range systems

Apart from conventional broiler production systems, several alternative and free-range systems for broiler production exist. For instance, under the Dutch 'Beter Leven' standards, described earlier in the section on genetics, systems that receive one star have to provide their birds with access to a covered veranda. Ruis et al. (2004) found that these verandas are well-used, with about 13% of the flock present in the veranda on average, and they were mainly used for active behaviours, such as foraging and dust bathing, while indoor levels of resting increased. This same pattern of active behaviour outside and passive behaviour inside the house was observed by Fanatico et al. (2016) in a study focusing on free-range broilers. Outdoor access in free-range systems offers many extra opportunities to the birds that may be difficult to offer in an indoor system: opportunities for foraging on fresh plant and animal material, a large space allowance (4 m^2 per bird according to European legislation), fresh air and daylight. It can be questioned whether the space allowance of 4 m^2 per bird is meaningful for the birds, as this seems very large compared with space inside the broiler house. Furthermore, broilers will often stay close to the house and use only a part of the outdoor run, especially when limited cover is provided in the outdoor run. Outdoor access also presents a risk for broiler health and may result in high mortality rates due to predation. Health risks mainly consist of contracting infectious diseases, such as avian influenza, due to increased contact with wild birds or their droppings in the outdoor run, but can also consist of parasitic infections (for laying hens, see Jansson et al., 2010), food-borne pathogens such as *Campylobacter* (Rodenburg et al., 2004) or enteric infections (Van Overbeke et al., 2006). This makes that free-range and organic systems are generally more popular in areas of Europe that are less densely populated and with relatively low concentrations of poultry production. Regarding mortality due to predators, foxes can be kept out by proper fencing. Mortality due to birds of prey is more difficult to control; however, providing shelter seems to play an important role. Dal Bosco et al. (2014) showed that by providing free-range broilers with high vegetation, almost no mortality due to predation was recorded (0.4%), while on average 2.6% mortality due to predation was recorded in a barren free range. Providing cover in the free range also seems to be key to attract a large percentage of the flock to use the free range (Dawkins et al., 2003). Furthermore, the type of cover seems to matter, for example, artificial or natural cover. Stadig et al. (2016) compared the use of the outdoor run between broilers with artificial, wooden shelters and broilers with access to short-rotation coppice with willows (Fig. 2). Broilers in the short-rotation coppice had a higher use of the outdoor run than broilers with artificial shelters (43% vs 35% of the flock). Percentages were relatively high compared with other studies, which may be explained by the fact that broilers were housed in small, mobile houses placed in the free range. Fanatico et al. (2016) reported a much lower average percentage of range use of 12.9% in broilers kept in more conventional floor

Figure 2 Slow-growing broilers using the short-rotation coppice of willows for shelter in the outdoor run (photo courtesy of Lisanne Stadig, ILVO, Melle, Belgium).

housing. They compared range use in broilers with and without structural enrichments and showed that they were able to draw birds further away from the house by offering enrichments (13 vs 6% of birds venturing 15–30 m from the house).

6 Summary: how research can contribute to enhanced and sustainable broiler production

Both case studies illustrate how research can contribute to sustainable broiler production. The various systems that allow the chicks access to feed and water directly after hatching are advantageous because chicks that hatch early do not have to wait in the hatcher without access to feed and water until all chicks have hatched. The long fasting period in conventional hatcheries is increasingly criticised by animal protection organisations in Europe. Furthermore, early access to feed seems to benefit early muscle development (Powell et al., 2016a), and to limit development of adipose tissue (Powell et al., 2016b) in comparison with 24-hour feed restriction. Early access to feed will also favour an early colonisation of the gut, which may favour gut health and immunity. This may also be the reason why broilers from this type of systems seem to require fewer antibiotic treatments. For the system where the broilers hatch on the farm, an added advantage is that chicks are not subjected to the stress of processing at the hatchery or transporting to the farm. Although long-term, negative effects of transport of broiler chicks seem limited (Jacobs et al., 2016), the situation where handling and transport are avoided seems much more natural and could have positive effects on welfare by reducing stress. As indicated previously, of course this type of system where the chicks hatch at the broiler farm requires more from the management skills of the farmer. Temperature of the house and floor becomes more critical if a producer chooses for on-farm hatching compared to placement of day-old chicks. Broiler farmers need to be trained in managing the hatching process and the early life phase. However, first results of this type of broiler production system are promising and the system seems to benefit broiler health, welfare and productivity as well as societal perception. The second case study focused on alternative and free-range broiler production system. As indicated previously, this type of middle-market systems is gaining importance in the European Union. Systems that offer a covered veranda or free-range access are perceived as more welfare-friendly than conventional broiler production systems and are supported by animal protection organisations. One of the major benefits of these systems is that slower-growing genotypes are used compared to the fast-growing broiler used in conventional broiler production. This seems to result in fewer health and welfare problems and lower mortality rates in these alternative systems. Interestingly, these middle-market systems seem to outperform conventional and organic systems not only regarding their welfare score, but also regarding cost-efficiency (Gocsik et al., 2016). In that way, these middle-market systems may allow producers to satisfy demands from society, retailers and consumers, while they are also able to produce high-quality broiler meat in a cost-effective way.

Of course, the focus on broiler welfare varies widely around the globe. While broiler welfare is an important issue in the European Union, countries like the United States, Brazil or China have been less focused on welfare (Robins and Phillips, 2011). Having said that, also food-producing companies in those countries are under pressure from NGOs and society to develop more sustainable methods of meat production. The focus for broiler production around the world should be to breed broilers that can be kept without major health and welfare problems and with low mortality rates, in line with the recommendation of the SCAHAW in 2000 and 2012. Furthermore, broiler management focused on maintaining a good litter quality should be promoted, as this was proven to be pivotal in maintaining good broiler welfare (Dawkins et al., 2004).

7 Future trends in research

Research on innovative hatching environments and on early feeding will likely continue to receive a lot of attention in future years, especially given the fact that benefits are both in the area of broiler health and welfare and human health (through reduced use of antibiotics). This type of research seems key to continue to develop the broiler industry into a more sustainable industry, which uses broilers that are more resilient to environmental changes and that are less susceptible to the health and welfare problems reported in conventional broilers in the past (SCAHAW, 2000; de Jong et al., 2012). Also, more research is needed on alternative and free-range systems. As described previously in this chapter, these middle-market systems seem very promising from both an animal welfare and an economic perspective. It would be interesting to investigate which attributes are most important for broiler welfare: slower-growing genotype, extra space, litter material or outdoor access. When it is clear which attribute or attributes are most important, it could also be investigated how one or more of these attributes could be applied to conventional broiler production systems. The slower-growing genotype seems to play a key role in increasing broiler activity and improving health and welfare. However, this genotype is per definition less efficient in growth than the conventional fast-growing broiler. For future broiler production, it should be investigated how efficiency and welfare can be balanced (Marchant Forde and Rodenburg, 2016). Despite the stabilising and even declining demand for meat in the Western world, demand for animal products is expected to double in the near future, mainly due to increased demand from Asia, the Middle East and Africa. The question is if and how we can meet this expected increased demand for poultry meat in a sustainable way? Should this be by intensifying existing poultry production facilities or by increasing the number of farms worldwide? If mega-farms develop, can we still guarantee good broiler health and welfare on those farms? Scientists should play a central role in this debate and inform the various stakeholders involved on the pros and cons of the various options for broiler health and welfare.

8 Where to look for further information

For further information on broiler welfare, the report 'The welfare of chickens kept for meat production (broilers)' by SCAHAW remains a very good starting point (SCAHAW, 2000). It describes modern broiler production, specific welfare problems in broilers, environmental factors linked to those welfare problems, the role of nutrition and feeding, welfare issues in broiler breeders (not covered in this chapter) and socio-economic aspects. The report was updated in 2012 (de Jong et al., 2012). A second useful resource is *Measuring and Auditing Broiler Welfare* (Weeks and Butterworth, 2003). This edited book first focuses on welfare problems in broilers and then explores how these welfare problems can be measured and audited and how they are linked to environmental factors. Also the 2006 review paper of Professor Werner Bessei on welfare of broilers is a useful source of information, focusing both on genetic and environmental aspects of broiler welfare issues (Bessei, 2006). Regarding the effects of stocking density on broiler welfare, the paper by Professor Marian Dawkins and colleagues is a must-read, also for being one of the few *Nature* papers in the field of animal welfare. For practical welfare assessment of broilers also the Welfare Quality Assessment Protocol for Poultry is very useful, presenting one method of measuring a number of animal- and environment-based welfare indicators and aggregating those into a welfare score that can be compared between flocks and farms (Welfare Quality Assessment Protocol for Poultry, 2009).

9 References

Ask, B. 2010. Genetic variation of contact dermatitis in broilers. *Poultry Science*, 89, 866–75.

Bessei, W. 2006. Welfare of broilers: a review. *World's Poultry Science Journal*, 62, 455–66.

Dal Bosco, A., Mugnai, C., Rosati, A., Paoletti, A., Caporali, S. and Castellini, C. 2014. Effect of range enrichment on performance, behavior and forage intake of free-range chickens. *Journal of Applied Poultry Research*, 23, 137–45.

Dawkins, M. S., Cook, P. A., Whittingham, M. J., Mansell, K. A. and Harper, A. E. 2003. What makes free-range broiler chickens range? In situ measurement of habitat preference. *Animal Behaviour*, 66, 151–60.

Dawkins, M. S., Donnelly, C. A. and Jones, T. A. 2004. Chicken welfare is influenced more by housing conditions than by stocking density. *Nature*, 427, 342–4.

Dawkins, M. S. and Layton, R. 2012. Breeding for better welfare: genetic goals for broiler chickens and their parents. *Animal Welfare*, 22, 147–55.

De Jong, I., Berg, C., Butterworth, A. and Estevéz, I. 2012. Scientific report updating the EFSA opinions on the welfare of broilers and broiler breeders. *EFSA Supporting Publications*, 9, 295E-n/a.

De Jong, I. C., Lourens, A. and Van Harn, J. 2015. Effect of hatch location and diet density on footpad dermatitis and growth performance in broiler chickens. *Journal of Applied Poultry Research*, 24, 105–114.

Fanatico, A. C., Mench, J. A., Archer, G. S., Liang, Y., Gunsaulis, V. B. B., Owens, C. M. and Donoghue, A. M. 2016. Effect of outdoor structural enrichments on the performance, use of range area, and behavior of organic meat chickens. *Poultry Science*, 95, 1980–8.

Gocsik, E., Brooshooft, S. D., de Jong, I. C. and Saatkamp, H. W. 2016. Cost-efficiency of animal welfare in broiler production systems: a pilot study using the Welfare Quality (R) assessment protocol. *Agricultural Systems*, 146, 55–69.

Hocking, P. M. 2014. Unexpected consequences of genetic selection in broilers and turkeys: problems and solutions. *British Poultry Science*, 55, 1–12.

Jacobs, L., Delezie, E., Duchateau, L., Goethals, K., Ampe, B., Lambrecht, E., Gellynck, X. and Tuyttens, F. A. M. 2016. Effect of post-hatch transportation duration and parental age on broiler chicken quality, welfare, and productivity. *Poultry Science*, 95, 1973–9.

Jansson, D. S., Nyman, A., Vagsholm, I., Christensson, D., Goransson, M., Fossum, O. and Hoglund, J. 2010. Ascarid infections in laying hens kept in different housing systems. *Avian Pathology*, 39, 525–32.

Julian, R. J. 1998. Rapid growth problems: Ascites and skeletal deformities in broilers. *Poultry Science*, 77, 1773–80.

Kapell, D. N. R. G., Hill, W. G., Neeteson, A.-M., Mcadam, J., Koerhuis, A. N. M. and Avendaño, S. 2012a. Genetic parameters of foot-pad dermatitis and body weight in purebred broiler lines in 2 contrasting environments. *Poultry Science*, 91, 565–74.

Kapell, D. N. R. G., Hill, W. G., Neeteson, A. M., Mcadam, J., Koerhuis, A. N. M. and Avendano, S. 2012b. Twenty-five years of selection for improved leg health in purebred broiler lines and underlying genetic parameters. *Poultry Science*, 91, 3032–43.

Knowles, T. G., Kestin, S. C., Haslam, S. M., Brown, S. N., Green, L. E., Butterworth, A., Pope, S. J., Pfeiffer, D. and Nicol, C. J. 2008. Leg Disorders in Broiler Chickens: Prevalence, Risk Factors and Prevention. *PLoS ONE*, 3, e1545.

Lamot, D. M., Van de Linde, I. B., Molenaar, R., van der Pol, C. W., Wijtten, P. J. A., Kemp, B. and van den , H. 2014. Effects of moment of hatch and feed access on chicken development. *Poultry Science*, 93, 2604–14.

Marchant Forde, J. N. and Rodenburg, T. B. 2016. Future directions for applied ethology. In Brown, J. A., Seddon, Y. and Appleby, M. C. (Eds), *Animals and Us: 50 years and more of Applied Ethology*. Wageningen, The Netherlands: Wageningen Academic Publishers, pp. 297–317.

Marchewka, J., Watanabe, T. T. N., Ferrante, V. and Estevez, I. 2013. Welfare assessment in broiler farms: Transect walks versus individual scoring. *Poultry Science*, 92, 2588–99.

Pakdel, A., Bijma, P., Ducro, B. J. and Bovenhuis, H. 2005. Selection strategies for body weight and reduced ascites susceptibility in broilers. *Poultry Science*, 84, 528–35.

Powell, D. J., Velleman, S. G., Cowieson, A. J., Singh, M. and Muir, W. I. 2016a. Influence of chick hatch time and access to feed on broiler muscle development. *Poultry Science*, 95, 1433–48.

Powell, D. J., Velleman, S. G., Cowieson, A. J., Singh, M. and Muir, W. I. 2016b. Influence of hatch time and access to feed on intramuscular adipose tissue deposition in broilers. *Poultry Science*, 95, 1449–56.

Rauw, W. M., Kanis, E., Noordhuizen-Stassen, E. N. and Grommers, F. J. 1998. Undesirable side effects of selection for high production efficiency in farm animals: a review. *Livestock Production Science*, 56, 15–33.

Rekaya, R., Sapp, R. L., Wing, T. and Aggrey, S. E. 2013. Genetic evaluation for growth, body composition, feed efficiency, and leg soundness. *Poultry Science*, 92, 923–9.

Robins, A. and Phillips, C. J. C. 2011. International approaches to the welfare of meat chickens. *Worlds Poultry Science Journal*, 67, 351–69.

Rodenburg, T. B., van der Hulst-van Arkel, M. C. and Kwakkel, R. P. 2004. Campylobacter and Salmonella infections on organic broiler farms. *NJAS Wageningen Journal of Life Sciences*, 52, 101–8.

Ruis, M. A. W., Coenen, E., van Harn, J., Lenskens, P. and Rodenburg, T. B. 2004. Effect of an outdoor run and natural light on welfare of fast growing broilers, In: Hanninen, L., Valros, A. (Eds), Proceedings of the 38th Interantional Congress of the ISAE, Helsinki, Finnland, p. 255.

Sandoe, P., Nielsen, B. L., Christensen, L. G. and Sorensen, P. 1999. Staying good while playing god – The ethics of breeding farm animals. *Animal Welfare*, 8, 313–28.

SCAHAW 2000. The welfare of chickens kept for meat production (broilers). European Commission. 149pp. http://ec.europa.eu/food/fs/sc/scah/out39_en.pdf.

Stadig, L. M., Rodenburg, T. B., Reubens, B., Aerts, J., Duquenne, B. and Tuyttens, F. A. M. 2016. Effects of free-range access on production parameters and meat quality, composition and taste in slow-growing broiler chickens. *Poultry Science*, 95, 2971–8.

Thiruvenkadan, A. K., Prabakaran, R. and Panneerselvam, S. 2011. Broiler breeding strategies over the decades: an overview. *World's Poultry Science Journal*, 67, 309–36.

van de Ven, L. J. F., van Wagenberg, A. V., Decuypere, E., Kemp, B. and van den Brand, H. 2013. Perinatal broiler physiology between hatching and chick collection in 2 hatching systems. *Poultry Science*, 92, 1050–61.

van de Ven, L. J. F., van Wagenberg, A. V., Groot Koerkamp, P. W. G., Kemp, B. and van den Brand, H. 2009. Effects of a combined hatching and brooding system on hatchability, chick weight, and mortality in broilers. *Poultry Science*, 88, 2273–9.

van Horne, P. L. M., van Harn, J. J. and Rodenburg, T. B. 2004. Slower growing broilers: performance, mortality and welfare. In Yalcin, S. (Ed.), XXII World's Poulty Congress – Abstract Book, Istanbul, Turkey, 326.

van Overbeke, I., Duchateau, L., de Zutter, L., Albers, G. and Ducatelle, R. 2006. A Comparison Survey of Organic and Conventional Broiler Chickens for Infectious Agents Affecting Health and Food Safety. *Avian Diseases*, 50, 196–200.

Weeks, C. A. and Butterworth, A. 2003. *Measuring and Auditing Broiler Welfare*, Wallingford, United Kingdom, CABI Publishing.

Welfare Quality 2009. Welfare Quality Assessment Protocol for Poultry (broilers, laying hens). Welfare Quality Consortium, Lelystad, The Netherlands. p. 111, http://edepot.wur.nl/233471.

Zuidhof, M. J., Schneider, B. L., Carney, V. L., Korver, D. R. and Robinson, F. E. 2014. Growth, efficiency, and yield of commercial broilers from 1957, 1978, and 2005. *Poultry Science*, 93, 1–13.

Chapter 11

Broiler breeding flocks: management and animal welfare

Ingrid C. de Jong and Rick A. van Emous, Wageningen Livestock Research, The Netherlands

1 Introduction

This chapter discusses the management of broiler breeders, that is, the parent stock of broiler chickens, and welfare issues related to the different life stages of broiler breeders. Where commercial broilers are the product of a cross of four lines, broiler parent stock (broiler breeders) are the product of a cross of two lines; both the male and female broiler breeder is the product of a cross of a specific paternal line and a specific maternal line (Hiemstra and Ten Napel, 2013). The actual figures on the number of broiler breeders worldwide are lacking; as an indication, in Europe the number of broiler breeders is estimated to be 44 million (Horne and Bondt, 2014). Management and housing of grandparents and great grandparents (but not pedigree stock) is to a large extent similar to that of broiler breeders (EFSA, 2010; Hiemstra and Ten Napel, 2013) and not described in this chapter. Nowadays, three companies dominate the world market for broiler breeding stock: Aviagen Broiler Breeders, Cobb-Vantress and Hubbard.

The majority of the broiler breeders worldwide are the parent stock of the so-called standard or fast growing broilers, which reach a body weight of 2.5 kg in 42 days or

http://dx.doi.org/10.19103/AS.2016.0011.26

less (EFSA, 2010). Dwarf parental females are used to produce broilers of intermediate (2.2 kg in 56–63 days of age) or slow growth rate (2.2 kg in 70–80 days of age) (De Jong and Swalander, 2013). Although worldwide only a small percentage of parent stock for intermediate or slow growing broilers is housed, in some countries a larger proportion of the total number of broiler breeders produce intermediate or slower growing broiler strains. For example, in France, the majority of the parent stock –85% according to De Jong and Guemene (2011) – are parents of intermediate or slower growing broiler strains (EFSA, 2010). In Europe, about 8% of the broiler breeders are estimated to be parent stock of intermediate or slower growing broilers (Horne and Bondt, 2014).

We provide a short, general description of housing and management of broiler breeders during both the rearing and the production period in the next paragraph. In addition, we focus on (major) welfare issues related to the management of broiler breeders and the current state-of-the-art research related to these welfare issues.

2 Housing conditions and management in the rearing period

In the rearing period, which comprises the age between 0 and 18–22 weeks, broiler breeder pullets are prepared for the production phase. The aim of the rearing period is to produce birds of ideal weight, uniformity, condition and stage of sexual maturity when they enter the production house (Leeson and Summers, 2000). Body weight and flock uniformity are important production indicators during the rearing period (Zuidhof et al., 2015).

Although there is worldwide variation in housing conditions during the rearing period due to differences in legislation, climatic conditions and labour costs, layout of rearing houses does not vary to a large extent. During the rearing period, broiler breeders are usually housed on a fully littered floor, and litter types vary between regions due to availability. In Europe, wood shavings, peat and straw are commonly used. Cage rearing is not common (De Jong and Swalander, 2013; EFSA, 2010). Feed is either distributed via pan feeders, chain feeders or spin feeders (systems that spread the feed in the litter) and can be provided as mash, crumb or pellet. Pelleted feed should be used in combination with spin feeders. Parent stock of fast growing broiler chickens is subjected to feed restriction during the rearing period (De Jong and Guemene, 2011; EFSA, 2010). In addition, water provision may be restricted (Hocking et al., 1993). Feed can be either provided daily or a skip-a-day feeding regime may be applied (EFSA, 2010). Skip-a-day feeding regimes can be in the form of 6/1, 5/2 or 4/3 feeding programmes (1, 2 or 3 days without feed each week and a larger portion on the feeding days). In Europe, usually daily feeding applied as legislation does not allow skip-a-day feeding programmes. Either raised platforms or perches are provided, usually from about 6 weeks of age onwards, to train the birds to use the higher levels in a house during the production period.

Depending on the region, climate and legislation, houses can be fully climate controlled and without daylight entrance or with windows, or either open-sided with mesh curtains. Males and females are reared separately so that different feeding programmes can be applied in rearing. Stocking density also varies between countries, depending on the climatic conditions and also on country-specific legislation. In European countries, stocking densities between 7 and 10 birds/m^2 (females) or 4 and 8 birds/m^2 (males) are generally applied, with lower stocking densities in open-sided houses (EFSA, 2010). After

Figure 1 Example of a broiler breeder rearing house in northwest Europe (photo provided by N. Katier).

a few days, usually a light programme with 8 h light per day is applied with a light intensity between 10 and 20 lux, although lower light intensities may be applied in case of injurious pecking or to reduce birds' activity (EFSA, 2010). The management guides of the breeding companies are usually taken as the basis for stocking density, light programme, feeding programme and so on (e.g. Aviagen, 2013; Cobb, n.d.; Hubbard, n.d.). Figure 1 shows an example of a rearing house in northwest Europe.

3 Housing conditions and management in the production period

The main goal in broiler breeder production is to provide fertilised eggs to produce a maximum number of healthy and robust day-old broiler chicks (Zuidhof et al., 2007). Important in relation to the management of adult breeders is maintaining the health status of the flock while keeping the egg production at a high level. Major criteria for monitoring birds for management purposes include body weight, body condition, egg production and hatching, hatchability and infertility and egg weight (Leeson and Summers, 2000).

The transition from the rearing period to the production period involves transportation to the production house, usually at a separate production farm. The production period starts between 18 and 22 weeks and lasts until 60–65 weeks, depending on the performance of a breeder flock. Males and females are reared separately and mixed at the beginning of the production period, and in Europe usually arrive on the same day at the production farm (EFSA, 2010; Van Tuijl, Aviagen, 2016, pers. comm.). The majority of the breeders are housed in floor house systems during the production period. Houses may be artificially lit, with or without windows or open-sided with curtains, which is dependent on legislation, region and climate. Artificial light in addition to natural light can be used to stimulate reproduction. Although less common, broiler breeders can also be housed in multi-tier colony cages during the production period. These are furnished cages with laying nests, perches and also a small litter area (De Jong and Swalander, 2013). Aviaries are not used for broiler breeders. During the production period broiler breeders can also be housed in conventional cages where artificial insemination is applied; this is uncommon in Europe, but may be found in other parts of the world (EFSA, 2010).

Floor house systems consist of a litter area and a certain proportion of slatted floors from which the nests can be accessed. The proportion of littered floor versus raised slatted floor (plastic or wooden slats) may differ between countries and regions, and the layout is also regionally dependent. Water is usually supplied on the slatted area and feed may be provided on the litter and/or on the slatted area. Litter types may vary between countries, for example, wood shavings, straw or peat. Feed can be provided in various forms (mash, pellet and crumbs) via feeder tracks or pans containing a male exclusion system. Male breeders are usually fed via trough feeding or feeder pans near the walls of the house, at such a height that the female breeders cannot reach there. Water can be provided via nipples, bell drinkers or cups. Feed restriction is applied during the production period to control the body weight, but the restriction level is much less severe as compared to the rearing period (De Jong and Jones, 2006). Water restriction may also be applied (EFSA, 2010). Figure 2 shows an example of a production house in northwest Europe.

Stocking density during the production period varies between 5 and 7.5 birds/m^2 depending on the legislation and region. Stocking density is usually lower in open-sided houses (EFSA, 2010). The percentage of males at the start of the production period is between 8 and 11%, and this decreases due to selection and mortality of males. At the start of the production around 23 weeks of age, 7.5–9% males are present in a flock (EFSA, 2010). The selection criterion of male breeders includes absence of mating activity and health problems (e.g. leg problems). About 15–25% of the males are selected during the production period. In some countries, 'spiking' of males is common practice. Inactive males are removed from the flock and replaced by younger and more active males to maintain the production of fertile eggs at a high level (Leeson and Summers, 2000). However, spiking involves the risk of introduction of pathogens and it may be stressful

Figure 2 Example of a broiler breeder production house. Laying nests and slatted floors are present in the middle of the house (left side of the picture) and the litter area is at both sides of the slatted area near the walls. This house does not have windows (Photo: R. van Emous).

to the birds because male aggression may increase (EFSA, 2010). Sometimes intra-spiking (swapping older males at the same farm from one house to the next) is used as a method to increase male activity and thus fertility, which has a much lower biosecurity risk (Casanovas, 2000).

4 Welfare issues: restricted feeding and water restriction

4.1 Restricted feeding

Parent stock of the so-called fast growing broilers (broilers that will reach a body weight of 2.5 kg in about 6 weeks) is subjected to restricted feeding programmes so that the birds will grow according to the growth curve as indicated by the breeding company. If broiler breeders are fed unrestricted, they will grow very fast, reach high body weights and as a result develop health (e.g. leg problems and increased mortality) (Heck et al., 2004; Mench, 2002) and reproduction problems. Low egg production associated with multiple ovulations, accelerated sexual maturity, low egg quality and low persistency of lay have been reported in unrestricted fed broiler breeders (Heck et al., 2004; Hocking et al., 1989), which is unfavourable with respect to the sustainability of the breeding flock. The high growth rate when fed unrestricted is due to the genetic selection for fast and efficient growth of standard, fast growing broiler chickens (Dawkins and Layton, 2012; Renema et al., 2007). Parent stock of so-called slower growing broiler strains are usually subjected to different, less restricted or unrestricted feeding programmes, as these birds have (much) slower growth rates due to a different genetic background (De Jong and Guemene, 2011). Parent stock of these so-called slower or intermediate growing broilers may either be a combination of a slower growing, usually dwarf female and a standard, fast growing male (that is subjected to a restricted feeding programme) or a combination of both a slower growing male and female.

Researchers in the early 1990s described the effect of the restricted feeding programme in broiler breeders on the welfare of the birds. Restricted fed broiler breeders show behavioural signs of stress and frustration, that is, redirected oral behaviours resulting in stereotypic object pecking (and if directed at the drinker resulting in overdrinking), hyperactivity, pacing (stereotypic walking) (De Jong et al., 2002; Hocking, 1993; Hocking, Maxwell and Mitchell, 1993; Hocking et al., 1996; Savory and Maros, 1993; Savory et al., 1992, 1993, 1996), increased aggression (Jones et al., 2004) and increased feeding motivation (Savory and Lariviere, 2000). It has also been reported that physiological indicators of (chronic) stress were observed, such as increased plasma corticosterone levels and increased heterophil:lymphocyte ratios in the blood (De Jong et al., 2003; Hocking, 1993; Hocking, Maxwell and Mitchell, 1996; Savory and Mann, 1997). However, more recently it has been debated if these physiological indicators indeed reflect stress, frustration and/or hunger in the birds or merely reflect the metabolic state (D'Eath et al., 2009; De Jong et al., 2003). Research continued with the aim to find reliable methods to measure hunger and motivation for appetitive behaviour in broiler breeders. A test was developed to measure appetitive behaviour (Dixon et al., 2014), but thus far no reliable behavioural test to assess hunger has been developed (Buckley et al., 2011a,b, 2015). Nevertheless, there is little doubt in research that animals undergoing quantitative feed restriction are hungry and that this negatively affects the welfare of the bird (D'Eath et al., 2009).

During the first 2–3 weeks of the rearing period, feed is provided unrestricted and thereafter a restricted feeding programme is started. The period in which the most severe restriction is applied starts around 6–7 weeks of age and runs to 15–16 weeks of age (De Jong and Jones, 2006). Restriction levels were estimated to be 25–33% of the intake of broiler breeders fed *ad libitum* (De Jong et al., 2002), but there are no recent data on relative restriction level. It can be expected that due to the continuing selection for efficient growth of the progeny, the relative restriction level has been increased since then (Zuidhof et al., 2014). By the end of the rearing period, that is, from 16 weeks of age onwards, the daily feed allowance slightly increases to prepare the birds for the production period. In addition, during the production period restricted feeding is applied, but restriction levels are much less severe as compared to the rearing period. Restriction levels of 45–80% of the *ad libitum* intake are applied until the peak of lay (Bruggeman et al., 1999) and restriction levels to about 80% of *ad libitum* intake are applied after peak of lay (Hocking et al., 2002), although also here no recent data are available.

The majority of research with respect to welfare of broiler breeders is related to the effects of restricted feeding on welfare. More recently, research focused on various methods to reduce the negative effects of restricted feeding such as using alternative feed composition or alternative feeding methods. The results of these studies are discussed below.

4.1.1 Fibre diluted diets or low protein diets in the rearing and the production period

It has been shown that increasing the fibre contents of the diet or decreasing the energy and protein contents of the diet can have positive effects on the welfare of broiler breeders, as measured by a reduction in time spent on behaviours indicative of frustration, stress or hunger. Hocking et al. (2004) applied diets diluted with 50, 100 and 200 g/kg ground oat hulls, ground unmolassed sugar beet pulp or sunflower meal and showed that diets with the highest concentration of oat hulls and sugar beet pulp had positive effects on behaviour during the rearing period (decreased prevalence of damaging pecking behaviour). The diets diluted with sugar beet pulp were associated with higher water content in the gastrointestinal tract and the authors suggested that these diets improved satiety and thus had the largest effect on the welfare of broiler breeders (Hocking et al., 2004).

De Jong et al. (2005a) applied diets with increased fibre contents and various fibre types in the rearing and the production period (8.4 MJ/kg (standard diet) vs 9.2 and 10.4 MJ/kg diets in the rearing period, and 11.7 MJ/kg (standard diet) vs 9.2 and 10.5 MJ/kg during the production period). They observed that the diet of 8.4 MJ/kg had some positive effects on the behaviour during the first half of the rearing period (decreased prevalence of stereotypic object pecking), but not during the second half of the rearing period, and even seemed to increase stress during the production period (indicated by a higher heterophil:lymphocyte ratio in the blood in the production period). Sandilands et al. (2005) and Tolkamp et al. (2005) fed broiler breeders *ad libitum* with diets diluted with oat hulls but in addition used an appetite suppressant (400 g/kg oat hulls or 400 g/kg oat hulls and 24–110 g/kg calcium propionate added to the standard diet, depending on bird's age) or used calcium propionate in combination with low protein content and provided these diets *ad libitum* (150 g/kg crude protein vs 200 g/kg crude protein in the standard diet vs diets diluted with oat hulls or oat hulls in combination with calcium propionate (Sandilands et al., 2006)). Stereotypic pecking was virtually absent, time spent sitting significantly increased and feeding motivation was reduced when applying diets diluted with oat hulls and including an appetite suppressant, indicating that welfare of broiler breeders during the rearing period might be improved (Sandilands et al., 2005; Tolkamp et al., 2005). The same effects on behaviour during the rearing period were found when appetite suppressants were used in combination with low protein diets (Sandilands et al., 2006). Morrissey et al. (2014b) showed that diets with increased fibre content (soy bean hulls, 400 g/kg) and an appetite suppressant (calcium propionate, 10–50 g/kg) had a positive effect on the behaviour of broiler breeders in the rearing period. However, it can be questioned if the use of appetite suppressants will be accepted by society; in addition, researchers questioned if the effects of appetite suppressants are caused by the fact that birds feel ill (Hocking and Bernard, 1993), which is unfavourable in terms of animal welfare.

Nielsen et al. (2011) tested the effects of diets with a high proportion of soluble or insoluble fibre. Three diets were applied, a commercial diet with non-starch polysaccharides (NSP) as fibre content (80% insoluble fibres) and two diets with twice the fibre content and a higher (89%) or lower (71%) proportion of insoluble fibres. They found that the experimental diet with twice the fibre amount and the high proportion of insoluble fibres had a positive effect on the behaviour of broiler breeders in the rearing period; stereotypic pecking was absent and more dust bathing, comfort behaviour and foraging was observed in these birds as compared to the control diet and the diet with soluble fibres. The diet with the low proportion of insoluble fibres negatively affected the litter quality (increased moisture level as compared to the other treatment groups) and birds fed these diets showed behavioural signs of discomfort. Thus, this diet was not preferred in terms of broiler breeder welfare (Nielsen et al., 2011). Others tested diets diluted with insoluble and soluble fibres (cellulose, wheat bran and cottonseed meal) during the production period and found that not only eating time increased, but that for the diets with insoluble fibres (cellulose and wheat bran) egg production also increased and plasma corticosterone concentrations decreased as compared to the diet diluted with cottonseed meal (Moradi et al., 2013), indicating a positive effect of these types of diluted diets on broiler breeder welfare and production.

Van Emous et al. (2014, 2015a) investigated the effects of three dietary protein levels and different growth patterns (2400 g vs 2200 g at 20 weeks of age) on indicators of welfare during the rearing and the production period. A positive effect on welfare indicators was found for the low dietary protein level diet in the rearing period, but not of the high growth pattern; stereotypic object pecking in the rearing period was significantly reduced in the low dietary protein treatment as compared to the other diets (van Emous et al., 2014, 2015a). When these diets in the rearing period were followed by a low, standard or high energy diet in the production period, it was shown that the low energy diet had a slightly positive effect on behavioural indicators of stress during the production period as compared to the medium or high energy diet, although both the low energy and low protein diet had negative effects on the quality of the feather cover. The authors concluded that increasing the energy:protein ratio had a positive effect on broiler breeder welfare as indicated by the behaviour of the pullets, especially in the rearing period (van Emous, 2015a).

To conclude, modifications of the diet composition by increasing the fibre level and/or increasing the energy:protein ratio are promising to alleviate the effects of feed restriction on broiler breeder welfare, especially during the rearing period. This research area needs further attention to promote application of these types of feed in practice.

4.1.2 Feeding programmes, feeding frequencies and feeding methods

In Europe daily feeding is commonly applied, whereas in North America skip-a-day feeding programmes are applied as these improve flock uniformity (de Beer and Coon, 2007). A few studies focused on the effects of daily or skip-a-day feeding programmes during the rearing period on broiler breeder welfare. Morrissey et al. (2014b) did not find evidence that either skip-a-day feeding (feeding on alternate days) or daily feeding differed in the extent to which broiler breeders were hungry or frustrated during the rearing period, although during the production period there seemed to be a positive effect of skip-a-day feeding over daily feeding (improved feather condition ascribed to less feather pecking) (Morrissey et al., 2014a). Skinner-Noble and Teeter (2009, cited in (EFSA, 2010)) reported no differences in stress levels in birds fed either daily or as per the skip-a-day feeding programme.

De Jong et al. (2005b) tested whether increasing the feeding frequency from once to twice a day, or scattering feed in the litter would reduce stress or hunger in the rearing period. They did not find any positive effects of both these methods on the behaviour and physiological indicators of stress. In general, it is advised to use mash feed instead of pellets to increase feeding time in broiler breeders, although this has not any effect on feelings of hunger. Furthermore, it is important to provide sufficient space at the feeders to prevent aggression in the birds at the time of feeding, as aggression around feeding may result in increased levels of feather and skin damage. Finally, the speed to which feed is distributed is very important to promote equal feed consumption between individuals, to promote flock uniformity and to prevent aggression between birds.

4.1.3 Feeding of males

Studies on the effect of feed restriction on broiler breeder welfare usually focus on the females, as there are numerically much more females present than male broiler breeders. According to Renema et al. (2007), males are subjected to a less severely restricted feeding regime during the rearing period as compared to females, but males are relatively more restricted during the production period as compared to females (EFSA, 2010). It is generally considered that feed restriction affects male welfare to a similar extent as the females; however, research is needed to confirm this hypothesis. De Jong et al. (2011) observed male and female broiler breeder behaviour during the rearing period when housed at two stocking densities (standard vs reduced stocking density), and observed that general behaviour of the males differed from female behaviour at both stocking densities. Object pecking was less common in male breeders as compared to female breeders during the rearing period, and males showed more standing and walking behaviour as compared to females. It is not clear if these differences in general behaviour imply a different stress level in males and females or just express differences in general behaviour between both sexes.

4.1.4 Parent stock of slower growing broilers

Substitution of the standard broiler breeders with dwarf or slower growing broiler breeders may be a strategy to improve welfare of broiler breeders during the rearing period (Decuypere et al., 2006). Jones et al. (2004) showed that replacing standard broiler breeders with dwarf female breeders favourably changed the behaviour (e.g. less stereotypic pecking), because the degree of food restriction during the rearing period was much less in the dwarf breeders as compared to the standard breeders. The dwarf breeders concern only one sex (only females), but numerically many more broiler breeder females are subjected to feed restriction as compared to males, and thus using dwarf females may be a solution for a large part of the broiler breeder population to improve welfare. However, using slower growing or dwarf females corresponds only to a specific market demand and in practice these bird types are not used on a large scale (De Jong and Guemene, 2011).

4.2 Water restriction

Farmers may apply water restriction during the rearing and the production period (starting from a few weeks of age onwards), although in some countries this is forbidden (EFSA, 2010). Water restriction is usually applied to prevent overdrinking in restricted fed broiler breeders, which may result in wet litter and, in severe cases, in polydipsia (Hocking et al., 1993). When water restriction is applied, water is usually provided around feeding and during a few hours after feeding, and possibly on other occasions during the day (EFSA, 2010). Water restriction is generally considered unfavourable with respect to welfare, although Hocking et al. (1993) showed that limited access to water during the rearing period of broiler breeders did not affect welfare indicators.

5 Welfare issues: excisions, mating behaviour and quality of feather cover

5.1 Excisions (beak trimming, toe clipping and de-spurring)

Depending on country-specific legislation, excisions can be carried out in broiler breeders. These are beak trimming (which can be applied in both males and females), toe clipping and de-spurring. EFSA (2010) reported that comb dubbing is only practised in less than 10% of the breeder population and not recommended by the breeders. Excisions can be carried out at the hatchery (de-spurring, toe clipping or beak trimming (applying hot blade) or infrared beak treatment (Gentle and McKeegan, 2007) or on-farm (hot blade beak trimming). Although toe clipping is commonly applied (usually the backward or inner toe is trimmed), de-spurring is not applied in all breeds (EFSA, 2010) and beak trimming is prohibited in some EU countries (Fiks-van Niekerk et al., 2009).

Both de-spurring and toe clipping are carried out to prevent feather and skin damage in the females due to mating behaviour of the males. Sharp and long spurs and toes may damage the feathers and subsequently the skin, and in flocks with non-mutilated males this can result in severe wounds in the females (and as a consequence increased mortality and a negative effect on egg production and fertility). This has especially been reported for males with intact toes (De Jong, 2016, pers. comm.). Beak trimming is performed in female and male broiler breeders to prevent feather and skin damage due to injurious pecking behaviour. Beak trimming in males also reduces feather damage at the back of the head, where the male grabs the female with his beak while mating (Gentle and McKeegan,

2007). However, beak trimming in males and females is not practised in all countries and in general injurious pecking (feather pecking and resulting cannibalism) is not very common in broiler breeder flocks with intact beaks. Specific management might be required, such as providing pecking substrates (bales, pecking blocks) or, in case of an outbreak of injurious pecking, managing the light intensity.

In a pilot study on a commercial broiler breeder farm, it was found that females (Ross 308) with intact beaks had better performance during the rearing period than infrared beak-trimmed females. This was due to the fact that there was no negative effect of beak treatment on feed and water intake during the first days on the rearing farm, resulting in improved growth and uniformity and less mortality in the non-beak-trimmed as compared to the beak-trimmed flocks. Feather scoring during the rearing and the production period showed that in general there was no negative effect of non-beak trimming on the feather cover in flocks with intact beaks (De Jong et al., 2013). A pilot study in commercial flocks with non-beak-trimmed and beak-trimmed males (Ross 308; both housed with non-beak-trimmed Ross 308 females) showed that non-beak trimming of male broiler breeders also did not result in any negative effects on feather damage in females and males, nor was there an effect on the technical results of these flocks (De Jong, Gunnink and Van Emous, 2016, pers. comm.). It is currently unknown if there are differences between broiler breeder lines in the propensity to develop injurious pecking behaviour.

5.2 Mating behaviour

Mating behaviour of both males and females is important with respect to maximising the production of fertile eggs. However, broiler breeder males may show rough, aggressive behaviour towards females during mating. This rough male behaviour may lead to feather and skin damage and fearfulness in females. As a result, females may hide in the nests and fertility will be negatively affected (Millman et al., 2000). One of the reasons for male aggression may be that males reach maturity earlier than females, leading to forced copulations and distress in the females, which in turn results in females hiding on the slats and in the nests (Leone et al., 2007). However, also with proper management in relation to sexual development of males and females, rough and aggressive mating behaviour of the males has been observed. In addition, it was observed that courtship behaviour in males was virtually absent (De Jong et al., 2009; Hocking and Bernard, 2000; Jones and Prescott, 2000; Millman et al., 2000). It has also been observed that females do not seem to respond properly (with crouching behaviour) to male approach (De Jong et al., 2009).

It is unknown why male broiler breeders show this rough behaviour towards females during mating. Rough mating behaviour was not related to aggressive behaviour per se (Millman and Duncan, 2000b) and feed restriction in the rearing period was not related to rough mating behaviour (Millman and Duncan, 2000a). Genetic background may play a role, as males from laying strains responded less aggressive during mating than broiler breeder males (Millman and Duncan, 2000b). Reducing the stocking density, thus, providing males with more space to perform courtship behaviour, indeed had a positive effect on the quality of the mating behaviour (more courtship behaviour and more successful matings) and also resulted in a higher egg production, more fertile eggs and a higher number of chicks per hen (De Jong et al., 2011). Jones et al. (2001) showed that enriching the light conditions with UV_A improved the transmission of sexual signals and thus the quality of the mating behaviour.

Figure 3 Quality Time® broiler breeder house. Males are penned in a separate area during feeding (central area on the picture) (Photo: R.A. van Emous).

It has been estimated that mating frequency is 5–10 times higher in a broiler breeder house as compared to flocks of chickens housed under natural conditions (Van Emous, 2010). This relatively high frequency of mating may affect the relationship between the males and females, resulting in females avoiding males as hypothesised by Fontana et al. (1992). Over-mating might be avoided by separating females and males temporarily during the day. Based on this hypothesis, a new housing system for broiler breeders, called the Quality Time® Concept (QTC), has been developed (Fig. 3) (Van Emous, 2010). Males are separated from females during 5 hours/day, using a separate feeding system and a moving fence. After a successful pilot experiment, two on-farm experiments were carried out in a broiler breeder house with 15000 birds. The house was divided in six compartments. In the QTC compartments more voluntary and successful matings were observed. In addition, quality of the sexual behaviour improved, which resulted in an improved feather cover between 37 and 48 weeks of age in the QTC compartments as compared to the control compartments. Separating males from females did not increase aggressive behaviour between the males in the male pen (Van Emous, 2010).

As indicated in the current paragraph, management practices to improve mating behaviour and thus welfare of broiler breeders exist, with concurrently improved performance of the breeder flocks. These methods need to be further developed for practical application.

5.3 Quality of the feather cover

Quality of the feather cover, especially during the production period, is important with respect to the prevention of skin damage in females resulting from male mating behaviour and with respect to thermoregulation of the bird. However, quality of the feather cover has decreased during the past 10–20 years due to unknown reasons (Van Emous and De Jong, 2013). Figure 4 shows examples of a deteriorated feather cover in females in the second half of the production period.

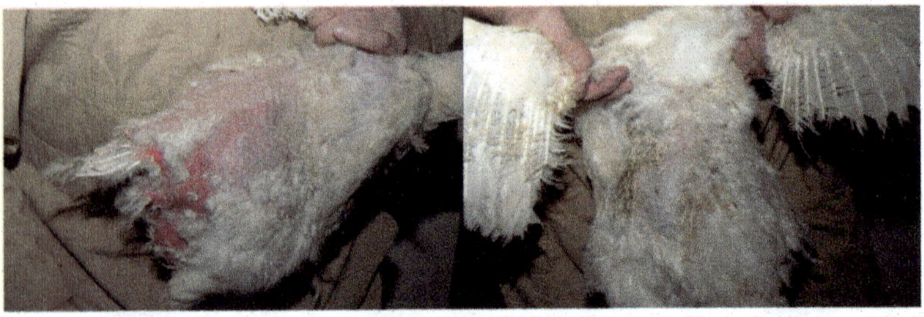

Figure 4 Examples of a deteriorated feather cover in female broiler breeders in the second half of the production period (Photo: R.A. van Emous).

Although not performing mutilations in broiler breeders implies an increased risk for a deteriorated feather cover (and skin damage), as indicated earlier, non-beak trimming of males and females did not seem to have a negative effect on the feather cover of both males and females (De Jong et al., 2013; De Jong, Gunnink and Van Emous, 2016, pers. comm.). Reduction of the stocking density improved the quality of the feather cover both in the rearing and in the production periods, although in general females at both standard and reduced stocking density still had a deteriorated feather cover at the end of the production period (De Jong et al., 2011).

Van Emous et al. (2014, 2015a) studied the effect of diets with different protein content in the rearing period and different energy content in the production period on the quality of the feather cover. They concluded that a low daily protein intake during the rearing period and during first phase of the production period resulted in an inferior feather cover as compared to diets with medium or high protein content at these ages (see Van Emous et al., 2014, 2015a for diet composition tables). This indicates that specific amino acids levels for feather development were deficient. The authors also suggested that a low daily protein intake between 2 and 6 weeks of age showed a more pronounced effect on feather cover than a low daily protein intake between 6 and 15 weeks of age.

6 Environmental enrichment

It is well known that chickens prefer a high and safe resting place to roost at night-time, due to the motivation to protect themselves from predators. Usually in the rearing period, raised slatted areas (~1 m^2/1000 birds) are provided from six weeks onwards to train the birds to go to the higher level, to prepare them to easily enter the nesting area during the production period. However, it is yet unclear if raised slatted areas might function as high and safe resting places or if breeders prefer perches. In addition, how these perches should be provided, including material and shape, is yet unclear. Most studies on perch design and perch use are carried out in laying hens, but it needs to be studied if results from the studies regarding laying hens are also valid for broiler breeders. A Swedish study, comparing wooden A-shaped perches with platforms, found a high incidence of breast blisters in the broiler breeders on perches which was not found in birds provided with platforms (Wachenfelt and Berndson, 2014). This could be due to the material and shape of the perches. In addition, they observed a preference of the birds to rest on the platforms, but this could have been influenced by birds experiencing pain due to the breast blisters and thus not using the perches (Wachenfelt and Berndson, 2014). Other studies only recorded that perches were used by broiler breeders, with increasing use with age (Hocking and Jones, 2006; Van Emous, 2016, pers. comm.). Further study is required to find whether perches or platforms are preferred by broiler breeders and which design should be applied when using perches. It is known from the studies regarding laying hens that a high proportion of birds in non-cage systems may suffer from keel bone fractures and deformities, caused by collisions with housing structures or prolonged pressure on the keel bone during perching (Heerkens et al., 2016). It is not known if broiler breeders suffer from keel bone fractures and keel bone deviations due to the design of perches and platforms.

Hocking and Jones (2006) studied if environmental enrichment, by providing plastic-coated bales of wood shavings or bunches of string, could decrease aggressive pecking and feather damage in broiler breeders in the rearing period. It turned out that plastic-coated bales of wood shavings, but not strings, were attractive for the breeders. However, neither behavioural changes nor improved feather cover were observed in the groups where enrichment was provided as compared to the control groups without enrichment.

Vertically placed cover panels in the litter area in the production house can be used to control excessive mating problems in commercial farms (Estevez, 1999) and thus may improve female broiler breeder welfare during the production period. In addition, cover panels improved reproductive performance in broiler breeder flocks, probably by attracting females to the litter floor and reducing male–male competition for females and over-mating (Leone and Estevez, 2008).

In general, environmental enrichment is not commonly applied in broiler breeder houses. Because environmental enrichment may improve the behavioural opportunities and reduce problematic behaviours such as injurious pecking (Estevez, 2009), further research to find appropriate enrichment for broiler breeders both in the rearing and in the production periods is needed.

7 Vaccinations

Maintaining flock health at a high level is an important issue during the rearing and the production period. Farmers usually apply a strict hygiene regime and disease prevention is an important issue with respect to flock management (EFSA, 2010). The applied vaccination schedule is dependent on country, region, individual farmers or advisers and integrations. However, during the rearing period broiler breeders can be subjected to a series of vaccinations starting at the hatchery or even *in ovo*. Examples are Marek's disease and infectious bronchitis (IB) vaccination at the hatchery, and vaccinations against Newcastle disease, IB, infectious laryngotracheitis (ILT), infectious rhinotracheitis, Gumboro, *Salmonella* and *Escherichia coli* at various ages in the rearing period. In the production period, broiler breeders are subjected to less frequent vaccinations (e.g. against IB), although this is also farmer or integration dependent. Vaccinations can be provided in various forms, such as injections, spray/aerosol, via drinking water and eye drops. Despite the need to prevent diseases in the breeder flock, (multiple) vaccinations are a challenge for the birds, resulting in depressed feeding (and thus growth) or local inflammation (Steentjes, 2016, pers. comm.).

8 Transgenerational effects

A relatively unknown area is how conditions during the life of broiler breeders affect the welfare and technical performance of the progeny. It is well known that age of the breeders is related to egg quality and chick quality, and first week mortality is higher in broilers from young breeding flocks (Yassin et al., 2009). However, it is important to consider that other conditions, for example, stress or diseases, in the life of the breeders may also affect the performance and welfare of the progeny. From laying hen studies it is known that feather pecking and anxiety in the layer breeding flock is related to the prevalence of feather pecking and anxiety in the progeny, although these effects were dependent on the genotype of the birds. The underlying mechanism may act through the deposition of corticosterone in the yolk of the eggs (de Haas et al., 2014).

Van der Waaij et al. (2011) studied the progeny of 60-week-old female broiler breeders that were either fed the recommended amount of feed or fed *ad libitum* during a 5-week period. The offspring of breeders fed *ad libitum* was heavier than the offspring of restricted fed breeders. In addition, female offspring of restricted fed breeders were lighter at hatching, but were heavier and had more abdominal fat at six weeks of age than female offspring of the *ad libitum* fed breeders. These results suggested a possible transgenerational effect of the feeding regime in breeders on the broilers. However, more research is needed to further explore this relationship, as reviewed by Berghof et al. (2013), who focused on possible transgenerational effects of stress in breeders on innate immunity in broiler chickens. They suggested that the mismatch between the breeder and broiler environment (e.g. by diet composition or microbiota) may negatively affect the innate immunity in the broiler chickens, but that more research is needed to further unravel the underlying mechanisms.

With respect to the effect of feed composition in the production phase of the breeders and its influence on the progeny, there is some evidence that low-density diets (21% lower nutrient density as compared to the standard diet) positively affect egg composition (higher egg white proportion) and hatchability, especially in young breeders (Enting et al., 2007b). The same authors found that these low-density broiler breeder diets improved growth rate of the progeny, reduced broiler mortality and either reduced or increased immune responses, depending on the breeder's age and egg weight (Enting et al., 2007a). It is not clear if these effects are due to transmission of nutrients from the breeder to the egg and the progeny, or to reduced stress in the breeders, or both.

Van Emous et al. (2015b) also studied the combined effects of low protein diets and different growth patterns in the rearing period on offspring performance. In general, they only found marginal effects of the different protein contents on progeny performance. However, male broilers of breeders fed the low protein diet during the rearing period had higher breast meat yield as compared to broilers of broiler breeders fed the medium or high protein diet – see Van Emous et al. (2015b) for diet composition tables. The effects of different daily protein intake of breeders during the rearing and the production period on progeny performance were observed in a consecutive experiment (van Emous et al., 2015a). A lower mortality and improved technical performance were observed in the progeny of 53-week-old breeders fed a low daily protein intake during the second phase of the laying period.

To conclude, there is evidence for transgenerational effects of management of broiler breeders and their progeny. This area needs further exploration to improve welfare and performance of both breeders and broilers.

9 Concluding remarks

The majority of the broiler breeders worldwide are parent stock of standard, fast growing broiler chickens. Because of the genetic selection for fast growth in the progeny, broiler breeders need to be subjected to restricted feeding programmes, especially during the rearing period. This is, despite a number of studies focusing on the methods to alleviate the negative effects of feed restriction on the welfare, still one of the most important welfare issues in broiler breeders. However, using feeds with higher insoluble fibre contents or lower protein content in the rearing period as compared to the standard commercial diets seems promising in terms of alleviating the effect of feed restriction on welfare. Another alternative is the use of slower growing or dwarf female broiler breeders that do not need to be fed restricted. Apart from feed restriction, other areas that need more attention in research are the relationship between management in the production period and mating behaviour, provision of environmental enrichment and how breeder management affects offspring performance.

10 Where to look for further information

Further information on housing and management can be found in management guides of the breeding companies that are updated on a regular basis (Aviagen, 2013; Cobb, n.d., Hubbard, n.d.) and textbooks (Leeson and Summers, 2000). Specific welfare issues have been discussed in various review papers (e.g. Decuypere et al., 2006, 2010; De Jong and Guemené, 2011) and the EFSA Scientific Opinion on welfare aspects of the management and housing of the grandparent and parent stocks raised and kept for breeding purposes (EFSA, 2010).

11 References

Aviagen. 2013. Ross parent stock management handbook. http://en.aviagen.com/assets/Tech_Center/Ross_PS/Ross_PS_Handbook_2013_i-r1.pdf (Accessed date 24 March 2016).

Berghof, T. V. L., H. K. Parmentier and A. Lammers. 2013. Transgenerational epigenetic effects on innate immunity in broilers: An underestimated field to be explored? *Poultry Science* 92:2904–13.

Bruggeman, V., O. Onagbesan, E. D'Hondt, N. Buys, M. Safi, D. Vanmontfort, L. Berghman, F. Vandesande and E. Decuypere. 1999. Effects of timing and duration of feed restriction during rearing on reproductive characteristics in broiler breeder females. *Poultry Science* 78:1424–34.

Buckley, L. A., L. M. McMillan, V. Sandilands, B. J. Tolkamp, P. M. Hocking and R. B. D'Eath. 2011a. Too hungry to learn? Hungry broiler breeders fail to learn a Y-maze food quantity discrimination task. *Animal Welfare* 20:469–81.

Buckley, L. A., V. Sandilands, P. M. Hocking, B. J. Tolkamp and R. B. D'Eath. 2015. Feed-restricted broiler breeders: State-dependent learning as a novel welfare assessment tool to evaluate their hunger state? *Applied Animal Behaviour Science* 165:124–32.

Buckley, L. A., V. Sandilands, B. J. Tolkamp and R. B. D'Eath. 2011b. Quantifying hungry broiler breeder dietary preferences using a closed economy T-maze task. *Applied Animal Behaviour Science* 133:216–27.

Casanovas, P. 2000. Management techniques to improve male mating activity and compensate for the age-related decline in broiler breeder fertility: Intra-spiking. *Cobb Technical News* 7:1–8.

Cobb, undated. Cobb500FF Breeder Mangement Supplement. http://www.cobb-vantress.com/docs/default-source/cobb-500-guides/cobb500ff-breeder-mangement-supplement---(english)3CA513F5C35A.pdf (Accessed date 24 March 2016).

D'Eath, R. B., B. J. Tolkamp, I. Kyriazakis and A. B. Lawrence. 2009. 'Freedom from hunger' and preventing obesity: the animal welfare implications of reducing food quantity or quality. *Animal Behaviour* 77:275–88.

Dawkins, M. S. and R. Layton. 2012. Breeding for better welfare: genetic goals for broiler chickens and their parents. *Animal Welfare* 21:147–55.

de Beer, M. and C. N. Coon. 2007. The effect of different feed restriction programs on reproductive performance, efficiency, frame size, and uniformity in broiler breeder hens. *Poultry Science* 86:1927–39.

Decuypere, E., V. Bruggeman, N. Li Everaert, R. Yue Boonen, J. De Tavernier, S. Janssens and N. Buys. 2010. The broiler breeder paradox: ethical, genetic and physiological perspectives, and suggestions for solutions. *British Poultry Science* 51(5): 569–79.

Decuypere, E., P.M. Hocking, K. Tona, O. Onagbesan, V. Bruggeman, E.K.M. Jones, S. Cassy, N. Rideau, S. Metayer, Y. Jego, J. Putterflam, S. Tesseraud, A. Collin, M. Duclos, J. J. Trevidy and J. Williams. 2006. Broiler breeder paradox: a project report. *World's Poultry Science Journal* 62:443–53.

de Haas, E. N., J. E. Bolhuis, B. Kemp, T. G. G. Groothuis and T. B. Rodenburg. 2014. Parents and early life environment affect behavioral development of laying hen chickens. *Plos ONE* 9. doi 10.1371/journal.pone.0090577.

De Jong, I. C., H. Enting, S. Van Voorst, E. W. Ruesink and H. J. Blokhuis. 2005a. Do low density diets improve broiler breeder welfare during rearing and laying? *Poultry Science* 84:194–203.

De Jong, I. C., M. Fillerup and H. J. Blokhuis. 2005b. Effect of scattered feeding and feeding twice a day during rearing on parameters of hunger and frustration in broiler breeders. *Applied Animal Behaviour Science* 92:61–76.

De Jong, I. C. and D. Guemene. 2011. Major welfare issues in broiler breeders. *World's Poultry Science Journal* 67:73–81.

De Jong, I. C., H. Gunnink and R. A. Van Emous. 2013. Monitoring van onbehandelde vleeskuikenmoederdieren op een praktijkbedrijf. 21 pp Wageningen UR Livestock Research, Lelystad, Report 716.

De Jong, I. C. and R. B. Jones. 2006. Feed restriction and welfare in domestic birds. In V. Bels (ed.), *Feeding in Domestic Vertebrates*, pp. 120–35. CABI publishing, Wallingford, UK.

De Jong, I. C., A. Lourens, H. Gunnink and R. A. Van Emous. 2011. Effect of stocking density on (the development of) sexual behaviour and technical performance in broiler breeders. 61 p. Wageningen UR Livestock Research, Lelystad, Report 457.

De Jong, I. C. and M. Swalander. 2013. Housing and management of broiler breeders and turkey breeders. In V. Sandilands and P. M. Hocking (eds), *Alternative Systems for Poultry: Health, Welfare and Productivity*, pp. 225–49. CABI publishing, Wallingford, UK.

De Jong, I. C., S. Van Voorst and H. J. Blokhuis. 2003. Parameters for quantification of hunger in broiler breeders. *Physiology & Behavior* 78:773–83.

De Jong, I. C., S. Van Voorst, D. A. Ehlhardt and H. J. Blokhuis. 2002. Effects of restricted feeding on physiological stress parameters in growing broiler breeders. *British Poultry Science* 43:157–68.

De Jong, I. C., M. Wolthuis-Fillerup and R. A. Van Emous. 2009. Development of sexual behaviour in commercially-housed broiler breeders after mixing. *British Poultry Science* 50:151–60.

Dixon, L. M., S. Brocklehurst, V. Sandilands, M. Bateson, B. J. Tolkamp and R. B. D'Eath. 2014. Measuring motivation for appetitive behaviour: food-restricted broiler breeder chickens cross a water barrier to forage in an area of wood shavings without food. *Plos ONE* 9. doi 10.1371/journal.pone.0102322.

EFSA. 2010. Scientific Opinion on welfare aspects of the management and housing of the grand-parent and parent stocks raised and kept for breeding purposes. *The EFSA Journal* 8:81. doi doi:10.2903/j.efsa.2010.1667.

Enting, H., W. J. A. Boersma, J. B. W. J. Cornelissen, S. C. L. van Winden, M. W. A. Verstegen and P. J. van der Aar. 2007a. The effect of low-density broiler breeder diets on performance and immune status of their offspring. *Poultry Science* 86:282–90.

Enting, H., T. A. M. Kruip, M. W. A. Verstegen and P. J. van der Aar. 2007b. The effect of low-density diets on broiler breeder performance during the laying period and on embryonic development of their offspring. *Poultry Science* 86:850–6.

Estevez, I. 1999. Cover panels for chickens: a cheap tool that can help you. In *Poultry Perspectives*, pp. 4–6.

Estevez, I. 2009. Behaviour and environmental enrichment in broiler breeders. In *Biology of Breeding Poultry*, pp. 261–83. CABI publishing, Wallingford, UK.

Fiks-van Niekerk, T. G. C. M., I. C. de Jong, T. Veldkamp, M. M. Van Krimpen and R. A. Van Emous. 2009. Mutliations in poultry. 46 p. Wageningen UR Livestock Research, Lelystad, Report 255 .

Fontana, W. A., A. D. Weaver and H. P. Van Krey. 1992. intermittend periods of infertility identified in naturally mated broiler breeder hens. *Journal of Applied Poultry Research* 1:190–3.

Gentle, M. J. and D. E. F. McKeegan. 2007. Evaluation of the effects of infrared beak trimming in broiler breeder chicks. *Veterinary Record* 160:145–8.

Heck, A., O. Onagbesan, K. Tona, S. Metayer, J. Putterflam, Y. Jego, J. J. Trevidy, E. Decuypere, J. Williams, M. Picard and V. Bruggeman. 2004. Effects of ad libitum feeding on performance of different strains of broiler breeders. *British Poultry Science* 45:695–703.

Heerkens, J. L. T., E. Delezie, T. B. Rodenburg, I. Kempen, J. Zoons, B. Ampe and F.A.M. Tuyttens. 2016. Risk factors associated with keel bone and foot pad disorders in laying hens housed in aviary systems. *Poultry Science* 95: 482–8.

Hiemstra, S. J. and J. Ten Napel. 2013. Study of the impact of genetic selection on the welfare of chickens bred and kept for meat production. 118 p. http://ec.europa.eu/food/animals/docs/aw_practice_farm_broilers_653020_final-report_en.pdf.

Hocking, P. M. 1993. Welfare of broiler breeder and layer females subjected to food and water control during rearing: quantifying the degree of restriction. *British Poultry Science* 343:53–64.

Hocking, P. M. and R. Bernard. 1993. Evaluation of putative appetite suppressants in the domestic fowl. *British Poultry Science* 34:393–404.

Hocking, P. M. and R. Bernard. 2000. Effects of the age of male and female broiler breeders on sexual behaviour, fertility and hatchability of eggs. *British Poultry Science* 41:370–6.

Hocking, P. M. and E. K. M. Jones. 2006. On-farm assessment of environmental enrichment for broiler breeders. *British Poultry Science* 47:418–25.

Hocking, P. M., M. H. Maxwell and M. A. Mitchell. 1993. Welfare assessment of broiler breeder and layer females subjected to food restriction and limited access to water during rearing. *British Poultry Science* 34:443–58.

Hocking, P. M., M. H. Maxwell and M. A. Mitchell. 1996. Relationships between the degree of food restriction and welfare indices in broiler breeder females. *British Poultry Science* 37:263–78.

Hocking, P. M., M. H. Maxwell, G. W. Robertson and M. A. Mitchell. 2002. Welfare assessment of broiler breeders that are food restricted after peak of lay. *British Poultry Science* 43:5–15.

Hocking, P. M., D. Waddington, M. A. Walker and A. B. Gilbert. 1989. Control of the development of the ovarian follicular hierarchy in broiler breeder pullets by food restriction during rearing. *British Poultry Science* 30:161–74.

Hocking, P. M., V. Zaczek, E. K. M. Jones and M. G. McLeod. 2004. Different concentrations and sources of dietary fibre may improve the welfare of female broiler breeders. *British Poultry Science* 45:9–19.

Horne, P. L. M. and N. Bondt. 2014. Competitiveness of the EU poultry meat sector. 46 p. LEI Wageningen UR, Wageningen, Report nr LEI 2014–038.

Hubbard, undated. Guide parent stock. http://www.hubbardbreeders.com/media/guide__ps__classic__08__01__2016_imp__051137500_1634_22012016.pdf (Accessed date 24 March 2016).

Jones, E. K. M. and N. B. Prescott. 2000. Visual cues used in the choice of mate by fowl and their potential importance for the breeder industry. *World's Poultry Science Journal* 56:127–38.

Jones, E. K. M., N. B. Prescott, P. Cook, R. P. White and C. M. Wathes. 2001. Ultraviolet light and mating behaviour in domestic broiler breeders. *British Poultry Science* 42:23–32.

Jones, E. K. M., V. Zaczek, M. McLeod and P. M. Hocking. 2004. Genotype, dietary manipulation and food allocation affect indices of welfare in broiler breeders. *British Poultry Science* 45:725–37.

Leeson, S. and J. D. Summers. 2000. *Broiler Breeder Production*. Unviersity Books, Guelph, Canada.

Leone, E. H. and I. Estevez. 2008. Economic and welfare benefits of environmental enrichment for broiler breeders. *Poultry Science* 87:14–21.

Leone, E. H., I. Estevez and M. C. Christman. 2007. Environmental complexity and group size: Immediate effects on use of space by domestic fowl. *Applied Animal Behaviour Science* 102:39–52.

Mench, J. A. 2002. Broiler breeders: feed restriction and welfare. *World's Poultry Science Journal* 58:23–30.

Millman, S. T. and I. J. H. Duncan. 2000a. Effect of male-to-male aggressiveness and feed-restriction during rearing on sexual behaviour and aggressiveness towards females by male domestic fowl. *Applied Animal Behaviour Science* 70:63–82.

Millman, S. T. and I. J. H. Duncan. 2000b. Strain differences in aggressiveness of male domestic fowl in response to a male model. *Applied Animal Behaviour Science* 66:217–33.

Millman, S. T., I. J. H. Duncan and T. M. Widowski. 2000. Male broiler breeder fowl display high levels of aggression toward females. *Poultry Science* 79:1233–41.

Moradi, S., M. Zaghari, M. Shivazad, R. Osfoori and M. Mardi. 2013. Response of female broiler breeders to qualitative feed restriction with inclusion of soluble and insoluble fiber sources. *Journal of Applied Poultry Research* 22:370–81.

Morrissey, K. L. H., T. Widowski, S. Leeson, V. Sandilands, A. Arnone and S. Torrey. 2014a. The effect of dietary alterations during rearing on feather condition in broiler breeder females. *Poultry Science* 93:1636–43.

Morrissey, K. L. H., T. Widowski, S. Leeson, V. Sandilands, A. Arnone and S. Torrey. 2014b. The effect of dietary alterations during rearing on growth, productivity, and behavior in broiler breeder females. *Poultry Science* 93:285–95.

Nielsen, B. L., K. Thodberg, J. Malmkvist and S. Steenfeldt. 2011. Proportion of insoluble fibre in the diet affects behaviour and hunger in broiler breeders growing at similar rates. *Animal* 5:1247–58.

Renema, R. A., M. E. Rustad and F. E. Robinson. 2007. Implications of changes to commercial broiler and broiler breeder body weight targets over the past 30 years. *World's Poultry Science Journal* 63:457–72.

Sandilands, V., B. J. Tolkamp and I. Kyriazakis. 2005. Behaviour of food restricted broilers during rearing and lay – effects of an alternative feeding method. *Physiology & Behavior* 85:115–23.

Sandilands, V., B. J. Tolkamp, C. J. Savory and I. Kyriazakis. 2006. Behaviour and welfare of broiler breeders fed qualitatively restricted diets during rearing: Are there viable alternatives to quantitative restriction? *Applied Animal Behaviour Science* 96:53–67.

Savory, C. J. and J.-M. Lariviere. 2000. Effects of qualitative and quantitative food restriction treatments on feeding motivational state and general activity level of growing broiler breeders. *Applied Animal Behaviour Science* 69:135–47.

Savory, C. J. and J. S. Mann. 1997. Is there a role for corticosterone in expression of abnormal behaviour in restricted-fed fowls? *Physiology & Behavior* 62:7–13.

Savory, C. J. and K. Maros. 1993. Influence of degree of food restriction, age and time of day on behaviour of broiler breeder chickens. *Behavioural Processes* 29:179–90.

Savory, C. J., K. Maros and S. M. Rutter. 1993. Assessment of hunger in growing broiler breeders in relation to a commercial restricted feeding programme. *Animal Welfare* 2:131–52.

Savory, C. J., E. Seawright and A. Watson. 1992. Stereotyped behaviour in broiler breeders in relation to husbandry and opiod receptor blockade. *Applied Animal Behaviour Science* 32:349–60.

Savory, C. J., F. A. M. Tuyttens and J. S. Mann. 1996. Temporal patterning of oral stereotypies in restricted-fed fowls: 2. influece of meal frequency and meal size. *International Journal of Comparative Psychology* 9:140–58.

Tolkamp, B. J., V. Sandilands and I. Kyriazakis. 2005. Effects of qualitative feed restriction during rearing on the performance of broiler breeders during rearing and lay. *Poultry Science* 84:1286–93.

van der Waaij, E. H., H. van den Brand, J. A. M. van Arendonk and B. Kemp. 2011. Effect of match or mismatch of maternal-offspring nutritional environment on the development of offspring in broiler chickens. *Animal* 5:741–8.

Van Emous, R. A. and I. C. De Jong. 2013. Promising management measures to solve the major welfare problems in broiler breeders. *Proceedings 2nd International Poultry Meat Congress Antalya* (Turkey).

Van Emous, R. A. 2010. Quality Time®; an innovative housing concept for broiler breeders. In *Proceedings of of the 2nd International symposium.* Highlights in nutrition and welfare in poultry production. Wageningen, The Netherlands, pp. 37–44.

van Emous, R. A., R. Kwakkel, M. van Krimpen and W. Hendriks. 2014. Effects of growth pattern and dietary protein level during rearing on feed intake, eating time, eating rate, behavior, plasma corticosterone concentration, and feather cover in broiler breeder females during the rearing and laying period. *Applied Animal Behaviour Science* 150:44–54.

van Emous, R. A., R. Kwakkel, M. van Krimpen and W. Hendriks. 2015a. Effects of different dietary protein levels during rearing and different dietary energy levels during lay on behaviour and feather cover in broiler breeder females. *Applied Animal Behaviour Science* 168:45–55.

van Emous, R. A., R. P. Kwakkel, M. M. van Krimpen, H. van den Brand and W. H. Hendriks. 2015b. Effects of growth patterns and dietary protein levels during rearing of broiler breeders on fertility, hatchability, embryonic mortality, and offspring performance. *Poultry Science* 94:681–91.

Wachenfelt, E. and E. Berndson. 2014. Usge of perches in meat fowlsSveriges lantbruksuniversitet Fakulteten för landskapsarkitektur, trädgårds- och växtproduktionsvetenskap.

Yassin, H., A. G. J. Velthuis, M. Boerjan and J. van Riel. 2009. Field study on broilers' first-week mortality. *Poultry Science* 88:798–804.

Zuidhof, M. J., D. E. Holm, R. A. Renema, M. A. Jalal and F. E. Robinson. 2015. Effects of broiler breeder management on pullet body weight and carcass uniformity. *Poultry Science* 94:1389–97.

Zuidhof, M. J., R. A. Renema and F. E. Robinson. 2007. Reproductive efficiency and metabolism of female broiler breeders as affected by genotype, feed allocation, and age at photostimulation. 3. Reproductive efficiency. *Poultry Science* 86:2278–86.

Zuidhof, M. J., B. L. Schneider, V. L. Carney, D. R. Korver and F. E. Robinson. 2014. Growth, efficiency, and yield of commercial broilers from 1957, 1978, and 2005. *Poultry Science* 93:2970–82.

Chapter 12

The effect of incubation temperature on embryonic development in poultry

M. S. Lilburn and R. Shanmugasundaram, Ohio State University, USA

1 Introduction

The biological process of avian incubation is the culmination of species-specific behavioural and physiological traits that have been studied and documented for over two hundred years (Reaumur, 1749, as cited by Henderson, 1939). It was recognized by Dareste (1891), and later confirmed by Eycleshymer (1907), that the parental movement of eggs was necessary to prevent embryonic attachment to the shell, resulting in embryonic mortality. From a commercial perspective, similar observations were subsequently confirmed by Martin and Insko (1935) in turkeys and Kaltofen (1961) in chickens where low or reduced rate of egg turning increased early embryonic losses. It is now well recognized that numerous factors influence the overall process of successful embryonic development (Landauer, 1967).

Embryos from precocial species including chickens and turkeys transition from an ectothermy state (heat exchange with the environment) to endothermy (*de novo* heat production) during the course of incubation (Whittow and Tazawa, 1991; Nitchelmann and Tzschentke, 1997, 2002; Tzschentke and Rumpf, 2011). Parental incubation behaviours influence the regulation of temperature and are a major source of the variability that exists in natural, unregulated nesting environments (Drent, 1970). Huggins (1941) observed a 5.6°C difference in mean internal temperature between the peripheral and central eggs within a large clutch (n = 18) of mallard eggs (*Anas platyrhynchos*). Smaller but

http://dx.doi.org/10.19103/AS.2016.0011.27

still considerable differences (1–3°C) in shell temperature were noted by Kessler (1960) within a clutch of pheasant eggs (*n* = 9). Randall (1943) observed that at hatch, the body temperature of chicks was close to that of the incubator (~39°C), and reached near adult body temperature (~41°C) at approximately ten days post-hatch.

The biology of incubation is a fascinating process when one considers the unique differences among species, in particular egg/embryonic temperature regulation and correlated physiologic adaptations (Webb, 1987). The objective of this chapter will be to focus on the specific effects of exogenous and endogenous heat on embryonic growth during different stages of incubation with an emphasis on controlled incubation temperature studies. It has long been recognized that incubation temperature is only one of many factors that can influence day-old chick quality (Tona et al., 2005) and post-hatch performance (Decuypere, 1984). The latter point has increasingly become a focal point as it relates to carcass growth and development and the ontogeny of growth-associated anomalies (Lourens et al., 2005; Joseph et al., 2006; Hulet et al., 2007; Molenaar et al., 2011; Da Costa et al., 2016). It is recognized that within the same poultry species (i.e. chickens), commercial selection goals have resulted in very different commercial strains (i.e. layers and broilers) and their phenotypic differences post-hatch have translated to different embryonic metabolic characteristics as well (Janke et al., 2004).

2 Establishing a temperature recommendation

The early 1900s was a very productive period with respect to research that furthered our understanding of the optimal temperature for incubating eggs for commercial poultry. Romanoff (1936) is a frequently cited literature source for the widely accepted 'optimal' incubation temperature of 37.5°C (99.5°F) for continuously incubated eggs. The data from Romanoff (1936) is shown in Fig. 1 and, as can be seen, the recommended mean temperature decreases to 36°C after 16 days of incubation. The decrease in the recommended temperature during the latter stages of incubation confirmed earlier studies on turkeys

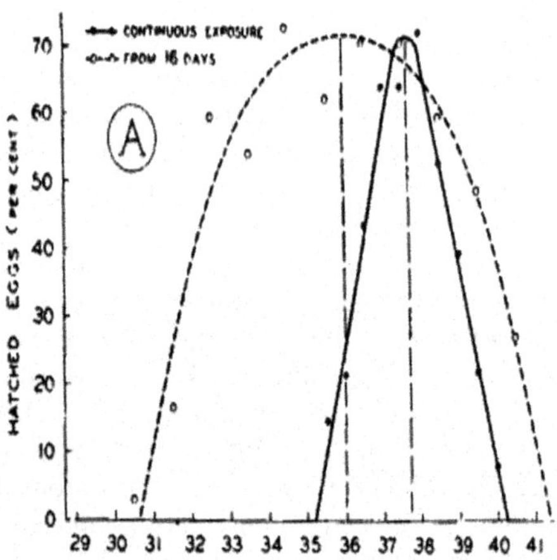

Figure 1 The effect of incubation temperature on hatchability (Romanoff, 1936). The vertical axis shows the percentage of eggs hatched; the horizontal axis shows temperature (°C).

with respect to incubation temperature and correlated effects on length of incubation and post-hatch growth (Romanoff, 1935). The results of Romanoff (1936) were in essence a confirmation of previously reported studies. Edwards (1901) had concluded, from previous literature at that time, including a report by Harvey (1651), that the optimum temperature for domestic fowl was 38°C with a range of 35° to 39°C. Eycleshymer (1907) observed that actual egg temperatures over the latter stages of incubation were 39.4–40°C. This is one of the first reports suggesting that during the latter stages of incubation, less artificial heat was needed. This was also one of the first published studies relating higher temperatures early in incubation to accelerated embryonic development and decreased hatch time. Philips (1917) reported that air temperatures of 38.3°–38.9°-39.4°C during each week of incubation were adequate for normal development. Philips and Brooks (1923) subsequently decreased the suggested optimal temperature range to 37.8°–39.4°C. These authors also noted that chicks produced heat during the course of incubation. Henderson and Brody (1927) and Henderson (1930) reported that the earliest stages of incubation were the most sensitive to temperature fluctuations and after 16 days endogenous embryonic heat was evident. These observations supported the conclusions by Romanoff and Faber (1933) and Romanoff (1936) that incubation temperature can actually be reduced by 2°–3°C during the latter stages of incubation to improve the thriftiness of hatched chicks. An interesting observation by Byerly (1938) was the high incidence of embryo malposition when eggs were incubated at high temperatures and this observation was subsequently reported in turkeys (French, 1994) and ducks (Kroesen and Lilburn, 2006).

3 Pre-incubation holding

It is often a necessary practice within the commercial poultry industry to store eggs for varying periods of time in controlled temperature holding rooms prior to setting. Decuypere and Michels (1992) published an older but still excellent review on the overall temperature management of eggs from holding through hatch. Edwards (1901) published one of the first papers related to establishing a 'physiological zero', an important concept with respect to fertile egg storage. He reported that there was no embryonic development below 20°C with some percentage of eggs showing development at 21°–22°C. Funk and Biellier (1944) concluded that approximately 27°C was the initiation point for blastoderm development and they provided some insight as to why they thought the observations by Edwards (1901) were low. In their introduction, Funk and Biellier (1944) referred to physiological threshold estimates of 28°–29°C by Prevost and Dumas (1821) and Kaestner (1895) as cited by Needham (1971) and these were more in line with their own observations. An approximate 27°C temperature minimum was later stated by Lundy (1969) and Wilson (1991) as the starting point for rapid embryonic development. Recent studies, however, suggest that temperatures as low as 14°–15°C will still allow for some degree of cellular development in chicken and turkey embryos, respectively (Eyal-Giladi and Kochav, 1976; Fasenko et al., 1992; Bakst and Gupta, 1997). These are the approximate holding temperatures currently recommended by Aviagen, a commercial broiler and turkey breeding company (Table 1). It has been recognized for many years that length of storage will delay hatch time (Mather and Laughlin, 1976) as well as negatively influence hatchability in chickens and turkeys (Scott, 1933; Asmundson, 1947; Kosin, 1950; Becker, 1964). Seven days of storage appears to be the point at which the decline in hatchability begins (Funk et al., 1950; Fasenko et al., 1992; Reijrink, 2008). The concept of advancing development prior to storage via pre-incubation warming is an interesting concept that has been in the literature for over 50 years (Kosin, 1956; Becker and Bearse, 1958; Bowling and Howarth, 1981). It is interesting that in the report by Becker and Bearse (1958), the storage times varied from 18 to 22 days and there were still noticeable improvements in hatchability due to some form of pre-storage warming or incubation. In a more recent study, Fasenko et al. (2001) incubated broiler eggs for variable periods up to 18 hours (37.5°C) followed by storage for 4 or 14 days. There were no pre-incubation effects after only 4 days of storage but the 6-hour treatment significantly improved hatchability after 14 days when compared with the control eggs (no pre-incubation). In chickens, Lourens (2006) reported that 3 to 9 hours of pre-incubation prior to a 14-day storage period resulted in significant (6.4–11.8%) improvements in hatchability compared with the control (no pre-incubation). Reijrink et al. (2010) suggest that a 24 h pre-warming period is better than a shorter, 4 h period (Table 2). Christensen

Table 1 Recommended pre-incubation storage temperatures for Aviagen turkeys and broilers

Turkeys[1]		Broilers[2]	
Storage (days)			
<5 d	17–18° C (63–65° F)	1–3 d	20–23° C (68–73° F)
5–10 d	14–15° C (57–59° F)	4–7 d	15–18° C (59–64° F)
>10 d	12–13° C (54–56° F)	>7 d	12–15° C (54–59° F)
		>13 d	12° C (54° F)

[1]Aviagen Turkeys, Inc, 31186 Midland Trail, East Lewisburg, W, 24901 USA
[2]Aviagen Group, 920 Explorer Boulevard NW, Huntsville, AL 35806

Table 2 Assessing pre-warming periods. Adapted from Reijrink et al. (2010)

	Experiment 1					Experiment 2		
Treatment	Fertility	Hatchability (%)		Embryo		Fertility	Hatchability (%)	
		Total	Fertile	Wt (g)	Lth (mm)		Total	Fertile
Storage (d)								
4	88.9	83.0	93.3	1.08	31.0	95.3	88.7	93.1
14 (Exp. 1); 13 (Exp. 2)	90.2	82.5	91.5	0.96	29.9	92.9	70.6	76.0
Pre-Incubation Warming (h)								
4	89.1	80.9	90.7	0.95	29.5	94.6	78.6	82.9
24	90.0	84.5	94.0	1.10	31.1	93.6	80.8	86.2
Probability								
Storage (d)	0.88	0.44	0.21	0.001	0.001	0.001	0.001	0.001
Pre-incubation (h)	0.99	0.99	0.99	0.001	0.001	0.11	0.04	0.002
Interaction	0.79	0.32	0.27	0.05	0.03	0.42	0.06	0.01

et al. (2003a,b) also observed benefits in turkey embryos by accelerating pre-incubation temperatures after 14 d of storage.

4 Early incubation temperature outcomes

As discussed above, Henderson and Brody (1927) and Henderson (1930) were among the first to observe that embryos are extremely sensitive to temperature variations during the early incubation period. Embryonic skeletal development is one system that can be influenced by both low and high incubation temperatures during early and late development (Romanoff, 1936; Oviedo-Rondon et al., 2008). Brookes and May (1972) showed that over the temperature range of 35–40°C, there was a linear increase in wet embryo weight (Fig. 2) after 10 days of incubation along with a correlated linear increase in femur length and calcification (Fig. 3). It should be emphasized that the data of Brookes and May (1972) was only through 10 days of incubation. As alluded to above, longer-term or late stage incubation at higher temperatures (>37.5°C) may have negative effects on overall skeletal health at hatch whether it be 'crippled chicks' at hatch as originally described by Romanoff (1936) or long bone growth and relative asymmetry (Oviedo-Rondon et al., 2008). Hammond et al. (2007) showed that these linear skeletal effects can be observed with as little as 3 days exposure (ED 4–7) to a one-degree increase in incubation temperature (37.5° vs 38.5°C; Fig. 4). More recent data (Shim and Pesti, 2011) confirmed the results of Hammond et al. (2007) and others with respect to a

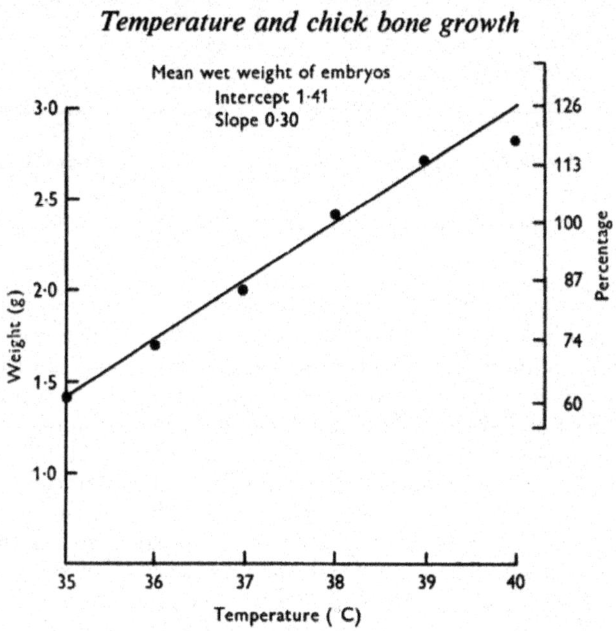

Figure 2 The relationship between temperature and embryo weight. Adapted from Brooks and May, 1972.

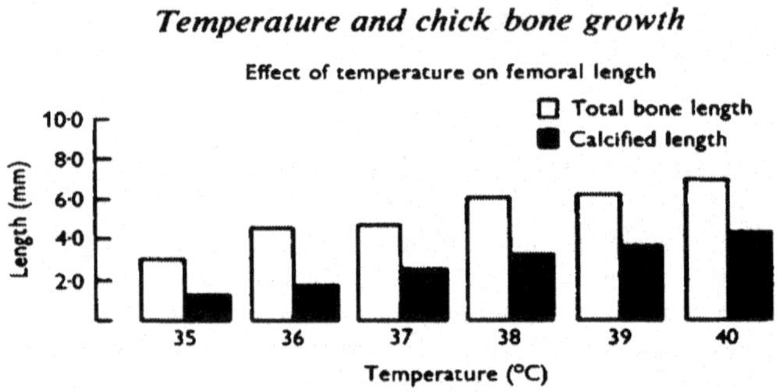

Figure 3 The effect of graded incubation temperatures (0-10 d) on embryonic skeletal growth (Brookes and May, 1972).

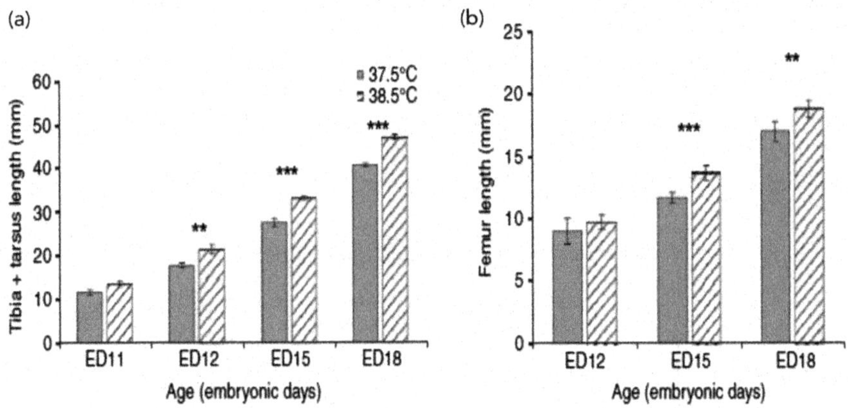

Figure 4 The effect of short term high temperature exposure (4-7 d) on embryonic skeletal development (Hammond et al., 2007).

short-term increase in temperature (ED 4–7) and a correlated decrease in the total length of incubation. These authors did not observe any carry-over effects of temperature on post-hatch skeletal development, however, and their data did not support conclusions by Oviedo-Rondon et al. (2009) that variable incubation temperatures can contribute to post-hatch skeletal issues, such as in broiler chicks. Shim and Pesti (2011) also correctly pointed out that differential temperature experiments need to consider the time of hatch when summarizing immediate post-hatch chick data. With respect to low incubation temperatures, Joseph et al. (2006) observed that a low eggshell temperature (EST 36.6°C) for the first 10 days of incubation resulted in lower hatchability and reduced body weights throughout rearing when compared with chicks from eggs incubated at the control EST (37.8°C). French (1994) observed that turkey eggs incubated at 36.5°C simply took longer to hatch compared with control poults incubated at 37.5°C, but there were no differences in embryonic mortality.

In a recent report by Krischek et al. (2016), broiler embryos were incubated at low (36.8°C), control (37.8°C) and high (38.8°C) temperatures between ED 7 and 10 or ED 10–13. The ED 7–10 embryos incubated at 38.8°C had increased body, liver and heart weights; mitochondrial respiratory activity; and lactate dehydrogenase and cytochrome oxidase activities compared with the embryos exposed to 36.8°C temperatures. Similarly, at ED 13, the increase in incubation temperature (38.8°C) showed similar effects when compared with the low-temperature treatment. These observations contribute to our understanding of some of the mechanisms associated with the temperature-induced acceleration of embryonic development.

DuRant et al. (2011) measured the respiration of wood duck embryos incubated at 35.0°, 35.9° and 37.0°C, temperatures that are within the normal temperature range of wood duck nests. Embryonic respiration was determined at 3-d intervals throughout incubation, at pipping and immediately post-hatch. The highest temperature resulted in accelerated embryonic growth and, not surprisingly, increased metabolic rates at most experimental timepoints. The embryos from eggs incubated at the lowest temperature, however, had considerably higher energy expenditures during hatching resulting in an overall increase in energy expenditure for the entire length of incubation. The authors concluded that this increased energy output coupled with slower embryonic development decreased the overall physiological status of these embryos compared with those from the higher temperature treatments. Recent work in Pekin duck embryos (Yan et al., 2016) showed that increasing incubation the temperature from 37.8°C to 38.8°C from ED 1 to 10 significantly increased the mRNA levels of target genes for selected enzymes associated with DNA methylation. They suggested that this could be one factor involved in 'thermal epigenetic' regulation of early embryonic development.

5 Late embryonic temperature outcomes

As alluded to previously, the early stages of incubation are significantly affected by differences in temperature, with higher temperatures decreasing the overall length of the incubation period (Romanoff and Faber, 1933; Henderson, 1939). Romanoff (1936) observed a linear decline in hatchability with each 1°C increase from 37.5° to 40.5°C, which was the basis for his suggested 36.5°C as the optimal temperature from 16 d to hatch in chickens. It has long been recognized that the temperature component of the microenvironment surrounding each egg is largely influenced by airflow (Sotherland, 1987) and incubator design is one source of variability in air velocity at different points within an incubator (Meijerhof and van Beek, 1993; Van Brecht et al., 2003, 2005). Within the published incubation literature, the control air temperature is often 37.5°–37.8°C and without a pre-planned temperature reduction protocol during incubation, egg temperatures will often exceed air temperature due to embryonic metabolic heat production (Tazawa and Rahn, 1987; Lourens, 2001).

Lourens et al. (2005) suggested that EST was the most accurate, non-destructive assessment of actual embryo temperature. However, even at a mean EST of 37.8°C, Lourens (2001) observed up to 5°C EST swings depending upon egg location within the incubator (single stage) and age of the embryo. Lourens et al. (2005) compared a control EST (37.8°C) with low (Week 1, 36.7°C) and high (Week 3, 38.9°C) temperatures during the first and last weeks of incubation. Both the low and high EST treatments resulted in reduced yolk-free body weight and embryo length at various stages of embryonic development ages (Table 3) and these same trends were observed for chick body weight at 7 d post-hatch. Elevated EST during week 3 had no effect on rectal temperatures at either day 1 or day 7, suggesting the overall thermoregulation was not compromised. Molenaar et al. (2011) imposed two EST treatments (37.8° vs 38.9°C) from ED 7 to ED 18 and reared the chicks at normal or cool brooding temperatures. They found that while total hatch was similar between incubation temperatures, yolk-free hatch weight and heart weights were significantly reduced by the higher EST and both increased EST and cooler brood treatments resulted in increased ascites-related mortality. The observations

Table 3 Embryo development in eggs set at low and high eggshell temperatures during the first and last week of incubation

	Embryo length (cm)			Yolk-free embryo weight (g)		
Days of incubation	14 d	18 d	21.5 d	14 d	18 d	21.5 d
Exp. 1 EST – Week 1						
36.7°	11.4^{b}	16.9	17.9^{b}	12.5^{b}	29.0	31.2^{b}
37.8°	11.9^{a}	17.6	18.8^{a}	13.6^{a}	30.4	34.6^{a}
Exp. 1 EST – Week 3						
37.8°	–	17.5	18.5^{a}	–	30.9	33.1
38.9°	–	17.0	18.2^{b}	–	28.5	32.7
Exp. 2 EST – Week 1						
36.7°	10.8^{b}	16.7	19.9^{b}	12.4^{b}	33.7	42.4^{b}
37.8°	11.4^{a}	17.0	20.3^{a}	13.4^{a}	34.9	42.9^{a}
Exp. 1 EST – Week 3						
37.8°	–	17.0	20.2^{a}	–	34.2	42.8^{a}
38.9°	–	16.7	19.9^{b}	–	34.5	42.5^{b}

Adapted from Lourens et al., 2005;
^{a,b} – Different superscripts represent significant differences, $P < .05$.

Table 4 Effects of egg shell temperature during and breeder flock age on post-hatch broiler growth (Adapted from Hulet et al., 2007)

	Egg shell temperature, C°			Breeder age (wk)		SEM
Age (d)	37.5°	38.6°	39.7°	29	57	
1	41.1^{c}	42.2^{b}	43.1^{a}	38.5^{b}	46.2^{a}	0.6
21	715.1^{a}	714.8^{a}	669.5^{b}	674.5^{b}	730.8^{a}	5.8
35	1722.5^{b}	1756.7^{a}	1663.6^{c}	1689.2^{b}	1745.7^{a}	8.9
44	2213.8^{b}	2263.3^{a}	2165.7^{c}	$2206.^{3}$	2225.7	9.8

$^{a\text{-}c}$ Different superscripts represent significant differences, $P < .05$.

on hatch weight and heart weight were similar to those made by Leksrisompong et al. (2007). Afsarian et al. (2016) recently reported that periodic exposure of eggs to a low incubation temperature (PLE; 15°C for one hr on ED 11,13,15, 17) compared with an EST of 37.8°C throughout incubation significantly improved body weight and feed efficiency at 42 d and significantly decreased mortality due to ascites. All chicks were reared in a 15°C environment from 14 to 42 d. Nangsuay et al. (2016) studied the interaction between breeder age, strain and EST from 7 d through hatch. They observed that breeder age more than strain had the biggest effect on yolk size and an increase in EST (38.9° vs 37.8°C) contributed to increased yolk-free body weight at 14 and 16 d of incubation but lower yolk-free body weight shortly after hatch. Across both ages and strains, the higher EST temperature significantly decreased the overall incubation period. Hulet et al. (2007) also incubated eggs from two breeder flocks (29, 57 wk) at three EST temperatures (36.5°, 37.6°, 38.7°) starting at 16 d. The eggs from older hens resulted in the highest body weight at hatch (Table 4). Across both breeder flock ages, the highest EST (38.7°C) increased body weight at hatch but decreased BW at all subsequent ages (Table 4).

Liu et al. (2015) increased the incubation temperature by 1°C between ED 11 and 20 and observed that out of 1370 genes identified, 1108 were upregulated and 262 were downregulated in leg muscles sampled from Pekin duck embryos.

6 Immunology of poultry

The fundamental mechanisms of the immune response are identical to those in avian species, mammals and most other vertebrates (Jeurissen et al., 1994). The thymus and bursa of Fabricius are primary lymphoid organs responsible for the maturation and acquisition of immunological competence of T and B lymphocytes in poultry (Rudrappa and Humphrey, 2007). Secondary lymphoid organs include the spleen and mucosal lymphoid tissues. The development of the immune system begins early in embryogenesis in poultry with bursal precursor cells being detected as early as ED 7 and cells expressing surface IgM, IgG and IgA at ED 10, ED 14 and ED 16, respectively (Sharma, 1999).

Oznurlu et al. (2010) reported that broiler embryos exposed to a 38.8°C incubation temperature (40.1–40.6°C EST) from 10 d through hatch had reduced thymus and bursa development as assessed by histological evaluation. It is still to be definitively determined, however, whether these observed histological effects have longer-term, practical

Table 5 Effect of incubation temperature on vaccine response in broiler chicks (IBDV = infectious bursal disease virus; NDV = Newcastle disease virus) (Santin et al., 2003)

		Vaccination					
		IBDV			NDV		
Incubation	T°	14 d	35 d	42 d	14 d	35 d	42 d
Control	(37.8° C)	2.81	3.51	3.66	1.58	2.10	2.03
High	(38.8° C)	2.79	3.62	3.81	1.54	2.07	1.84
Low	(36.8° C)	2.68	3.55	3.84	1.54	2.33	1.84
		NS	NS	NS	NS	NS	NS

implications. Santin et al. (2003) incubated broiler eggs at 36.8°, 37.8° or 38.8°C from ED 13 through hatch and observed no differences in the immune responses to infectious bursal disease or Newcastle disease at multiple ages post-hatch (Table 5). A recent report by Filho et al. (2015) suggests that incubation temperature warrants more attention as a potential tool to deal with food safety, specifically with respect to *Salmonella* infection in chicks immediately post-hatch. These authors used three temperatures (36.7°, 37.7° and 38.7°C) from ED 11 through hatch. Chicks were inoculated with *Salmonella* at 2 d post-hatch, and the chicks hatched at the highest incubation temperature had significantly reduced *Salmonella* colonization; the overall number of *Salmonella*-positive chicks was reduced. This treatment also resulted in increased mucin expression and decreased Hsp70 expression in inoculated chicks compared with the two lower incubation treatments.

7 Thermal manipulations and heat stress post-hatch

The induction of some degree of thermotolerance in young chicks by exposing them to a short-term heat stress shortly after hatch is not a new concept (Arijona et al., 1988, 1990; Yahav and Hurwitz, 1996). With respect to embryonic development, the ontogeny of thyroid hormone synthesis, secretion and regulatory control is established during the first half of embryonic development between ED 8 and ED 11 (Thommes, 1987; Reyns et al., 2002). This has prompted several authors to pursue the hypothesis that exposing embryos to a short-term thermal stress at various incubation ages could also induce post-hatch thermotolerance (Yahav et al., 2004; Collin et al., 2005; Piestun et al., 2008, 2009; Shinder et al., 2009). Yahav et al. (2004) exposed embryos to daily, 3-h temperature extremes (39.5°C, 41.0°C) during early (EE; ED 8–10) and late (LE; ED 16–18) incubation. At 3 d post-hatch, sampled chicks from each treatment were heat challenged (6 h at 41.0°C). These authors concluded that exposure to the 39.5°C but not 41.0°C thermal manipulation during both the EE and LE stages of incubation resulted in lower body temperatures at hatch. When compared with control chicks that were not thermally challenged as embryos, the embryonic thermal manipulations also resulted in improved heat tolerance. A

Table 6 Body weight and temperature of chicks at hatch and 3 d when exposed to an embryonic or post-hatch thermal challenge

	Embryonic thermal challenge (ED 16–18)					
Variables	Control	3 h	6 h	12 h	24 h	SEM
	Body weight and temperature at hatch					
BW (g)	45.49	45.53	45.74	45.52	45.67	0.27
Tb (C°)[1]	36.86[a]	36.34[b]	36.40[b]	36.43[b]	36.46[b]	0.01
	Body temperature (Tb) in naive and heat challenged chicks (3 d)					
Naive	41.00[a]	40.77[c]	40.85[bc]	40.75[c]	40.96[ab]	0.04
Challenged	42.60[a]	41.89[b]	42.01[b]	41.83[b]	41.93[b]	0.09

[1] Body temperature after hatching. (Adapted from Collin et al., 2005.)
[a–c] Different superscripts represent significant differences, $P < .05$.

subsequent study by Collin et al. (2005) used only one temperature (39.5°C) but increased the length of the exposure treatments (3 h, 6 h, 12 h, 24 h) and only applied them to embryos at ED 16–18. All chicks were subsequently heat stressed at 3 d post-hatch (6 h at 41.0°C). As can be seen in Table 6, there was a reduction in body temperature at hatch in thermally challenged embryos, similar to what was observed by Yahav et al. (2004) and a concomitant improvement in response to heat stress at 3 d post-hatch. Piestun et al. (2008, 2009) concluded that these thermal exposure effects are mediated via the hypothalamic–hypophyseal–thyroid and adrenal axis and circulating thyroid hormones and corticosterone.

As alluded to earlier in this chapter, there is considerable research on the adverse effects of high temperatures during the latter stages of incubation and fewer reports on low temperatures prior to and at transfer (ED 18–19). In one such study, Shinder et al. (2009) exposed embryos at ED 18–19 to 15°C for varying time periods (30, 60 min) and hatched chicks were subsequently given a cold challenge (3 h; 10°C) at 3 and 10 d post-hatch. There were no effects of incubation cold exposure on hatchability. Chicks derived from the 60-min cold exposure treatment during incubation were most capable of withstanding the cold challenge post-hatch as assessed by body temperature regulation. In a subsequent study by Willemsen et al. (2010), embryos were exposed to one of three incubation temperatures (34.6°, 37.6°, 40.6°C) from ED 16 to 18.5 and then returned to the control temperature (37.5°C). There were few differences between the low and control temperatures in yolk-free body weight and selected hormones and metabolites. The main effect of the low temperature was an observed delay in hatch time and this has been observed previously. In a recent study by Morita et al. (2016), broiler embryos exposed to a low incubation temperature during late embryonic stage showed lower plasma T3 and GH levels at hatch and reduced T3 level at 42 d post-hatch.

8 Future trends and conclusion

It is well recognized that during avian incubation, the effects of temperature alone (without consideration for the additive effects of humidity, CO_2 and O_2) can have significant overall effects on many aspects of embryonic physiology with potential carry-over effects on commercial post-hatch performance. These observations are intimately tied to parental behaviours in wild species and have been studied and recorded for over two hundred years. In commercial practice, the measurement of EST has been recommended as a practical and accurate reflection of embryonic heat status during all stages of incubation when compared with simply monitoring incubation temperature. Monitoring EST is a particularly effective tool for controlling the overheating of embryos during the latter stages of incubation when embryonic heat production is greatest. The standardization of incubation temperature profiles within large commercial incubators is less easily regulated in multi-stage (multiple egg ages) versus single-stage (single egg age) incubators and this can influence embryonic growth and development during both the early and late stages of incubation. Exposure to higher than expected temperatures during early incubation can accelerate embryonic growth, increase the length of the hatching period (i.e. hatch window) and predispose the earliest hatchlings to dehydration post-hatch. Elevated temperatures during the latter stages of incubation can likewise affect the overall physiologic homeostasis of hatchlings and an assessment of this via cloacal temperatures should be regularly incorporated into experimental protocols, particularly when post-hatch factors or variables will be determined.

It is now recognized that some physiological systems (i.e. thermoregulatory) are functional during embryonic development and this has been addressed in research as a potential for programming and improving the responses of chicks to post-hatch temperature challenges. There is also evidence that incubation temperature profile may have post-hatch consequences on factors such as footpad dermatitis (Da Costa et al., 2016) and 'gait scores' in older birds (Ipek and Sozcu, 2016), two increasingly important metrics related to overall animal welfare. The latter two areas, however, need considerably more research attention and carefully designed experimental protocols so as to eliminate the potential for post-hatch confounding effects. Similar to thermoregulatory development, selected aspects of the immune system are also functional during the latter stages of incubation, and this has been the basis for the successful implementation of *in ovo* vaccination. This is an area that also warrants more research attention, particularly given the new levels of production transparency and medication limitations that are enveloping commercial poultry production.

9 Where to look for further information

Readers of this chapter who want more information on this topic can refer to *Avian Incubation* by Charles Deeming (2002) and his most recent book co-authored with S. James Reynolds entitled *Nests, Eggs, and Incubation: New Ideas about Avian Incubation*. Individual scientific papers can be sourced via the PubMed and Google Scholar search engines.

10 References

Afsarian, O., M. H. Shahir, A. Akhlaghi, H. Lotfolahian, A. Hoseini and A. Lourens, 2016. Periodical low eggshell temperatures during incubation and post hatch dietary arginine supplementation: effects on performance and cold tolerance acquisition in broilers. *Poult. Sci.* 00:1–8 http://dx.doi.org/10.3382/ps/pew193.

Arjona, A. A., D. M. Denbow and W. D. Weaver, Jr., 1988. Effect of heat stress early in life on mortality of broilers exposed to high environmental temperatures just prior to marketing. *Poult. Sci.* 67:226–31.

Arjona, A. A., D. M. Denbow and W. D. Weaver, Jr., 1990. Neonatally induced thermotolerance: physiological responses. *Comp. Biochem. Physiol.* 95a:393–9.

Asmundson, V. S. 1947. Time held prior to incubation and hatchability of turkey eggs. *Poult. Sci.* 26:305–7.

Bakst, M. R., and S. K. Gupta. 1997. Preincubation storage of turkey eggs: impact on rate of early embryonic development. *Br. Poult. Sci.* 38:374–7.

Barott, H. G. 1937. Effect of temperature, humidity and other factors on hatch of hen's eggs and on energy metabolism of chick embryos. *U.S. D.A. Tech. Bull.* 553:1–46.

Becker, W. A. 1963. Length of preincubation storage of turkey eggs and its effects on body weight. *Poult. Sci.* 42:1356–9.

Becker, W. A., and G. E. Bearse. 1958. Pre-incubation warming and hatchability of chicken eggs. *Poult. Sci.* 37:944–8.

Bowling, J. A., and B. Howarth 1981. The effects of exposing broiler breeder eggs to high temperatures before storage on hatchability and subsequent performance of chicks. *Poult. Sci.* 60:2333–6.

Brookes, M., and K. U. May, 1972. The influence of temperature on bone growth in the chick. *J. Anat.* 111:351–63.

Byerly, T. C. 1938. Effect of different incubation temperatures on mortality of chick embryos. *Poult. Sci.* 17:200–5.

Christensen, V. L., J. L. Grimes, M. W. Wineland and G. S. Davis, 2003 a. Accelerating embryonic growth during incubation following prolonged egg storage. 1. Embryonic livability. *Poult. Sci.* 82:1863–8.

Christensen, V. L., J. L. Grimes, M. J. Wineland and G. S. Davis, 2003 b. Accelerating embryonic growth during incubation following prolonged egg storage 2. Embryonic growth and metabolism. *Poult. Sci.* 82:1869–78.

Collin, A., M. Picard and S. Yahav, 2005. The effect of duration of thermal manipulation during broiler chick embryogenesis on body weight and body temperature of post-hatched chicks. *Anim. Res.* 54:105–11.

Da Costa, M. J., E. O. Oviedo-Rondon, M. J. Wineland, K. Claassen and J. Osborne, 2016. Effects of incubation temperature and trace mineral sources on chicken live performance and footpad skin development. *Poult. Sci.* 95:749–59.

Dareste, C. 1891. Recherches sur la production artificielle des monstrosites ou essais de teratogenie experimentale. Paris (Reinwald), 2nd edn.

Ducuypere, E. 1984. Incubation temperature in relation to post-natal performance in chickens. *Arch. Exp. Vet. Med.* 38:439–49.

Decuypere, E. and H. Michels 1992. Incubation temperature as a management tool. *W. Poult. Sci. J.* 48:28–38.

Drent, R. H. 1970. Functional aspects of incubation in the Herring Gull. *Behaviour. Supplement* The Hearing Gull and Its Egg 17:1–132.

DuRant, S. E., W. A. Hopkins and G. R. Hepp 2011. Embryonic developmental patterns and menergy expenditure are affected by incubation temperature in wood ducks (Aix sponsa). *Physiol. Biochem. Zool.* 84:451–7.

Edwards, C. L. 1901. The physiological zero and the index of development for the egg of the domestic fowl. *Am. J. Physiol.* 6:351–97.

Eyal-Giladi, H., and S. Kochav, 1976. From cleavage to primitive streak formation: a complementarynormal table and a new look at the first stages of the development of the chick. *I. Gen. Morphol. Dev. Biol.* 49:321–37.

Eycleshymer, A. C. 1907. Some observations and experiments on the natural and artificial incubation of the Common Fowl. *Biol. Bull.* 12: 360–75.

Fasenko, G. M., F. E. Robinson, R. T. Hardin and J. L. Wilson, 1992. Variability in preincubation embryonic development in domestic fowl. 2. Effects of duration of egg storage period. *Poult. Sci.* 71:2129–32.

Fasenko, G. M. 2001. Pre-storage incubation of long term stored broiler breeder: 1. Effects on hatchability. *Poult. Sci.* 80:1406–11.

Filho, A. L. C. J. de Oliveira, H. B. de Oliveira, D. B. Campos, R. R. Guerra, F. G. Costa and P. E. Givisiez 2015. High incubation temperature and threonine dietary level improve ileum response against post-hatch Salmonella enteritidis inoculation in broiler chicks. *PLOS ONE* 10(7):e0131474.

French, N. 1994. Effect of incubation temperature on the gross pathology of turkey embryos. *Br. Poult. Sci.* 35:363–71.

Funk, E. M., and H. V. Biellier, 1944. The minimum temperature for embryonic development in the domestic fowl (*Gallus domesticus*). *Poult. Sci.* 23:538–40.

Funk, E. M., J. Forward and H. L. Kempster, 1950. Effect of holding temperature on hatchability of eggs. *Missouri Agr. Expt. Sta. Bull.* 539.

Hammond, C. L., B. H. Simbi and N. C. Stickland, 2007. In ovo temperature manipulation influences embryonic motility and growth of limb tissues in the chick (Gallus gallus). *J. Exp. Biol.* 210: 2667–75.

Harvey, W. 1651. Exercitationes de generatione animalium. London.

Henderson, E. W. 1930. The influence of temperature and breeding upon the rate of growth of chick embryos. *Missouri Agr. Exp. Station Res. Bull.* 149:1–47.

Henderson, E. W. 1939. Influence of temperature on length of artificial incubation period of gallus domesticus. Proceedings of VII World's Poultry Congress, pp. 180–4.

Henderson, E. W., and S. Brody 1927. The effect of temperature on the percentage – rate of growth of chick embryos. *Missouri Agr. Exp. Station Res. Bull.* 99.

Huggins, R. A. 1941. Egg temperatures of wild birds under natural conditions. *Ecology* 22:148–57.

Hulet, R., G. Gladys, D. Hill, R. Meijerhof and T. El-Shiekh, 2007. Influence of egg shell embryonic temperature and broiler breeder flock age on posthatch growth performance and carcass characteristics. *Poult. Sci.* 86:408–12.

Ipek, A., and A. Sozcu 2016. The effects of eggshell temperature fluctuations during incubation on welfare status and gait score of broilers. *Poult. Sci.* 95:1296–1303.

Janke, O., B. Tzschentke and M. Boerjan 2004. Comparative investigations of heat production and body temperature in embryos of modern chicken breeds. *Avian Poult. Biol. Rev.* 15:191–6.

Jeurissen, S. H. M., L. Vervelde and E. M. Janse. 1994. Structure and function of lymphoid tissues of the chicken. *Poult. Sci. Rev.* 5:183–207.

Joseph, N. S., A. Lourens and E. T. Moran, Jr. 2006. The effects of suboptimal eggshell temperature during incubation on broiler chick quality, live performance, and further processing yield. *Poult. Sci.* 85:932–8.

Kaestner, S. 1895. über künstliche KÄlteruhe von Hühnereiern im Verlauf der mBebrütung. *Arch. f. m Anat.*

Kaltofen, R. S. 1961. The effect of fertility rate and frequency of turning on embryonic mortality during artificial incubation. *Anim. Breed. Abstr.* 29:106–7.

Kessler, F. 1960. Egg temperatures of the Ring-necked Pheasant obtained with a self-recording potentiometer. -Auk 77: 330–6.

Kosin, I. L. 1950. A relationship between the length of storage and incubation periods in Broad Breasted Bronze eggs. *Poult. Sci.* 29:620–1.

Kosin, I. L. 1956. Studies on pre-incubation warming of chicken and turkey eggs. *Poult. Sci.* 35: 1384–92.

Krischek, C., K. Janisch, W. Naraballobh, R. Brunner, K. Wimmers and M. Wicke 2016. Alteration of the incubation temperature between embryonic day 7 and 13 influences weights and mitochondrial respiratory and enzyme activities in breast and leg muscles of broiler embryos. *Mol. Reprod. Dev.* 83:71–8.

Kroesen, K. A., and M. S. Lilburn 2006. Effect of breeder hen age and incubation temperature on embryonic temperature and development of White Pekin ducklings. *Poult. Sci.* 85:138.

Landauer, W. 1967. The hatchability of chicken eggs as influenced by environment and heredity. Storrs Agricultural Experiment Station. Monograph 1., Storrs, Ct.

Leksrisompong, N., H. Romero-Sanchez, P. W. Plumstead, K. E. Brannan and J. Brake 2007. Broiler incubation. 1. Effect of elevated temperature during late incubation on body weight and organs of chicks. *Poult. Sci.* 86:2685–91.

Liu, H., J. Liu, X. Yan, Q. Li, Y. Zhao, Y. Wang, R. Zhang, G. Wang, H. Wang, X. Li, C. Yang, L. Li, C. Han and J. Wang. 2015. Impact of thermal stress during incubation on gene expression in embryonic muscle of Peking ducks (Anasplatyrhynchos domestica). *J. Therm. Biol.* 53:80–9.

Lourens, A. 2001. The importance of air velocity in incubation. *World Poult.* 17:29–30.

Lourens, A., H. van den Brand, R. Meijerhof and B. Kemp, 2005. Effect of eggshell temperature during incubation on embryo development, hatchability, and posthatch development. *Poult. Sci.* 84:914–20.

Lourens, A. 2006. Heating eggs before storage increases hatchability. *World Poult.* 22:22–3.

Lundy, H. A. 1969. A review of the effects of temperature, humidity, turning and gaseous environment in the incubator on the hatchability of the hen's egg. In T. C. Carter and B. M. Freeman (eds), *The Fertility and Hatchability of the Hen's Egg*, Oliver and Boyd, Edinburgh, pp. 143–76.

Martin, J. H., and W. M. Insko 1935. The effect of temperature and position in the incubation of turkey eggs. *Poult Sci.* 14:152–5.

Mather, C. M., and K. F. Laughlin. 1976. Storage of hatching eggs: the effect on total incubation period. *Br. Poult. Sci.* 17:471–9.

Meijerhof, R., and G. van Beek 1993. Mathmetical modelling of temperature and moisture loss of hatching eggs. *J. Theor. Biol.* 165:27–41.

Molenaar, R., R. Hulet, R. Meijerhof, C. M. Maatjens, B. Kemp and H. van den Brand, 2011. High eggshell temperatures during incubation decrease growth performance and increase the incidence of ascites in broiler chickens. *Poult. Sci.* 90:624–32.

Morita, V. S., V. R. Almeida, J. B. Matos, T. I. Vicentini, H. van den Brand and I. C. Boleli, 2016. Incubation temperature alters thermal preference and response to heat stress of broiler chickens along the rearing phase. *Poult. Sci.* (2016): pew071.

Nangsuay, A., R. Meijerhof, I. van den Anker, M. J. W. Heetkamp, V. De Souza Morita, B. Kemp and H. van den Brand 2016. Effects of breeder age, broiler strain, and eggshell temperature on development and physiological status of embryos and hatchlings. *Poult. Sci.* 00:1–14, http://dx.doi.org/10.3382/ps/pew080.

Needham, J. 1931. *Chemical Embryology*, 3 Vols. New York: The Macmillan Company.

Nichelmann, M., and B. Tzschentke 1997. Ontogeny of thermoregulation during the prenatal period in birds. *Ann. N.Y Acad. Sci.* 813:78–86.

Nichelmann, M., and B. Tzschentke 2002. Ontogeny of thermoregulation in precocial birds. *Comp. Biochem. Physiol. Pt. A.* 131:751–63.

Oviedo-Rondon, E., J. Small, M. J. Wineland, V. L. Christensen, P. S. Mozdziack, M. D. Koci, S. V. L. Funderburk, D. T. Ort and K. M. Mann 2008. Broiler embryo bone development is influenced by incubator temperature, oxygen concentration and eggshell conductance at the plateau stage in oxygen consumption. *Br. Poult. Sci.* 49:666–76.

Oviedo-Rondon, E. O., M. J. Wineland, J. Small, H. Cutchin, A. McElroy, A. Barri and S. Martin 2009. Effect of incubation temperatures and chick transportation conditions on bone development and leg health. *J. Appl. Poult. Res.* 18:671–8.

Oznurlu, Y., I. Celik, T. Telatar and E. Sur 2010., Histochemical and histological evaluations of the effects of high incubation temperature on embryonic development of thymus and bursa of Fabricius in broiler chickens. *Br. Poult. Sci.* 51:43–51.

Philips, A. G. 1917. Temperature experiments in incubation. *Poult. Sci.* 9:65–6.

Philips, A. G., and F. D. Brooks 1923. Temperature experiments during the incubation of hens eggs. *Purdue University Agr. Exp. St. Bull.* 275:1–16.

Piestun, Y., D. Shinder, M. Ruzal, O. Halevy, J. Brake and S. Yahav 2008. Thermal manipulations during broiler embryogenesis. Effect on the acquisition of thermotolerance. *Poult. Sci.* 87:1516–25.

Piestun, Y., O. Halevy and S. Yahav 2009. Thermal manipulations of broiler embryos-The effect on thermoregulation and development during embryogenesis. *Poult. Sci.* 88:2677–88.

Prevost, J. L., and J. A. Dumas. *Examen du sang et de son action dans les divers phénomènes de la vie.* 1821.

Randall, W. C. 1943. Factors influencing the temperature regulation of birds. *Am. J. Physiol.* 139:56–63.

Reaumur, M. 1749. Pratique l'art de faire eclare d'elever toute saison des oiseaux deomestique de tautes especes soit par le moyen de la chaleur des fumier, soit par le moyen de celle du feu ordinaire. Paris.

Reijrink, I. A. M. R. Meijerhof, B. Kemp and H. van den Brand 2008.The chicken embryo and its micro environment during egg storage and early incubation. *W. Poult. Sci. J.* 64:581–98.

Reijrink, I. A. M., D. Berghmans, R. Meijerhof, B. Kemp and H. van den Brand 2010. Influence of egg age time and preincubation warming profile on embryonic development, hatchability, and chick quality. *Poult. Sci.* 89:1225–38.

Reyns, G. E., K. Venken, G. Morreale de Escobar, E. R. Kuhn and V. M. Darras 2003. Dynamics and regulation of intracellular thyroid hormone concentrations in embryonic chicken liver, kidney, brain and blood. *Gen. Comp. Physiol.* 134:80–7.

Romanoff, A. 1935. Influence of incubation temperature on the hatchability of eggs, post-natal growth and survival of turkeys. *J. Agri. Sci.* 25:318–25.

Romanoff, A. 1936. Effects of different tempertures in the incubator on the prenatal and postnatal development of the chick. *Poult. Sci.* 15:311–15.
Romanoff, A. L. 1939. Effect of temperature shock on development of chick embryo. In *Proceedings of the VIM World's Poultry Congress, Cleveland*, pp. 184–6.
Romanoff, A., and H. A. Faber 1933. Effect of temperature on the growth, fat and calcium metabolism and mortality of the chick embryo during the latter part of incubation. *J. Cell Comp. Physiol.* 2:457–66.
Rudrappa, S. G., and B. D. Humphrey 2007. Energy metabolism in developing chicken lymphocytes is altered during the embryonic to post hatch transition. *J. Nutr.* 137: 427–32.
Santin, E., A. Maiorka, W. J. C. Polveiro, A. C. Paulillo, A. C. Laurentiz, S. A. Borges and A. V. Fischer da Silva 2003. Effect of environmental temperature on immune response of broilers. *J. Appl. Poult. Res.* 12:247–50.
Scott, H. M. 1933. The effect of age and holding temperature on hatchability of turkey and chicken eggs. *Poult. Sci.* 12:49–54.
Sharma, J. M. 1999. Introduction to poultry vaccines and immunity. *Adv. Vet. Med.* 41:481–94.
Shim, M. Y., and G. M. Pesti 2011. Effects of incubation temperature on the bone development of broilers. *Poult. Sci.* 90:1867–77.
Shinder, D., M. Rusal, M. Giloh and S. Yahav 2009. Effect of repetitive acute cold exposures during the last phase of broiler embryogenesis on cold resistance through the life span. *Poult. Sci.* 88:636–46.
Sotherland, P. R., J. R. Spotila and C. V. Paganelli 1987. Avian eggs: barriers to the exchange of heat and mass. *J. Exp. Zool.* 1 (Suppl. 1):81–6.
Tazawa, H., and H. Rahn 1987. Temperature and metabolism of chick embryos and hatchlings after prolonged cooling. *J. Exp. Zool.* 1:105–9.
Thommes, R. C. 1987. Ontogenesis of thyroid function and regulation in the developing chick embryo. *J. Exp. Zool. Supp.* 1:273–9.
Tona, K., V. Bruggeman, O. Onagbesan, F. Bamelis, M. Gbeassor, K. Mertens and E. Decuypere 2005. Day-old chick quality: relationship to hatching egg quality, adequate incubation practice and prediction of broiler performance. *Avian Poult. Biol. Rev.* 16:109–19.
Tzschentke, B., and M. Rumpf 2011. Embryonic development of endothermy. *Resp. Phys. Neuro.* 178:97–107.
Van Brecht, A., J. M. Aerts, P. Degraeve and D. Berckmans 2003. Quantification and control of the spatio-temporal gradients of air speed and air temperature in an incubator. *Poult. Sci.* 1677–87.
Van Brecht, A., H. Hens, J. L. Lemaire, J. M. Aerts, P. Degraeve and D. Berckmans 2005. Quantification of the heat exchange of chicken eggs. *Poult. Sci.* 84:353–61.
Webb, D. R. 1987. Thermal tolerance of avian embryos: a review. *The Condor* 89:874–98.
Whittow, G. C., and H. Tazawa 1991. The early development in thermoregulation in birds. *Physiol. Zool.* 64:1371–90.
Willemsen, B. Kamers, F. Dahlke, H. Han, Z. Song, Z. Ansari Pirsaraei, K. Tona, E. Decuypere and N. Everaert, 1010. High and low temperature manipulation during late incubation: effects on embryonic development, the hatching process, and metabolism in broilers. *Poult. Sci.* 89:2678–90.
Wilson, H. R. 1991. Physiological requirements of the developing embryo: temperature and turning. In S. G. Tullet (ed.)*Avian Incubation*, London: Butterworths, pp. 145–56.
Yahav, S., and S. Hurwitz 1996. Induction thermotolerance in male broiler chickens by temperature conditioning at an early age. *Poult. Sci.* 75:402–6.
Yahav, S., A. Collin, D. Shinder and M. Picard 2004. Thermal manipulations during broiler chick embryogenesis: effects of timing and temperature. *Poult. Sci.* 83:1959–63.
Yan, X. P., H. H. Liu, J. Y. Liu, R. P. Zhang, G. S. Wang, Q. Q. Li, D. M. Wang, L. Li and J. W. Wang 2016. Evidence in duck for supporting alteration of incubation temperature may have influence on methylation of genomic DNA. *Poult. Sci.* 94:2537–45.

Chapter 13

The contribution of environmental enrichment to sustainable poultry production

Inma Estevez, Neiker-Tecnalia and Ikerbasque (The Basque Foundation for Science), Spain; and Ruth C. Newberry, Norwegian University of Life Sciences, Norway

1 Introduction

Extraordinary progress in poultry production efficiency has been achieved through advances in genetics, nutrition and animal health. New technology has permitted major improvements in control of the housing environment. Despite this success, the poultry industry faces new and important challenges that are no longer based solely on improving productive efficiency. Contemporary poultry production must still maintain competiveness but must also consider ethical and social responsibility factors that influence consumer acceptance such as food safety, environmental impact and animal welfare. These are key aspects to ensure the long-term sustainability of the poultry industry.

http://dx.doi.org/10.19103/AS.2016.0011.28

Any sustainable system must respond to consumer demands for farming methods that are based on the respect and ethical treatment of the production animals. Therefore, consideration of animal welfare is an inherent aspect of any sustainable production system. No less importantly, improving animal welfare can also contribute to the reduction of system inefficiencies by promoting healthier and more efficient animals that are better prepared to confront stress challenges and disease threats. These animals should also require less medication. Thus, better welfare can result in more efficient (higher production outputs with a similar use of resources or inputs) and socially responsible methods that are in line with consumers' ethical considerations and demands.

There have been notable advances in our knowledge about the behaviour and welfare of meat poultry, especially chickens (e.g. Dawkins, 2004; Estevez, 2007). A key advance has been in the area of environmental enrichment, a concept referring to additions and modifications to the production environment that facilitate the biological adaptation of the animals to their environment and improve their welfare (Newberry, 1995). A variety of welfare benefits can be targeted, such as better health, promotion of positive emotional states and increased stress resilience. Various studies have remarked on the positive effects of environmental enrichment strategies to control unwanted behaviours, promote adaptive behavioural responses, improve welfare and maximize productive efficiency in poultry (e.g. Cornetto and Estevez, 2001a,b; Leone and Estevez, 2008; Estevez, 2009; Ventura et al., 2010, 2012).

Despite these advances, the implementation of environmental enrichment at a commercial level has been limited. This response is driven, in part, by the assumption that addressing behavioural needs and animal welfare leads to higher production costs and less competitiveness. This may be the case in situations involving major structural modifications to the housing, or the banning of a production system as has happened in European Union countries due to legislative requirements (Directive 1999/74/EC). However, environmental enrichment strategies can be cost-effective and beneficial to both the birds and the producers. For example, Leone and Estevez (2008) demonstrated that a $400 investment to outfit a standard commercial broiler breeder house with short vertical 'cover panels' made with plastic netting resulted in a $6304 return on investment due to an increment of 4.5 more chicks produced per female in a single production flock. This is just one illustration of the potential of environmental enrichment.

The use of different forms of environmental enrichment has usually occurred in response to the need to reduce a health problem that has a negative economic impact. For example, a high incidence of leg problems in meat poultry is a serious welfare concern (Marchewcka et al., 2013, 2015) that causes major economic losses. The reduction in production efficiency not only relates to the mortality caused by leg problems, often in birds close to market weight, but also to increased downgrades and condemnations associated with lameness and immobility as demonstrated in turkeys (Marchewka et al., 2015). It is speculated that the low activity observed in broiler chickens (Newberry et al., 1985; Weeks et al., 2000) may influence the incidence of leg problems. Newberry (1999) showed that environmental enrichments that stimulate exploration increase the birds' motivation to move and use the available space. The provision of key elements in the environment that motivate the birds to move may help to reduce the incidence of leg problems, thus achieving more sustainable production in the longer term.

Feather pecking and cannibalism are unwanted behaviours that have a catastrophic impact on the welfare and performance of poultry. Limited opportunities to explore and forage early in life may lead to feather pecking and cannibalism, although it is recognized that these behaviours have a multifactorial origin (Newberry, 2004; Drake et al., 2010; Nicol et al., 2013; Estevez, 2015). They are more typically found in laying hens, especially in alternative egg production (Green et al., 2000), and are uncommon in broilers. However, they occur occasionally in slow-growing meat chicken strains (Nielsen et al., 2003), while turkeys (Marchewka et al., 2013), ducks and broiler breeders are often beak trimmed to prevent them from developing. Though beak trimming is administered to prevent major welfare problems, this practice raises ethical concerns to the point that it is banned in some European countries, and others envision it being banned for all poultry species in the near future. It is important to be able to provide the means to reduce the incidence of feather pecking and cannibalism through alternative strategies, such as environmental enrichment, because banning of beak trimming or unacceptance of this practice by consumers is likely to extend globally.

Environmental enrichment strategies incorporate elements into commercial housing that are key to promoting normal behavioural development and facilitating adaptation to the physical and social environment. By considering these aspects, it should be possible to minimize problems that are costly in terms of animal welfare but that also cause important losses in productive efficiency (e.g. recurrent panic attacks leading to crowding and suffocation at specific locations that may result in high mortality). For birds such as broiler breeders that are reared in a bi-dimensional environment and later moved to a three-dimensional production environment, facilitating their transition to the latter is particularly relevant (Estevez, 2009). There is a diversity of options to incorporate simple changes in the housing environment to help birds cope better with the physical and social environment. These approaches can spare birds unnecessary stress while also rewarding farmers with higher productive efficiency.

In this chapter, we review environmental enrichment strategies that are relevant to commercial meat poultry production, their benefits in terms of behaviour and welfare of the birds and their potential for contributing to more efficient and sustainable production models. Potential risks or problems that may arise when environmental enrichment is not appropriately implemented are also discussed, finishing with discussion of current gaps in knowledge and concluding remarks.

2 Structural complexity: an introduction

Natural environments offer a wide range of structures used by animals for different purposes. Environmental features such as cover provided by bushes, rocks and uneven terrain offer protection from adverse weather conditions, predators and aggressive conspecifics (Elton, 1939), thereby reducing the need for vigilance (Lazarus and Symonds, 1992). Elevated structures and tree branches are used by the domestic fowl under natural and semi-natural conditions to perch during the day and to roost at night (Wood-Gush et al., 1978; Blokhuis, 1984), suggesting that perching provides an increased perception of safety (Newberry et al., 2001). Environmental complexity also serves to reduce visual contact between conspecifics (Estep and Baker, 1991) which, in chickens, helps to prevent, or terminate, aggressive interactions (Cornetto et al., 2002) and coercive matings (Pizzari, 2016).

In contrast, commercial meat poultry are typically raised in large open houses with little option for moving to protected areas if feeling threatened, to rest undisturbed, or to ascend to an elevated structure to roost at night. We can speculate that because of the unlikelihood of predation occurring in the indoor production environment, there is no need to offer structurally complex indoor environments. However, the need for protection against predation has shaped behaviour in the ancestors of our current poultry species across thousands of years of evolutionary pressure. Although genetic selection has indeed altered the behaviour of poultry, these changes appear to relate more to quantitative than to qualitative aspects (Newberry, 2017a). Behavioural strategies that were critical for survival under natural conditions are especially likely to persist, even in highly selected genetic production lines (Pizzari, 2016). Behaviour such as perching (Newberry et al., 2001) and the use of cover (Newberry and Shackleton, 1997; Cornetto and Estevez, 2001a) to avoid predation, and variation in competitive behaviour for access to basic resources according to the resource distribution (e.g. food, Leone and Estevez, 2007; mating opportunities, Bilcik et al., 2005) are easily observed in commercial poultry. These behavioural responses are not problematic in themselves as long as birds have adequate environmental features to allow them to respond appropriately, thus minimizing the risk of problematic behaviours and costly stress responses (Jones, 1996).

The addition of different forms of enrichment that provide structural complexity (e.g. perches, ramps and cover panels) in commercial meat poultry houses can be cost-effective to resolve or minimize behaviour-related management and welfare problems. These structures create microenvironments within the production house facilitating the birds' adaptive behavioural responses. If birds are able to modulate their responses, or avoid overexposure to social contact, stress levels should be minimal, permitting birds to maximize their investment of resources in growth and reproduction.

3 Structural complexity: cover panels

3.1 Promoting uniform distribution of birds throughout the house

A mismatch between the birds' behaviour and the characteristics of the production environment may be the source of costly mortality due to panic reactions, even in apparently apathetic broiler flocks. The natural flock reaction when a danger is perceived is to seek cover. Thus, birds will run to the nearest (and only) cover, usually the end wall, where they cluster. In severe cases of panic, several thousand birds may die as a consequence of piling up against walls. Short vertical 'cover panels' distributed along the house can provide cover that shelters birds from the perceived danger (Newberry and Shackleton, 1997; Cornetto and Estevez, 2001a,b), stopping birds in smaller groups along the house before a large number reach the end of the house where they pile on top of each other, risking scratches from each other's claws, overheating and suffocation. Vertical cover panels can be constructed with a frame of wood, PVC piping or other materials, supporting a screen of plastic netting or similar material that partially obstructs the vision of the birds without impeding ventilation (Fig. 1).

A homogeneous bird distribution along the house is desirable to facilitate uniform use of resources and a homogeneous airflow among birds. However, birds tend to be found at relatively high density along the walls, while central areas may be relatively underused (Newberry and Hall, 1990). Although the effects of uneven flock distribution are difficult to quantify, they may have an important impact on sustainability. Uneven flock distribution may result in compromised welfare and lower productive efficiency and may even affect food quality (Cornetto and Estevez, 2001a). By limiting airflow around birds, higher than desirable bird density at specific locations increases heat stress in hot weather.

Figure 1 Broiler chickens resting by vertical mesh cover panels providing partial visual cover in the central area of the house. (a) Photo: I. Estevez, (b) Photo: R. C. Newberry.

Litter condition may deteriorate faster in highly used areas adjacent to walls, increasing the risk of footpad dermatitis (De Jong et al., 2014), one of the most clear indicators of compromised welfare in broiler chickens (Welfare Quality, 2009). As birds move in and out of highly populated areas, they may bump into or step on other birds rather than manoeuvring around them (Estevez, 1994), causing disturbance of resting birds (Cornetto et al., 2002). This behaviour not only disrupts resting periods that are important for physiological processes such as sleep, tissue restoration and growth (Blokhuis, 1984), but may also increase the risk of skin scratches and bruises to the birds that are disturbed. Under experimental conditions, the use of vertical cover panels to distribute birds more evenly has been found to reduce the frequency of disturbances and promote resting and preening within central areas of the house (Newberry and Shackleton, 1997; Cornetto and Estevez, 2001a,b; Cornetto et al., 2002; Fig. 2). Enrichment of the indoor housing of commercial slow-growing broilers with cover panels has also been effective in promoting a more even bird distribution, although no further benefits were detected in regard to behavioural changes or use of an outdoor area (Rodriguez-Aurrekoetxea et al., 2014; Rodriguez-Aurrekoetxea and Estevez, 2015).

Although cover panels are not typically used in commercial conditions, it is not uncommon to find the use of straw bales (Fig. 3) or wood shavings bales in chicken houses, especially in European countries. This simple enrichment has a similar function to cover panels in providing protection to the birds while also offering opportunities for the more audacious birds to perch and forage. The bales disintegrate into the litter at

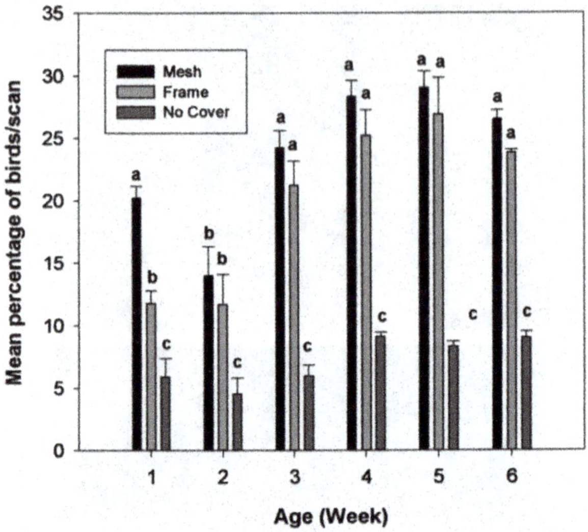

Figure 2 Mean percentage of birds using the central region in enclosures enriched with vertical mesh cover panels (black) or panel frames without mesh (grey) as compared to control (no cover) enclosures (dark grey). Treatments with different letters denote statistical differences ($P < 0.05$). From Cornetto and Estevez (2001a).

Figure 3 Broilers using straw bales for roosting, foraging and as cover. Photo: R. C. Newberry.

a rate that depends on whether or not they are wrapped in plastic that prolongs their integrity, and on the number of bales provided. The bales may help to maintain better litter condition, by adding dry material to the existing litter. Apart from collecting the plastic wrap, they do not require any additional cleaning or maintenance effort as new bales are added with the new flocks. However, in one of the few studies on the effects of straw bales under commercial conditions, Bailie et al. (2013) reported only minor increases in locomotion, standing and latency to lie down, which were interpreted as indicating better leg condition. The congregation of birds around the bales suggests that the bales were acting as a form of cover. Despite positive indications, benefits may be limited by the sparsity of bales typically provided relative to the dimensions of chicken houses and the large number of birds housed (Bailie et al., 2013), although Bailie and O'Connell (2014) did not detect benefits from adding more bales.

3.2 Benefits of cover panels in breeding flocks

Substantial female mortality occurs in broiler breeder flocks as a result of male harassment and forced matings (Fig. 4), especially in situations of high male–male competition for access to females (Estevez, 1999; Pizzari, 2016). Placing cover panels along the litter area successfully controlled ongoing female mortality in a problematic commercial flock by attracting a larger number of females down from the slatted area to the litter area (Fig. 5), thus reducing male–male competition for females (Estevez, 1999). The presence of cover panels (or structures such as straw or hay bales) may also act to reduce the visibility of females in the litter area, thereby diminishing mating pressure.

It might be speculated that, if mating pressure declines, the presence of structural complexity in the mating area would have a negative impact on fertility, which is, in itself, a relevant performance issue in broiler breeders (McGary et al., 2002). Follow-up studies conducted by Leone and Estevez (2008) in commercial breeder flocks showed that, on the contrary, females in houses enriched with cover panels laid 2% more eggs than females in unenriched houses, and both flock fertility and hatchability were improved in enriched flocks. Enriched and control flocks started to diverge at around 35 weeks of age, when

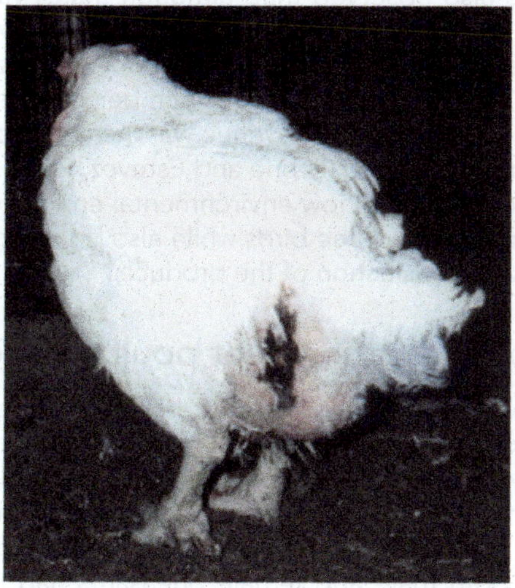

Figure 4 Female broiler breeder with injuries due to forced matings. Photo: I. Estevez.

Figure 5 (a) Broiler breeder males in the litter area, with a low number of females; (b) the same litter area 24 h after introduction of cover panels. Note the difference in the number of females. From Estevez (1999).

flock fertility started declining. The improved reproductive performance may be due to a combination of factors such as reduced stress, reduced male–male competition and lower mating interference. A higher proportion of successfully completed matings is one of the factors speculated to promote fertility in broiler breeder flocks (Bilcik and Estevez, 2005; Bilcik et al., 2005). It was also observed that, although males varied in the size of their home ranges, they generally used a higher proportion of the available space when the house contained cover panels (Leone and Estevez, 2008). Their greater roaming behaviour increased the possibility of mating with a larger number of different females along the house. As indicated above, the economic revenue was increased by over $6000 in a single flock with an initial investment of $400 (Leone and Estevez, 2008). This is probably one of the best examples for understanding how environmental enrichment techniques can help to improve the health and welfare of the birds while also boosting economic returns, and probably the degree of work satisfaction of the producer.

3.3 Findings on cover in other meat poultry species

Although there have been few studies on environmental enrichment in meat poultry species other than chickens, there are a couple of examples of the use of structures similar to cover panels. Sherwin et al. (1999) observed a decline in injurious pecking and a possible improvement in musculoskeletal function in male turkeys reared under experimental conditions when provided with structures offering cover. Deeming et al. (2011a,b) observed that both male and female pheasants had better feather condition at the end of the breeding season in pens with visual barriers as compared to controls. Deeming et al. (2011b) also showed that, although barriers did not affect the frequency of courtship and mating, they did reduce the frequency of aggressive pecks and chases in pheasants. Fertility was also significantly higher and persisted for longer in the pens with visual barriers as compared to the controls, similar to the findings of Leone and Estevez (2008) in broiler breeders.

The consistency of results across meat poultry species suggests that cover panels serving as partial visual barriers offer a simple and cost-effective type of enrichment that facilitates the birds' adaptation to commercial facilities according to their behavioural biology. The addition of cover panels results in fewer behavioural problems such as disturbances, aggressive interactions and sexual harassment while improving fertility and hatchability in breeding flocks. Other effects such as protection against fatalities resulting from panic reactions are difficult to test but likely to be beneficial. Thus, increasing the level of visual cover could have a significant impact in improving welfare and bird performance, thereby contributing to long-term sustainability.

4 Structural complexity: perches, barriers and ramps

Perches are provided to satisfy the natural motivation of domestic fowl, turkeys and pheasants to roost in safe, elevated locations (ancestrally, in trees) throughout the night and for shorter periods during the day (Collias and Collias, 1967; Wood-Gush et al., 1978; Blokhuis, 1984). Perches offer birds opportunities to jump on and off them, which is known to improve bone strength and reduce the risk of osteoporosis in laying hens (e.g. Tauson, 1984; Appleby et al., 1992; Leyendecker et al., 2005). From a functional standpoint, accessing the third dimension permits laying hens to avoid cannibalism, feather pecking (Huber-Eicher and Audige, 1999) and aggressive individuals (Cordiner and Savory, 2001). These benefits are likely to be similar in meat poultry in which feather pecking and aggressive interactions occur such as in turkeys (Marchewka et al., 2013). The use of perches appears to increase perceived safety (Newberry et al., 2001). When used from an early age, the experience manoeuvring in three dimensions improves the birds' spatial skills (Gunnarsson et al., 2000). It also increases the use of nest boxes, reducing the incidence of floor eggs in laying hens (Gunnarsson et al., 1999) and broiler breeder flocks (Brake, 1987; Appleby et al., 1988).

Because of the demonstrated benefits for the health and welfare of laying hens, current European legislation (Directive 1999/74/EC) requires that a minimum of 15 cm of perching space per hen is provided in commercial egg-producing systems for adult laying hens within EU countries. This legal requirement contrasts with the situation for commercial meat poultry where perches are not usually provided. This difference may relate to the low use of perches found in some studies on perching behaviour of young broilers (Fiscus Le Van et al., 2000; Martrenchar et al., 2000; Pettit-Riley and Estevez, 2001; Kaukonen et al., 2016). Additionally, broilers are often slaughtered prior to the age when jungle fowl would start to be brooded in trees under natural conditions, suggesting that they may not have a strong motivation to roost on perches. They also have relatively ample space to move around on the floor until the last week of the rearing period. These conditions may have contributed to the perception that perches are not necessary for broilers. Nevertheless, crossing obstacles, exploring in low undergrowth and sitting on low structures are components of their natural behaviour, and studies with different perch designs indicate that enrichment of broiler houses with perches may, in fact, be advantageous from health, welfare and performance points of view.

Several studies concerning potential benefits of perches for meat poultry have been conducted in broilers under simulated commercial management conditions. Perch use in these studies has varied depending on factors such as perch design (length, height, material, shape, disposition and location), bird age and environmental factors (Koene et al., 1999; Martrenchar et al., 2000; Su et al., 2000). For example, less than 1–7% of the flock perched on angled and 8.5-cm-high horizontal perches (Fiscus Le Van et al., 2000; Pettit-Riley and Estevez, 2001), and from 7 to 27% on perches set 15–30 cm above the floor (Hughes and Elson, 1977; Bizeray et al., 2000a,b; Ventura et al., 2012; Kiyma et al., 2016), although Norring et al. (2016) noted very low use of 10- and 30-cm-high perches. Peak perch use usually occurred at around 3–5 weeks of age, declining thereafter. However, Estevez et al. (2002) observed that the use of 15-cm-high perches increased from 4 to 6 weeks, while the use of 7.5-cm-high perches declined during this period. Their study was conducted in midsummer to test the effectiveness of cool perches (i.e. perches constructed from pipes refrigerated by cooled water circulating through them) to reduce the effects of heat stress. It is likely that the relatively high perch use was due to increased cooling when perched on the higher perches and, especially, on the cool perches.

Access to elevated platforms is an alternative to perches that may also act as a form of cover. Although the information available is limited, Norring et al. (2016) noted a higher use of platforms as compared to perches.

4.1 Benefits of using the third dimension

Even though the volume of research on perch use in broilers is not as ample as for layers, improvements in health and welfare as well as product quality and production efficiency have been identified. Bizeray et al. (2002a,b) used perches (100 cm long by 4 cm wide by 15 cm high) serving as barriers between feeders and drinkers to stimulate locomotion in broilers, with the aim of reducing the incidence of leg problems (Fig. 6). Although access to barrier perches did not have a detectable effect on gait score (assessed according to Kestin et al., 1992), bone quality or amount of walking, birds reared with barrier perches had wider tibias. This effect was probably related to perching behaviour as birds spent nearly 10% of the photoperiod perching at the expense of reduced time spent lying on

Figure 6 Four-week-old broilers using barrier perches (from Bizeray et al., 2002a, b).

the litter. Considering that broilers were observed to spend 60–80% of their time resting during the photoperiod (Cornetto and Estevez, 2001b), a 10% reduction in the time spent in contact with litter by resting on perches may be beneficial. In fact, follow-up studies by Ventura et al. (2010, 2012) and Kiyma et al. (2016) demonstrated that birds provided with barrier perches had a lower incidence of foot pad dermatitis. Furthermore, in commercial broilers raised with elevated platforms, Kaukonen et al. (2016) reported a reduced incidence and severity of tibial dyschondroplasia, and a lower mean percentage of birds with gait score 3 (moderate gait abnormality), though no differences in higher gait scores indicating severe leg problems.

Access to the third dimension provides some additional benefits. Ventura et al. (2012) found a lower frequency of resting disturbances (i.e. interruption of resting due to disruption caused by the action of another bird; Cornetto et al., 2002), especially at high rearing densities (Fig. 7). Reduced disturbances could contribute to bird health and production efficiency by minimizing the risk of skin scratches, thereby lowering the incidence of downgraded carcasses. Ventura et al. (2012) also observed higher use of the central areas in pens with barrier perches, indicating that they contributed to a more homogeneous bird distribution and therefore to a more uniform use of resources. Similarly to cover panels, barrier perches may also be helpful in controlling panic reactions.

One of the arguments against the implementation of perches in commercial conditions is the potentially added cost of higher feed consumption due to increased activity. However, there was no evidence of increased feed conversion ratios or reduced body weights in studies where broilers had access to perches (Pettit-Riley and Estevez, 2001; Bizeray et al., 2002b; Ventura et al., 2010; Kiyma et al., 2016), even when perching reached as high as 25% of the time budget (Ventura et al., 2012). Any increase in energy expended climbing on and off perches may be compensated by a reduction in running and aggressive interactions (Ventura et al., 2012). Availability of perches may be particularly beneficial at high densities when walking is obstructed, and lying on the floor is increasingly disrupted

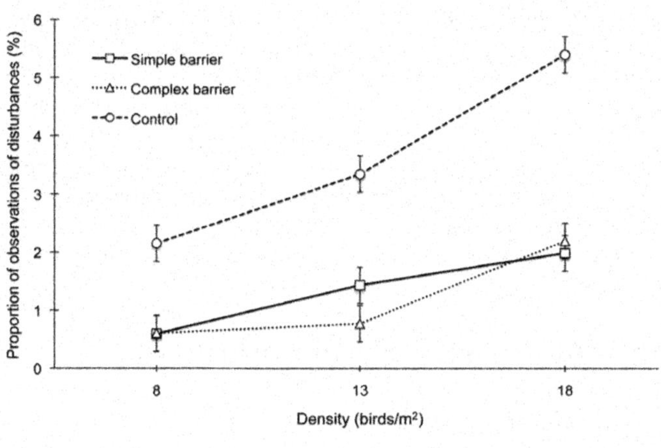

Figure 7 The effect of access to single (simple, continuous line) and double (complex, dotted line) barrier perches or no access to perches (control, dashed line) on the proportion of observations with disturbances (i.e. resting birds getting up due to jostling by active birds) at three bird densities. From Ventura et al. (2012).

(Buijs et al., 2011). Perches are not only effective at reducing the frequency of disturbances contributing to better quality rest (Ventura et al., 2012), but could reduce skin damage related to deteriorating litter condition at high densities (Estevez, 2007). A further potential benefit of perches may be reduced injuries during catching and shackling, considering that perch access resulted in less wing flapping by broilers when caught and suspended by the legs (Newberry and Blair, 1993).

The scientific literature for other meat poultry species in regard to the potential benefits of perches is limited. In turkeys, Martrenchar et al. (2001) reported a peak in perching in Week 5, with 10–13% of birds perching, declining to 0 by Week 10. They also found reduced injurious pecking in this treatment as compared to the control. However, besides perches, the pens were also enriched with bright pecking objects and straw, making it unclear whether the effects were due to the availability of perches, the diversion of attention towards the other enrichment elements or a combination of both. Bergmann et al. (2009) observed more use of perches and platforms by free-range turkeys during the night than during the day. In pheasants, those reared in pens with perches were more likely to use elevated roosting sites after release on hunting estates, and had lower juvenile mortality due to natural causes, than controls reared without perches (Whiteside et al., 2016).

4.2 Perches and the thermal environment

Any opportunities for the birds to get off the litter for extended periods by using perches or platforms gives birds more options for behavioural thermoregulation. Studies on laying hens and broiler breeder females (Muiruri and Harrison, 1991; Muiruri et al., 1991) and broiler chickens (Reilly et al., 1991; Estevez et al., 2002) have shown that the birds prefer to roost on cold perches when exposed to high environmental temperatures. These results indicate that birds actively used cool perches for thermoregulatory purposes, helping to reduce heat stress and thereby improving egg production and hatchability of broiler breeders (Muiruri and Harrison, 1991), and increasing final body weights and feed efficiency in broiler chickens (Reilly et al., 1991; Estevez et al., 2002). No panting was observed in broiler breeder hens with access to cool perches, even when temperatures reached as high as 40°C (Muiruri et al., 1991). This result suggests that cool perches are effective at controlling heat stress, probably at a lower energy cost and avoiding the inconvenience of wetting of litter caused by some cooling systems used in commercial broiler breeder houses. Although the number of experimental studies is limited, and no studies have been conducted under commercial conditions, the results suggest that perches that stimulate exercise in the third dimension may also provide a cost-effective alternative for optimizing the thermal environment through cooling, and also potentially warming, opportunities.

5 Other benefits of structural complexity

5.1 Acquiring navigational skills for the future

Research has shown that early life experience can have an important effect on behavioural and neuronal development, use of space and learning ability in mammals (Rosenzweig and Bennett, 1996; Chaloupkova et al., 2007) and in the domestic fowl (Gunnarsson et al., 2000). When birds are maintained in a different environment during rearing (typically a bi-dimensional environment) than during the reproductive period (typically a more complex, three-dimensional environment), the transition between environments can be stressful (Hocking et al., 2005) until the birds get acquainted with the physical and social features of the new environment. The same applies to preparation of young birds for success in free-range conditions. The difficulty adapting will be more severe the larger the disconnect between environments. Access to perches or any elements that confer young birds with the opportunity to experience the third dimension and to navigate in a more complex environment (e.g. cover panels) can help them to acquire essential skills that facilitate their transition between environments. For example, pheasants reared with perches made fewer errors in finding food than controls without perches when tested in an eight-arm radial maze, which was relevant to their subsequent survival when released outdoors (Whiteside et al., 2016).

One of the main differences between the rearing and the egg production environment is that, in the latter, feeders, drinkers and nest boxes are elevated above the floor. It may take some days for the birds to learn to locate and use them, especially considering that feeders and drinkers may be of different types than provided during rearing, and the birds have never previously been exposed to nest boxes. Due to the birds' failure to find and use nest boxes, it is quite common to have a high incidence of floor eggs in breeding flocks during early lay (Appleby et al., 1988; Estevez, personal observations). This problem is often managed by fencing the females on the slatted area for several days hoping that they learn to lay the eggs in the nest boxes. However, this practice can be stressful for the birds leading to reduced performance, and it requires additional work (and cost) for the farmer. Providing early access to perches, ramps and platforms should facilitate transition to the laying house without the need for confinement on the slats as birds will be better prepared to jump on and off heights and to explore in three dimensions.

Although limited data are available for meat poultry, both Brake (1987) and Appleby et al. (1988) reported that broiler breeders reared with perches had around 5% fewer floor eggs when reared with access to perches. Appleby et al. (1988) reported that 86% of control hens laid their first egg on the floor whereas for hens reared with access to perches this proportion was reduced to 21%. Differences in the rate of floor eggs between birds reared without perches or with access to perches starting within four weeks post-hatch were evident up to 35 weeks of age in layers (Gunnarsson et al., 1999) and appear to have persisted to 40–50 weeks in broiler breeders (Brake, 1987). These results illustrate the role of learning in the utilization of nest boxes and the long-lasting effect of providing birds with early access to perches or other elevated structures such as ramps or platforms.

Preparedness for the new environment does not require that young breeders are reared under exactly identical conditions to the ones they will experience later in life. It has been argued that providing young birds with the opportunity to explore an enriched, variable environment in which to develop their cognitive skills can also improve their welfare (Meehan and Mench, 2007). For example, the skills required to adapt to the breeding facility can be acquired when young birds are provided with opportunities to learn the spatial competencies necessary to navigate and access resources. These opportunities could include the provision of cover panels in the rearing facility (Newberry and Shackleton, 1997; Cornetto and Estevez, 2001a), potentially improving the birds' ability to get around obstacles. Research has also shown that poultry species use visual cues not only to locate food resources but also to exploit them more efficiently, especially in changing environments (Vallortiaga, 1996; Roper and Marples, 1997). It should be possible to use this learning capacity to our advantage to ease the transition of breeders from rearing to breeding facilities. How broiler breeders and other poultry are raised when young will have long-lasting effects on their behavioural flexibility and stress resilience that will be reflected in their health and performance during their reproductive life. Investment in enriching the rearing environment to prepare young breeders for their reproductive life could, thus, pay off from both the welfare and economic standpoints.

5.2 Dark brooders

Although fast-growing broiler chickens have a low propensity to feather peck, slow-growing broilers do show this behavioural problem occasionally (Nielsen et al., 2003). Other meat poultry, such as turkeys or broiler breeders are often beak trimmed to prevent injurious feather pecking. However, this management practice is stressful for the birds, even if performed within the ten first days of life, a reason why this management practice has already been banned or is expected to be banned in some European countries. Thus, non-invasive strategies must be developed to prevent injurious feather pecking and cannibalism.

Although most environmental enrichment practices will serve to reduce the incidence of feather pecking and cannibalism by, for example, reducing visual (cover panels) and physical contact (perches) with resting birds, dark brooders offer an additional approach for minimizing the risk. Early rearing with dark brooders as compared to regular heating lamps (Johnsen and Kristensen, 2001) gave excellent preliminary results in controlling severe feather pecking, improving feather condition and reducing mortality in layer pullets observed up to 23 weeks of age (Jensen et al., 2006). Similar results were obtained by Gilani et al. (2012) in a study of small commercial flocks of Columbian Blacktail chickens with intact beaks. The birds raised with dark brooders had a lower incidence of severe feather pecking, were less likely to have missing feathers and were slightly heavier. In addition to reducing early feather pecking and stimulating foraging and dustbathing, Riber and Guzman (2016) also obtained evidence suggesting that dark brooders reduced fearfulness in layer chicks. Although so far no studies have been conducted to assess potential application to turkeys, broiler breeders or other meat poultry species at risk of severe feather pecking and cannibalism when beaks are left intact, environmental enrichment with dark brooders may yield welfare and performance benefits in the meat poultry industry as well.

6 Visual enrichment through lighting

Poultry species, with the exception of emus and tinamous, have four cone types supporting colour vision across a broad spectrum extending to the near-ultraviolet range (Hart et al., 1999; Wright and Bowmaker, 2001). However, meat birds are usually kept at low light intensity under unvarying lighting conditions with the intention of minimizing the risk of cannibalism and feather pecking, and maximizing conversion of feed to growth as opposed to diversion into energetically costly behavioural activities. However, dim, artificial lighting impairs discrimination between colours (Olsson et al., 2015), ability to navigate between perches (Taylor et al., 2003) and social recognition (D'Eath and Stone, 1999), suggesting that welfare may be enhanced by providing a more varied and dynamic visual landscape.

Bizeray et al. (2002a) used a disco ball to project brightly coloured (red, green, blue and yellow) moving light patterns on the floor during four 1-h periods daily throughout rearing of broilers. It was rare for the birds to follow the moving light spots and no pecking at light spots was detected. Foraging and locomotion were unaffected, but the light treatment did increase time spent feeding by comparison with the control condition. Since feed intake and growth were unaffected (Bizeray et al. 2002b), the authors speculated that the increased time at the feeders may have been occupied in increased feed particle selection when eating.

Newberry et al. (1985) found that providing brighter spotlights that alternated from one side of the house to the other every 1–2 hours stimulated movement in male broilers, although without significantly reducing the incidence or severity of leg disorders. Between 4 and 7 weeks of age, birds of certain strains congregated under the lights, whereas for other strains the density was lower under the lights. Above seven weeks of age, the area under the lights had a similar or lower density of birds compared to the darker areas. It was noted that when males engaged each other in non-harmful sparring under the lights, the resulting disturbance cleared the area of other birds. In a study with male turkeys, Sherwin et al. (1999) provided two bright spotlights that switched on for 1–5 min at random times during the photoperiod as part of a complex enrichment treatment that included multiple pecking stimuli. Injurious pecking was reduced relative to the control, and the turkeys appeared to have stronger legs, although it is not known how the spotlights, specifically, may have contributed to these effects. Some welfare guidelines require that bright spots of light in the poultry house are avoided due to concerns that birds might smother when congregating in those spots (Prescott et al., 2003), although no smothering occurred in the above studies. It should also be noted that, when birds are raised at low light intensities, an uneven bird distribution caused by sunlight flooding through ventilation inlets and fans may have a negative impact on performance and welfare (see Section 3.1 above), especially if resting cycles are not synchronized by a long daily scotoperiod.

Some regulations and welfare assurance schemes, nevertheless, require that poultry have access to natural lighting. Brighter intensity lighting can be expected to increase activity levels (Newberry et al., 1988), and early exposure to the varied patterns of lighting resulting from changing weather condition and across time of day should help to habituate birds to such variations. Indeed, Bailie et al. (2013) observed that light intensities were higher in commercial broiler houses allowing natural light to enter through windows, and birds were more active. Importantly, they found that gait scores tended to be lower and latency to lie was higher in birds exposed to natural light through windows, suggesting that the windows were an effective source of visual enrichment. There may be concerns that exposure to natural light through windows would be a risk factor for feather pecking and cannibalism in adult flocks but this was not found to be the case in a study of 81 Swedish layer flocks (Yngvesson et al., 2011). For birds being reared for free-range production, early exposure to natural light is particularly recommended because lighting preference is influenced by experience (Gunnarsson et al., 2008).

7 Foraging enrichments

Foraging behaviour incorporates the travel time taken to find food, the handling time required to catch, extract and process food into a form that can be swallowed and the ingestion time involved in consuming the prepared food. Wild birds generally spend a large part of their day engaged in the travelling and handling components of foraging, which together comprise the appetitive phase of foraging, whereas only a small proportion of each foraging bout is devoted to the consummatory phase. The appetitive phase of foraging includes learning about new sources of food through exploratory and play behaviour, and attraction to food sources discovered by others (i.e. social learning). It also involves using prior learning to recall and navigate to locations where food has previously been found; recognize which items are edible; select appropriate items to meet nutritional needs; and apply skills developed to catch, extract and process them. Understanding the different aspects of the natural foraging behaviour of poultry species provides insights into the features of materials that can be expected to stimulate foraging behaviour. These include a degree of concealment requiring searching and extraction, particle variability that attracts investigation, a degree of movement, manipulability and destructibility providing opportunities to practice food handling skills and edibility providing a consummatory reward.

In modern poultry production, where nutritionally complete, processed feed can be easily found in predictable locations, the demands for appetitive behaviour are minimal. A majority of foraging time is devoted to consummatory behaviour and overall expenditure in foraging behaviour is low. Although domesticated poultry, and especially broilers (Lindqvist et al., 2006), invest less time in foraging than wild birds, they nevertheless exhibit some foraging. Chickens, for example, can be seen pecking and scratching in the litter despite the nearby presence of feeders, investigating novel objects, running into a familiar area accessible for a restricted period each day (Newberry, 1999; Fig. 8), engaging in playful 'worm running' with small objects found in the litter (Cloutier et al., 2004) and jumping to catch insects and seeds swaying in the breeze at the end of stalks. Ostriches given access to sand enriched with cabbage, conifer cones and sticks explored more, pecked more at a novel food source and tended to peck less at pen fixtures than control birds (Christensen and Nielsen, 2004; Fig. 8).

The emergence of an outbreak of cannibalism provides a prime example that modern poultry retain the motivation for individual and social learning about new food sources (Cloutier et al., 2002). A major economic incentive for investing in foraging enrichments relates, therefore, to their role in guiding appetitive foraging activities towards appropriate substrates and reducing investigation of conspecific body parts as potential food sources. This is important because even if only one bird in a flock learns through exploration that feathers, eggs, blood or internal organs represent rich food sources, others will be quick to join in, thereby having the opportunity to learn the undesirable behaviour as well (Cloutier et al., 2002). Although unusual in broilers, these redirected foraging behaviours (Newberry et al., 2007) can be a serious concern in turkeys, pheasants, ducks and geese with intact beaks, especially if kept on slatted flooring.

There is limited research on foraging enrichments for poultry species other than chickens but methods found effective for reducing feather pecking and cannibalism in laying hens may be beneficial. First, it is important to rear birds from day 1 with access to litter. Early access to particulate substrate on a solid surface has long-term value in reducing the risk of damaging feather pecking (Tahamtani et al., 2016). However, access to litter alone is insufficient to prevent the development of severe feather pecking in the birds that are most motivated to forage (Newberry et al., 2007). Long-cut straw is an effective foraging enrichment found to retain sustained interest as a foraging substrate and help to reduce the risk of unwanted behaviours in laying hens, as has provision of feed in mash rather than pelleted form (Aerni et al., 2000), and hanging bunches of thick string that can be pulled apart (McAdie et al., 2005). It is suggestive that straw was a component in combined enrichment treatments found to reduce pecking damage in turkeys (Sherwin et al., 1999; Martrenchar et al., 2001). However, Hocking and Jones (2006) detected no improvement

Figure 8 (a) Heavy broilers running into an area accessible for only 3 hours daily (see Newberry, 1999 for details). Photo: R. C. Newberry. (b) Ostrich foraging in sand. Photo: I. Estevez.

in the feather and skin condition of commercial broiler breeders provided with wood shavings bales and bunches of string.

Another goal in providing foraging enrichments is to stimulate movement in the hope of improving skeletal and cardiovascular health, especially in broilers. Bailie et al. (2013) observed that commercial broilers provided with straw bales performed more walking and running and less sitting than birds in houses lacking bales, and this was associated with longer latencies to lie down suggestive of improved leg health. Increasing the density of straw bales from 1 bale/44 m^2 to 1 bale/29 m^2 floor space did not result in added welfare benefits (Bailie and O'Connell, 2014). The addition of bunches of string to windowed broiler houses provided with straw bales did not have clear effects on leg health but the strings were frequently pecked (Bailie and O'Connell, 2015) suggesting that this added opportunity for foraging behaviour was appreciated by the birds. Bizeray et al. (2002a,b) found no benefit of scattering whole wheat in the litter as it did not appear to be eaten, had no impact on time spent foraging or other measured activities and made no difference to production or health parameters relative to the control condition. Live mealworms were effective in stimulating broiler activity and ground pecking and scratching (Pichova et al., 2016). Although probably not practical to supply as an enrichment for broilers kept indoors, this finding points to the enrichment value of access to insects under free-range conditions.

8 Comfort behaviour enrichments

8.1 Dust bathing substrates

Dustbathing is a vigorous behaviour that involves loosening the substrate through pecking and scratching, and then squatting and using rapid vertical wing shakes to toss dusty substrate into the feathers. The bird then lies for a period, sometimes on one side with stretched legs, while stale, dirty feather lipids are absorbed into the dust. The dust bath ends when the bird stands up and shakes off the dust. Later, birds engage in preening to apply fresh waterproofing lipids from their oil gland to their feathers (van Liere, 1992). Dustbathing helps to maintain the integument in good condition by removing excess feather lipids (Olsson and Keeling, 2005), and it is considered as one of the behavioural responses evolved to maintain the insulative and waterproofing value of the feathers (van Liere, 1992) and, possibly, to control ectoparasites in wild birds (Clayton et al., 2010).

Dustbathing has been demonstrated to be an effective mechanism to control northern fowl mites (*Ornithonyssus sylviarum*) and body lice (*Menacanthus stramineus*) in laying hens when provided with dust boxes containing diatomaceous earth and kaolin (Martin and Mullens, 2012). These benefits were acquired only by the birds that did dustbathe in the dust boxes. Dust boxes containing sand alone were not effective in removing ectoparasites (Velozzi et al., 2015). Although ectoparasites are not a major health or performance issue in meat poultry because of their relatively short lifespan, infestations may be a bigger issue in breeding poultry.

Besides, with the exception of waterfowl, meat poultry species have a high motivation to perform dustbathing behaviour. When denied access to appropriate substrates, they may repeatedly start dustbathing sequences but be unable to complete the full sequence that would lead to a functional outcome (van Liere, 1992). Furthermore, they typically perform the behaviour at high levels when a suitable substrate is provided after a period of deprivation. These observations suggest that dustbathing behaviour is pleasurable (Widowski and Duncan, 2000) and that thwarting of access to dusty material is stressful (Vestergaard et al., 1997).

It is often assumed that, as most meat poultry are raised on litter, their motivation to dustbathe is satisfied. However, broilers (Shields et al., 2004, Toghyani et al., 2010, Villagrá et al., 2014) and other poultry (van Liere et al., 1990; Olsson and Keeling, 2005) have a strong preference to dustbathe in loose, fine-grained material such as sand and, especially, peat moss (Petherick and Duncan, 1989; Fig. 9). Such materials are preferred over the coarser materials that are most commonly provided as litter (e.g. wood shavings, rice hulls and chopped straw), which are not the best materials in which to perform dustbathing behaviour. Furthermore, the litter quality deteriorates over time, especially if environmental conditions are not well controlled, leading to caked litter that is unsuitable for dustbathing, sometimes very early on.

Because access to a good dustbathing substrate is important for the welfare of poultry (Duncan et al., 1998), provision of sand or peat moss patches through the house should be an effective enrichment. Peat moss is, in fact, currently used as an enrichment in certain commercial broiler marketing schemes. Additionally, considering the relevance of dustbathing for maintaining optimal feathering, perhaps access to a high-quality dustbathing substrate would also carry some health and performance benefits. Some authors have hypothesised that, given the wide range of leg and wing movements

Figure 9 Broiler dustbathing in peat moss. Photo: R. C. Newberry.

performed during dustbathing bouts, stimulating dustbathing may be a potential avenue to increase exercise level and improve leg condition in meat poultry (Shields et al., 2004). Vestergaard and Sanotra (1999) were unable to determine if higher rates of dustbathing were protective against lameness in broilers but they did observe a negative association between lameness score and dustbathing frequency. Potential health and performance benefits from enrichment with attractive dustbathing materials have not yet been evaluated under commercial conditions for meat poultry, but possibly birds with longer production cycles would benefit the most.

8.2 Water baths

Whereas most meat poultry species use dust as their bathing substrate, waterfowl use water. Research on Pekin ducks showed that, when they were given access to water only from nipple drinkers or narrow-lipped bell drinkers, they were not always able to properly clean their eyes, nostrils and feathers (Jones et al., 2009; Driscoll and Broom, 2011). These results suggest that ducks need to be able to dip their head in water and splash water over their feathers to fully clean these areas. Access to water from showers, water baths and sufficiently wide troughs was effective in meeting these needs and in avoiding elevated water-directed behaviour towards a water bath at seven weeks of age as was observed in ducks reared with nipple drinkers only (Jones et al., 2009). On the other hand, Schenk et al. (2016) found that ducks had dirtier eyes, nostrils and feathers when reared with water troughs than nipple drinkers, along with lower feed intake, higher plasma corticosterone levels, higher mortality and more condemnations. Reasons for different results across studies are unresolved, although dirtiness of birds reared with nipple drinkers could result from insufficient water flow or nipple access (Schenk et al., 2016). It is clear that, when using open water sources, great care is needed to ensure that the water is clean and to avoid wet litter conditions that can have adverse effects on foot and leg health, and lead to elevated ammonia levels. Jones et al. (2009) observed that ducks spent more time near open water sources than nipple drinkers, which is important to note given the challenge of avoiding wet litter in these areas.

9 Enrichment and use of outdoor areas

Access to an outdoor area by commercial poultry is often perceived as a guarantee of animal welfare because of the wider possibilities for the birds to express natural behaviour including locomotory behaviours (Keppler and Folsch, 2000), foraging, playing and sunbathing (Fig. 10). In broilers, the extra space provided and the higher possibilities for exercise may have additional positive effects in regard to leg health (Fanatico et al., 2005), reduced risk of contact dermatitis and improved feather condition.

However, when access is voluntary, use of the outdoor area is often low and birds that do go outdoors show a strong tendency to remain within the immediate proximity of the house, presumably because the outdoor open area is somewhat intimidating for birds raised indoors. For commercial free-range broilers, Dawkins et al. (2003) reported a maximum of 15% of the flock using the range, and Rodriguez-Aurrekoetxea et al. (2014) indicated an average use of 5% in a slow-growing strain, with nearly 64% of the birds never observed outdoors. In adult laying hens, the frequency of use of the outdoor area is reported to range from 8 to 32% (Hegelund et al., 2005, 2006; Gilani et al., 2014; Rodriguez-Aurrekoetxea et al., 2015), with higher rates probably related to greater time to develop experience using the outdoor space as compared to meat poultry.

Many factors affect the use of the outdoor area in commercial free-range poultry, including climatic conditions (Hegelund et al., 2005; Richards et al., 2011, 2012), the size of the flock (Gebhardt-Henrich et al., 2014) and experience. Research in laying hens suggests that early experience in accessing the outdoor space is important for encouraging high use later on (Rodriguez-Aurrekoetxea and Estevez, 2016). Furthermore, this research indicated that individuals that used the outdoors more often had larger home ranges,

Figure 10 Slow-growing broilers with outdoor access: (a) exploring close to dense cover, (b) play fighting and (c) sunbathing. Photos: R. C. Newberry.

moved further from the house, had better plumage condition and had a lower incidence of footpad dermatitis. Thus, early management of the birds in free-range systems to promote contact with the outdoors appears to be critical to ensure frequent and more uniform use of the outdoor area. Because the rearing periods for meat poultry are shorter, early management to promote outdoor use is especially relevant, as long as the climatic conditions are suitable considering the birds' age and feathering.

Rodriguez-Aurrekoetxea and Estevez (2014) and Fanatico et al. (2016) used environmental enrichments in an attempt to encourage early use of the outdoor area by slow-growing broilers. The enrichments, comprising perches (Fig. 11) and different forms of cover panels and shades, failed to attract more chickens to go outdoors in either study, but Fanatico et al. (2016) observed that chickens in enriched outdoor areas moved further away from the chicken house as compared to controls. When provided with a high number of protective wooden A-frames or short rotation coppice with willow, 35% and 43% of the flock, respectively, used the outdoor range (Stadig et al., 2016), suggesting that a high degree of protection is necessary in the outdoor area for the birds to make a good use of it. However, despite the relatively high use of the outdoor range, access to range was only moderately associated with reduced hock dermatitis and resistance to tonic immobility induction (an indicator of reduced fear), possibly because data were collated at the flock rather than individual level (Stadig et al., 2016). Given wide inter-individual variation in use

Figure 11 Slow-growing meat chickens in (a) standard outdoor area and (b) area enriched with perches. From A. Rodriguez-Aurrekoetxea (2016).

of the outdoor areas, it would be desirable to assess health and welfare effects according to the duration of range use by individual birds.

The availability of a high degree of cover allows chickens the option to hide in the cover when perceiving a danger such as a potential predator (Zeltner and Hirt, 2008), rather than running inside the house, and can actually protect birds from being caught by predators. The degree of shelter from sun, rain and wind provided outdoors is also relevant. The shelter provided by trees planted within 20–25 m of the house was found to be effective in drawing laying hens outdoors (Bright et al., 2011).

10 Discussion

10.1 Design considerations and constraints on enrichment

Although enrichments, by definition, have positive effects, details of their design and implementation are crucial to achieving the desired effects without unintended adverse side effects. For instance, indoor cover panels should be porous to avoid disrupting the ventilation but without holes so large that the birds could get their head or other body parts caught in them. They should be narrow to avoid serving as obstacles that could result in piling and short so that they do not disrupt the farmers' ability to check the condition of the flock. Evidence suggests that they should provide partial rather than solid visual cover, allowing for a 'Venetian blinds' effect, whereby the cover offers a sense of security through concealment while allowing birds to monitor events on the other side (Newberry and Shackleton, 1997). Furthermore, whereas vertical cover panels were effective in promoting vulnerable behaviours such as rest and preening, indicating their security-inducing effect, horizontal overhead panels of the same design did not have this effect (Newberry and Shackleton, 1987 and unpublished observations).

When considering perches, ramps and platforms, the design needs to take into account the large size and relative lack of agility of meat poultry birds. Unless well occupied, they may have negative effects by taking up floor space, thus effectively reducing the usable space per bird. The increased crowding could result in greater stress and reduced immunocompetence (Heckert et al., 2001). A challenge is deciding how much space to allocate to such structures given that, if birds are attracted to rest on them or around their base, their presence blocks access by other birds and may make it hard for birds located on top of them to get off without disturbing birds below. Well-used structures must be designed and located in a manner that does not lead to excessive accumulation of manure on top or underneath them, or manure raining upon birds below. The materials used in construction, and the profile, should allow birds to grip easily while also minimizing the risk of breast blisters (Nielsen, 2004) and bumblefoot, and avoiding the harbouring of mites. Ruiz-Feria et al. (2014) found that triangular ramps placed between feeders and drinkers reduced leg tendon breaking strength and growth rate of broilers reared to 49 days, suggesting that the design of those structures was difficult for heavy broilers to negotiate. The durability and ease of cleaning are also factors to consider when designing structural enrichments. The investment cost declines when items can be reused for many flocks but this advantage is lost if the labour cost to clean them is high.

When considering dustbathing and foraging materials, there can be concerns about the use of peat moss from the perspective of environmental sustainability. Sand is heavy to handle and abrasive on equipment. The use of straw, hay and other forages as enrichment can be constrained by concerns about biosecurity when bringing them into houses and by the potential for introducing unwanted seeds when used litter is spread on fields. Wood shavings bales may disintegrate within only a few days, limiting their durability as an enrichment. Because of their size, it would be challenging to bring in new ones once the birds are older and take up more floor space. Bale wrap can extend bale integrity but the plastic materials currently used are not environmentally sustainable. Care must be taken that materials are non-toxic and safe if ingested, will not cause gastrointestinal impaction and will biodegrade and not become pollutants in used litter.

When buildings incorporate windows to allow natural light, blinds that diffuse the light may be necessary to avoid panic reactions from lightning flashes or from patterns of moving light across the ceiling created by car lights on passing vehicles. Also, adjustments to the amount of light coming in through different windows may be needed if variation in lighting across the house leads to consistently higher densities in certain areas of the house. Diffuse light may also be needed to minimize the risk of feather pecking and cannibalism in flocks prone to such behaviour. Whereas dark brooders may be useful enrichments in houses where heating is supplied by brooders, they may not be effective in houses with underfloor heating where there is no incentive to seek brooder heat, considering that broilers do not appear to be particularly attracted to rest in dark enclosed spaces without supplementary heating (Newberry, personal observation).

There are multiple challenges to be considered when introducing access to the outdoors as an enrichment strategy (Newberry, 2017b). A well-drained substrate free of contaminants (e.g. dioxins and heavy metals) is essential. Birds should have clean feet when entering the house to avoid tracking in mud harbouring endoparasites. Although free-range broilers are attracted out of the house to drink from puddles and water troughs after rain showers (Jones et al., 2007), suggesting an enriching effect, open water that attracts wild waterfowl serves as a risk factor for the spread of avian influenza. Shade is necessary to avoid sun burn and birds must be protected from predators and from humans who may introduce garbage, inappropriate foodstuffs and even unwanted pet poultry. Finally, there can be challenges managing the diet when there is individual variation in consumption of forage, and crop impaction with fibrous vegetation can be a concern.

10.2 Future research

Many enrichment ideas presented in this chapter require further evaluation under commercial conditions. There is a need to determine optimum amounts of the different types of enrichment according to flock size and density, and to assess how different genetic lines and management conditions may affect results. Investigations should also evaluate how different forms of enrichment can be combined to maximize benefits towards different enrichment goals. When results are obtained across a representative number of flocks, it will be possible to evaluate how different enrichment programmes can be used to target problems detected based on flock culling and mortality data and on processing plant data such as incidence of footpad lesion scores and different reasons for condemnation and downgrading.

It is also clear that such research needs to be extended beyond broilers and broiler breeders to other meat poultry species such as turkeys and ducks, for which very limited data currently exist. Species- and age-appropriate enrichment strategies need to be developed taking into consideration the evolutionary background that has shaped the behavioural repertoire of each species. Such research will be useful when developing welfare assurance standards that currently incorporate vague enrichment requirements or incorporate a level of specificity that has not been established empirically (e.g. perch space standards for turkeys).

We have emphasized benefits that can be achieved through structural (e.g. perches, cover panels), spatial (e.g. access to free-range), foraging (e.g. straw bales) and bathing enrichments. Sensory enrichment is another enrichment concept that has received little attention to date, where the focus is on exposure to a variety of sights, sounds, smells, tastes and textures that stimulate different sensory modalities. Practical benefits could come from providing opportunities for voluntary exploration and habituation to a wide range of sensory stimuli, potentially reducing the risk of a panic reaction when unexpectedly encountering novel stimuli, such as during catching and transport. Cognitive enrichment is a related concept that is useful to consider when developing enrichment strategies, where the emphasis is on addressing opportunities for the birds to develop different skills that promote behavioural flexibility and stress resilience.

Research would also be warranted in the area of developing measures of quality of life that extend beyond reducing negative impacts on health into the realm of promoting positive psychological states such as pleasurable anticipation and joyfulness. The incorporation of enrichments into welfare assurance schemes is based on the perception that consumers are willing to pay more for products that contribute positively to the birds' quality of life. The science is lagging behind these marketing initiatives. There is a need for practical on-farm methods for quantifying positive aspects of quality of life that have a value beyond avoiding suffering.

Future lines of research should also place a high emphasis on considering the notion of enrichment while developing the basic housing elements. For example, platforms and perches can be designed to incorporate cooling and heating, enabling birds to select a comfortable microenvironment according to their individual needs. Automatic feeder lines could be designed in such a way that feed pans permit a larger number of birds to feed together, with the pipes in between feeders designed as perches, thus gaining added benefits from standard equipment. Enclosed verandahs attached to houses could be used to offer varied sensory and cognitive enrichments while retaining a high degree of control over nutrition and biosecurity. Such developments could lead, eventually, to a future when the concept of enrichment is no longer relevant because enrichment has become standard, corresponding with a high level of sustainability from the combined perspectives of economic efficiency and social responsibility.

11 Conclusions

In this chapter, we have provided examples of how environmental enrichment can enhance the biological adaptation of poultry beyond the level obtained from providing basic resources such as feed and water. The main emphasis is to improve welfare by promoting species-typical behaviour, thereby reducing the risk of abnormal and damaging behaviours such as panic responses and feather pecking. Other potential benefits include safe, effective use of space and resources, increased ability to cope with change and promotion of positive psychological states.

Too often, enrichment is equated to the provision of a few simple 'toys' with no biological relevance to the birds. These toys may elicit a brief response to their novelty but are soon ignored. It is, thus, essential for a good enrichment programme to first determine the objectives of environmental intervention, then to design the required features of the enrichment to achieve the proposed goals, and finally to quantify and validate both positive and negative effects to assess overall net benefits. It is not sufficient simply to add an object and assume that it has an enriching effect; to categorize the modification as an enrichment requires documenting actual improvements in the welfare of the birds relative to results in its absence (Newberry, 1995). For enrichment programmes to produce a meaningful long-term benefit for the birds' welfare, they must have a sustained effect (e.g. through providing perches where birds can roost, straw bales that stimulate foraging and cover panels to reduce the sense of exposure). Thus, achievement of sustained beneficial effects can be seen as the key to sustainability in the development of effective environmental enrichment strategies.

12 References

Aerni, V., El-Lethey, H. and Wechsler, B. (2000), 'Effect of foraging material and food form on feather pecking in laying hens', *Brit. Poultry Sci.*, 41(1), 16–21. http://dx.doi.org/10.1080/00071660086349.

Appleby, M. C., Duncan, I. J. H. and McRae, H. E. (1988), 'Perching and floor laying by domestic hens: experimental results and their commercial application', *Brit. Poultry Sci.*, 29, 351–7. http://dx.doi.org/10.1080/00071668808417060.

Appleby, M. C., Smith, S. F. and Hughes, B. O. (1992), 'Individual perching behavior of laying hens and its effects in cages', *Brit. Poultry Sci.*, 33, 227–38. http://dx.doi.org/10.1080/00071669208417462.

Bailie, C. L., Ball, M. E. E. and O'Connell, N. E. (2013), 'Influence of the provision of natural light and straw bales on activity levels and leg health in commercial broiler chickens', *Animal*, 7(4), 618–26. http://dx.doi.org/10.1017/S1751731112002108.

Bailie, C. L. and O'Connell, N. E. (2014), 'The effect of level of straw bale provision on the behaviour and leg health of commercial broiler chickens', *Animal*, 8, 1715–21. http://dx.doi.org/10.1017/S1751731114001529.

Bailie, C. L. and O'Connell, N. E. (2015), 'The influence of providing perches and string on activity levels, fearfulness and leg health in commercial broiler chickens', *Animal*, 9, 660–8. http://dx.doi.org/10.1017/S1751731114002821.

Bergmann, S., Platz, S., Heyn, E., Schweizer, C., Strassmeier, P. and Erhard, M. (2009), 'Turkey fattening under extensive rearing conditions – Selected parameters concerning health, performance and behavior', *Berl. Münch. Tierärztl.*, 122(7–8), 284–91. http://dx.doi.org/10.2376/0005-9366-122-284.

Bilcik, B. and Estevez, I. (2005), 'Impact of male-male competition and morphological traits on mating strategies and reproductive success in broiler breeders', *Appl. Anim. Behav. Sci.*, 92, 307–23. http://dx.doi.org/10.1016/j.applanim.2004.11.007.

Bilcik, B., Estevez, I. and Russek-Cohen, E. (2005), 'Reproductive success of broiler breeders in natural mating systems: The effect of male-male competition, sperm quality and morphological characteristics', *Poultry Sci.*, 84, 1453–62. http://dx.doi.org/10.1093/ps/84.9.1453.

Bizeray, D., Estevez, I., Leterrier, C. and Faure, J. M. (2002a), 'Influence of increased environmental complexity on leg condition, performance and level of fearfulness in broilers', *Poultry Sci.*, 81, 767–73. http://dx.doi.org/10.1093/ps/81.6.767.

Bizeray, D., Estevez, I., Leterrier, C. and Faure, J. M. (2002b), 'Effects of increasing environmental complexity on the physical activity of broiler chickens', *Appl. Anim. Behav. Sci.*, 79, 27–41. http://dx.doi.org/10.1016/S0168-1591(02)00083-7.

Blokhuis, H. J. (1984), 'Rest in poultry', *Appl. Anim. Behav. Sci.*, 12, 289–303. http://dx.doi.org/10.1016/0168-1591(84)90121-7.

Brake, J. (1987), 'Influence of presence of perches during rearing on incidence of floor laying in broiler breeders', *Poultry Sci.*, 66, 1587–9. http://dx.doi.org/10.3382/ps.0661587.

Bright, A., Brass, D., Clachan, J., Drake, K. A. and Joret, A. D. (2011), 'Canopy cover is correlated with reduced injurious feather pecking in commercial flocks of free-range laying hens', *Anim. Welf.*, 20, 329–38.

Buijs, S., Keeling, L. J., Vangestel, C., Baert, J. and Tuyttens F. A. M. (2011), 'Neighbourhood analysis as an indicator of spatial requirements of broiler chickens', *Appl. Anim. Behav. Sci.*, 129, 111–20. http://dx.doi.org/10.1016/j.applanim.2010.11.017.

Chaloupkova, H., Illmann, G., Bartos, L. and Spinka, M. (2007), 'The effect of pre-weaning housing on the play and agonistic of domestic pigs', *Appl. Anim. Behav. Sci.*, 103, 25–34. http://dx.doi.org/10.1016/j.applanim.2006.04.020.

Christensen, J. W. and Nielsen, B. L. (2004), 'Environmental enrichment for ostrich, Struthio camelus, chicks', *Anim. Welf.*, 13, 119–24.

Clayton, D. H., Koop, J. A., Harbison, C. W., Moyer, B. R. and Bush, S. E. (2010), 'How birds combat ectoparasites', *Open Ornithol. J.*, 3, 41–71. http://dx.doi.org/10.2174/1874453201003010041.

Cloutier, S., Newberry, R. C. and Honda, K. (2004), 'Comparison of social ranks based on worm-running and aggressive behaviour in young domestic fowl', *Behav. Proc.*, 65, 79–86. http://dx.doi.org/10.1016/j.beproc.2003.07.001.

Cloutier, S., Newberry, R. C., Honda, K. and Alldredge, J. R. (2002), 'Cannibalistic behaviour spread by social learning', *Anim. Behav.*, 63, 1153–62. http://dx.doi.org/10.1006/anbe.2002.3017.

Collias, N. E. and Collias, E. C. (1967), 'A field study of the Red Jungle Fowl in North-Central India', *Condor*, 69, 360–86. http://dx.doi.org/10.2307/1366199.

Cordiner, L. S. and Savory, C. J. (2001), 'Use of perches and nestboxes by laying hens in relation to social status, based on examination of consistency of ranking orders and frequency of interaction', *Appl. Anim. Behav. Sci.*, 71, 305–17. http://dx.doi.org/10.1016/S0168-1591(00)00186-6.

Cornetto, T. L. and Estevez, I. (2001a), 'Influence of vertical panels on use of space by domestic fowl', *Appl. Anim. Behav. Sci.*, 71, 141–53. http://dx.doi.org/10.1016/S0168-1591(00)00171-4.

Cornetto, T. L. and Estevez, I. (2001b), 'Behavior of the domestic fowl in presence of vertical panels', *Poultry Sci.*, 80, 1455–62. http://dx.doi.org/10.1093/ps/80.10.1455.

Cornetto, T. L., Estevez, I. and Douglass L. (2002), 'Using artificial cover to reduce aggression and disturbances in domestic fowl', *Appl. Anim. Behav. Sci.*, 75, 325–36. http://dx.doi.org/10.1016/S0168-1591(01)00195-2.

Dawkins, M. S., Cook, P. A., Whittingham, M. J., Mansell, K. A. and Harper, A. E. (2003), 'What makes free-range broiler chickens range? In situ measurement of habitat preference', *Anim. Behav.*, 66, 151–60. http://dx.doi.org/10.1006/anbe.2003.2172.

Dawkins, M. S., Donnelly, C. A. and Jones, T. A. (2004), 'Chicken welfare is influenced more by housing conditions than by stocking density', *Nature*, 427, 342–4. http://dx.doi.org/10.1038/nature02226.

D'Eath, R. B. and Stone, R. J. (1999), 'Chickens use visual cues in social discrimination: an experiment with coloured lighting', *Appl. Anim. Behav. Sci.*, 62, 233–42. http://dx.doi.org/10.1016/S0168-1591(98)00216-0.

De Jong, I. C., Gunnik, H. and Van Harn, J. (2014), 'Wet litter not only induces footpad dermatitis but also reduces overall welfare, technical performance, and carcass yield in broiler chickens', *J. Appl. Poultry Res.*, 23, 51–8. http://dx.doi.org/10.3382/japr.2013-00803.

Deeming, D. C., Hodges, H. R. and Cooper, J. J. (2011a), 'Effect of sight barriers in pens of breeding ring-necked pheasants (Phasianus colchicus) I. Behaviour and welfare', *Brit. Poultry Sci.*, 52, 403–14. http://dx.doi.org/10.1080/00071668.2011.590796.

Deeming, D. C., Hodges, H. R. and Cooper, J. J. (2011b), 'Effect of sight barriers in pens of breeding ring-necked pheasants (Phasianus colchicus) II. Reproductive parameters', *Brit. Poultry Sci.*, 52, 415–22. http://dx.doi.org/10.1080/00071668.2011.587182.

Dixon, L. M., Duncan, I. J. H. and Mason, G. J. (2010), 'The effects of four types of enrichment on feather-pecking behaviour in laying hens housed in barren environments', *Anim. Welf.*, 19, 429–35.

Drake, K. A., Donnelly, C. A. and Dawkins, M. S. (2010), 'Influence of rearing and lay risk factors on propensity for feather damage in laying hens', *Brit. Poultry Sci.*, 51, 725–33. http://dx.doi.org/10.1080/00071668.2010.528751.

Driscoll, K. K. M. and Broom, D. M. (2011), 'Does access to open water affect the health of Pekin ducks (Anas platyrhynchos)?', *Poultry Sci.*, 80, 299–307. http://dx.doi.org/10.3382/ps.2010-00883.

Duncan, I. J. H., Widowski, T. M., Malleau, A. E., Lindberg, A. C. and Petherick, J. C. (1998), 'External factors and causation of dustbathing in domestic hens', *Behav. Proc.*, 43, 219–28. http://dx.doi.org/10.1016/S0376-6357(98)00017-5.

Elton, C. (1939), 'On the nature of cover', *J. Wildlife Manag.*, 3, 332–8.

Estep, D. Q. and Baker, S. C. (1991), 'The effects of temporary cover on the behavior of socially housed stump-tailed macaques (Macaca arctoides)', *Zoo Biol.*, 10, 465–72. http://dx.doi.org/10.1002/zoo.1430100605.

Estevez, I., Tablante, N. L., Pettit-Riley, R. and Carr, L. (2002), 'Use of cool perches by broiler chickens', *Poultry Sci.*, 81, 62–9. http://dx.doi.org/10.1093/ps/81.1.62.

Estevez, I. (1999), 'Cover panels for chickens: A cheap tool that can help you', *Poult. Perspect.*, 1(2), 4–6.

Estevez, I. (2009), Behavior and environmental enrichment in broiler breeders, pp. 261–83. In *Biology of Breeding Poultry*, 29, Chapter 15, P. M. Hocking (Ed.). CAB international, British Library of London, UK. ISBN: 13: 978-1-84593-375-3.

Estevez, I. (2007), 'Density allowances for broilers; Where to set the limits?', *Poultry Sci.*, 86, 1265–72. http://dx.doi.org/10.1093/ps/86.6.1265.

Estevez, I. (2015), *Análisis multifactorial del picaje en avicultura*. LII Simposio Científico de Avicultura, Málaga, Spain, 28–30 October, pp. 67–80.

Fanatico, A. C., Mench, J. A., Archer, G. S., Liang, Y., Brewer Gunsaulis, V. B., Owens C. M. and Donoghue A. M. (2016), 'Effect of outdoor structural enrichments on the performance, use of range area, and behavior of organic meat chickens', *Poultry Sci.*, 95, 1980–8. http://dx.doi.org/10.3382/ps/pew196.

Fanatico, A. C., Pillai, P. B., Cavitt, L. C., Owens, C. M. and Emmert, J. L. (2005), 'Evaluation of slower-growing broiler genotypes grown with and without outdoor access: growth performance and carcass yield', *Poultry Sci.* 84, 1321–7. http://dx.doi.org/10.1093/ps/84.8.1321.

Fiscus-LeVan, N. J., Estevez, I. and Stricklin, R. W. (2000), 'Use of horizontal and angled perches by broiler chickens', *Appl. Anim. Behav. Sci.*, 65, 349–65. http://dx.doi.org/10.1016/S0168-1591(99)00059-3.

Gebhardt-Henrich, S. G., Toscano, M. J. and Fröhlich E. K. F. (2014), 'Use of outdoor ranges by laying hens in different sized flocks', *Appl. Anim. Behav. Sci.*, 155, 74–81. http://dx.doi.org/10.1016/j.applanim.2014.03.010.

Gilani, A. M., Knowles, T. G. and Nicol C. J. (2012), 'The effect of dark brooders on feather pecking on commercial farms', *Appl. Anim. Behav. Sci.*, 142, 42–50. http://dx.doi.org/10.1016/j.applanim.2012.09.006.

Gilani, A. M., Knowles, T. G. and Nicol, C. J. (2014), 'Factors affecting ranging behaviour in young and adult laying hens', *Brit. Poultry Sci.*, 55, 127–35. http://dx.doi.org/10.1080/00071668.2014.889279.

Green, L. E., Lewis, K., Kimpton, A. and Nicol, C. J. (2000), 'A cross sectional study of the prevalence of feather pecking in laying hens in alternative systems and its associations with management and disease', *Vet. Rec.*, 147, 233–8. http://dx.doi.org/10.1136/vr.147.9.233.

Gunnarsson, S., Heikkilä, M., Hultgren, J. and Valros, A. (2008), 'A note on light preference in layer pullets reared in incandescent or natural light', *Appl. Anim. Behav. Sci.*, 112, 395–9. http://dx.doi.org/10.1016/j.applanim.2007.09.004.

Gunnarsson, S., Keeling, L. J. and Svedberg, J. (1999), 'Effect of rearing factors on the prevalence of floor eggs, cloacal cannibalism and feather pecking in commercial flocks of loose housed laying hens', *Brit. Poultry Sci.*, 40, 12–18. http://dx.doi.org/10.1080/00071669987773.

Gunnarsson, S., Yngvesson, J., Keeling, L. J. and Forkman, B. (2000), 'Rearing without early access to perches impairs the spatial skills of laying hens', *Appl. Anim. Behav. Sci.*, 67, 217–28. http://dx.doi.org/10.1016/S0168-1591(99)00125-2.

Hart, N. S., Partridge, J.C. and Cuthill, I. C. (1999), 'Visual pigments, cone oil droplets, ocular media and predicted spectral sensitivity in the domestic turkey (Meleagris gallopavo)', *Vision Res.*, 39, 3321–8. http://dx.doi.org/10.1016/S0042-6989(99)00071-1.

Heckert, R. A., Estevez, I., Russek-Cohen, E. and Pettit-Riley, R. (2001), 'Effects of density and perch availability on the immune status of broilers', *Poultry Sci.*, 81(4), 451–7. http://dx.doi.org/10.1093/ps/81.4.451.

Hegelund, L., Sørensen, J. T., Kjær, J. B. and Kristensen, I. S. (2005), 'Use of the range area in organic egg production systems: effect of climatic factors, flock size, age and artificial cover', *Brit. Poult. Sci.*, 46, 1–8. http://dx.doi.org/10.1080/00071660400023813.

Hegelund, L., Sørensen, J. T. and Hermansen, J. E. (2006), 'Welfare and productivity of laying hens in commercial organic egg production systems in Denmark', *NJAS – Wagen. J. Life Sci.*, 54(2), 147–55. http://dx.doi.org/10.1016/S1573-5214(06)80018-7.

Hocking, P. M. and Jones, E. K. M. (2006), 'On-farm assessment of environmental enrichment for broiler breeders', *Brit. Poultry Sci.*, 47(4), 418–25. http://dx.doi.org/10.1080/00071660600825074.

Huber-Eicher, B. and Audige, L. (1999), 'Analysis of risk factors for the occurrence of feather pecking in laying hen growers', *Brit. Poultry Sci.*, 40, 599–604. http://dx.doi.org/10.1080/00071669986963.

Hughes, B. O. and Elson, H. A. (1977), 'The use of perches by broilers in floor pens', *Brit. Poultry Sci.*, 18, 715–22. http://dx.doi.org/10.1080/00071667708416426.

Jensen, A. B., Palme, R. and Forkman, B. (2006), 'Effect of brooders on feather pecking and cannibalism in domestic fowl (Gallus gallus domesticus)', *Appl. Anim. Behav. Sci.*, 99, 287–300. http://dx.doi.org/10.1016/j.applanim.2005.10.017.

Johnsen, P. F. and Kristensen, H. H. (2001), Effect of brooder quality on the early development of feather pecking behaviour in domestic chicks, pp. 209–12. In *Proceedings of the 6th European Symposium on Poultry Welfare*, H. Oester and C. Wyss (Eds). Zollikofen, Switzerland.

Jones, R. B. (1996), 'Fear and adaptability in poultry: insights, implications and imperatives', *World Poult. Sci. J.*, 52, 131–74. http://dx.doi.org/10.1079/WPS19960013.

Jones, T., Feber, R., Hemery, G., Cook, P., James, K., Lamberth, C. and Dawkins, M. (2007), 'Welfare and environmental benefits of integrating commercially viable free-range broiler chickens into newly planted woodland: A UK case study', *Agr. Syst.*, 94, 177–88. http://dx.doi.org/10.1016/j.agsy.2006.08.007.

Jones, T. A., Waitt, C. D. and Dawkins, M. S. (2009), 'Water off a duck's back: Showers and troughs match ponds for improving duck welfare', *Appl. Anim. Behav. Sci.*, 116, 52–7. http://dx.doi.org/10.1016/j.applanim.2008.07.008.

Kaukonen, E., Norring, M. and Valros, A. (2016), 'Perches and elevated platforms in commercial broiler farms: use and effect on walking ability, incidence of tibial dyschondroplasia and bone mineral content', *Animal*, 11(5), 864–71. http://dx.doi.org/10.1017/S1751731116002160.

Kells, A., Dawkins, M. S., Borja, M. C. and Cortina, M. (2001), 'The effect of a 'freedom food' enrichment on the behaviour of broilers on commercial farms', *Anim. Welf.*, 10, 347–56.

Keppler, C. and Fölsch, D. W. (2000), 'Locomotive behaviour of hens and cocks (Gallus gallus f. dorn.) – Implication for housing systems', *Arch. Tierz., Dummerstorf*, 43, 184–8. http://archtierz.fbn-dummerstorf.de/pdf/2000/at00si2p184.pdf.

Kestin, S. C., Knowles, T. G., Tinch, A. E. and Gregory, N. G. (1992), 'Prevalence of leg weakness in broiler chickens and its relationship with genotype', *Vet. Rec.*, 131, 190–4. http://dx.doi.org/10.1136/vr.131.9.190.

Kiyma, Z., Küçükyılmaz, K. and Orojpour, A. (2016), 'Effects of perch availability on performance, carcass characteristics, and footpad lesions in broilers', *Arch. Anim. Breed.*, 59, 19–25. http://dx.doi.org/10.5194/aab-59-19-2016.

Koene, P., Ruiten, S. V. and Bokkers, E. (1999), The effect of increasing broiler behaviour possibilities by giving extra furniture and a slimmer body: the effect of perches and feed restriction, p. 136. In *Proceedings of the 33rd International Congress of the International Society for Applied Ethology, Lillehammer, Norway, 17–21 August 1999*, K. E. Boe, M. Bakken and B. O. Braastad (Eds). Agricultural University of Norway, Ås, Norway.

Lazarus, J. and Symonds, M. (1992), 'Contrasting effects of protective and obstructive cover on avian vigilance', *Anim. Behav.*, 43, 519–21. http://dx.doi.org/10.1016/S0003-3472(05)80110-1.
Leone, E. N. H. and Estevez, I. (2007), 'Patch discrimination as a function of handling time', *Poult. Sci.*, 86, 447–52. http://dx.doi.org/10.1093/ps/86.3.447.
Leone, E. H. and Estevez, I. (2008), 'Economic and welfare benefits of environmental enrichment for broiler breeders', *Poult. Sci.*, 87, 14–21. http://dx.doi.org/10.3382/ps.2007-00154.
Leyendecker, M., Hamann, H., Hartung, J., Kamphues, J., Neumann, U., Sürie, C. and Distl, O. (2005), 'Keeping laying hens in furnished cages and an aviary housing system enhances their bone stability', *Brit. Poult. Sci.*, 46, 536–44. http://dx.doi.org/10.1080/00071660500273094.
Lindqvist, C., Zimmerman, P. and Jensen, P. (2006), 'A note on contrafreeloading in broilers compared to layer chicks', *Appl. Anim. Behav. Sci.*, 101(1–2), 161–6. http://dx.doi.org/10.1016/j.applanim.2006.01.006.
Marchewka, J., Estevez, I., Vezzoli, G., Ferrante, V. and Makagon, M. (2015), 'The transect method: a novel approach to on-farm welfare assessment of commercial turkeys', *Poult. Sci.*, 94, 7–16. http://dx.doi.org/10.3382/ps/peu026.
Marchewka, J., Watanabe, T. T. N., Ferrante, V. and Estevez, I. (2013), 'Social and environmental factors affecting the behavior and welfare of turkeys (Meleagris gallopavo)', *Poult. Sci.*, 92, 1467–73. http://dx.doi.org/10.3382/ps.2012-02943.
Martin, C. D. and Mullens, B. A. (2012), 'Housing and dustbathing effects on northern fowl mites (Ornithonyssus sylviarum) and chicken body lice (Menacanthus stramineus) on hens', *Med. Vet. Entomol.*, 26(3), 323–33. http://dx.doi.org/10.1111/j.1365-2915.2011.00997.x.
Martrenchar, A., Huonnig, D., Cotte, J. P., Boilletot, E. and Morisse, J. P. (2000), 'Influence of stocking density, artificial dusk and group size on the perching behaviour of broilers', *Brit. Poult. Sci.* 41, 125–30. http://dx.doi.org/10.1080/713654921.
Martrenchar, A., Huonnig, D. and Cotte, J. P. (2001), 'Influence of environmental enrichments on injurious pecking and perching behaviour in young turkeys', *Brit. Poult. Sci.*, 42(2), 161–70. http://dx.doi.org/10.1080/00071660120048393.
McAdie, T. M., Keeling, L. J., Blokhuis, H. J. and Jones, R. B. (2005), 'Reduction in feather pecking and improvement of feather condition with the presentation of a string device to chickens', *Appl. Anim. Behav. Sci.*, 93(1–2), 67–80. http://dx.doi.org/10.1016/j.applanim.2004.09.004.
McGary, S., Estevez, I., Bakst, M. and Pollock D. (2002), 'Phenotypic traits as reliable indicators of fertility in male broiler breeders', *Poult. Sci.*, 81, 102–11. http://dx.doi.org/10.1093/ps/81.1.102.
Meehan, C. L. and Mench, J. A. (2007), 'The challenge of challenge: can problem solving opportunities enhance animal welfare?', *Appl. Anim. Behav. Sci.*, 102, 246–61. http://dx.doi.org/10.1016/j.applanim.2006.05.031.
Muiruri, H. K. and Harrison, P. C. (1991), 'Effect of roost temperature on performance of chickens in hot ambient environments', *Poult. Sci.*, 70, 2253–8. http://dx.doi.org/10.3382/ps.0702253.
Muiruri, H. K., Harrison, P. C. and Gonyou, H. W. (1991), 'The use of water-cooled roosts by hens for thermoregulation', *Appl. Anim. Behav. Sci.*, 28, 333–9. http://dx.doi.org/10.1016/0168-1591(91)90165-T.
Nielsen, B. L. (2004), 'Breast blisters in groups of slow-growing broilers in relation to strain and the availability and use of perches', *Brit. Poult. Sci.*, 45(3), 306–15. http://dx.doi.org/10.1080/00071660410001730798.
Newberry, R. C. (1995), 'Environmental enrichment: increasing the biological relevance of captive environments', *Appl. Anim. Behav. Sci.*, 44, 229–43. http://dx.doi.org/10.1016/0168-1591(95)00616-Z.
Newberry, R. C. (1999), 'Exploratory behaviour of young domestic fowl', *Appl. Anim. Behav. Sci.*, 63, 311–21. http://dx.doi.org/10.1016/S0168-1591(99)00016-7.
Newberry, R. C. (2004), Cannibalism, pp. 239–57. In *Welfare of the Laying Hen*, G. C. Perry (Ed.). CAB International Wallingford, UK. http://dx.doi.org/10.1079/9780851998138.0239.
Newberry, R. C. (2017a), Behavioral, emotional, and cognitive effects of domestication, pp. 315–29. In *APA Handbook of Comparative Psychology: Vol. 1. Basic Concepts, Methods, Neural Substrate, and Behavior*, J. Call, G. M. Burghardt, I. M. Pepperberg, C. T. Snowdon and T. R.

Zentall (Eds). American Psychological Association, APA Books, Washington DC. http://dx.doi.org/10.1037/0000011-016.

Newberry R. C. (2017b), Commercial free-range egg production practices, pp. 89–102. In *Egg Innovation and Strategies for Improvement*, P. Y. Hester (Ed.). Academic Press, Oxford, UK. http://dx.doi.org/10.1016/B978-0-12-800879-9.00009-3.

Newberry, R. C. and Blair, R. (1993), Perches ease handling of broiler chickens, pp. 270–1. In *Proceedings 4th Europ. Symp. Poultry Welfare, Edinburgh, Scotland*, C. J. Savory and B. O. Hughes (Eds). Universities Federation for Animal Welfare, Potters Bar, UK.

Newberry, R. C., Estevez, I. and Keeling, L. J. (2001), 'Group size and perching behaviour in young domestic fowl', *Appl. Anim. Behav. Sci.*, 73, 117–29. http://dx.doi.org/10.1016/S0168-1591(01)00135-6.

Newberry, R. C. and Hall, J. W. (1990), 'Use of pen space by broiler chickens: Effects of age and pen size', *Appl. Anim. Behav. Sci.*, 25, 125–36. http://dx.doi.org/10.1016/0168-1591(90)90075-O.

Newberry, R. C., Hunt, J. R. and Gardiner, E. E. (1985), 'Effect of alternating lights and strain on behaviour and leg disorders of roaster chickens', *Poultry Sci.*, 64, 1863–8. http://dx.doi.org/10.3382/ps.0641863.

Newberry, R. C., Hunt, J. R. and Gardiner, E. E. (1988), 'Influence of light intensity on the behavior and performance of broiler chickens', *Poultry Sci.*, 67, 1020–5. http://dx.doi.org/10.3382/ps.0671020.

Newberry, R. C., Keeling, L. J., Estevez, I. and Bilcik, B. (2007), 'Behaviour when young as a predictor of severe feather pecking in adult laying hens: The redirected foraging hypothesis revisited', *Appl. Anim. Behav. Sci.*, 107, 262–74. http://dx.doi.org/10.1016/j.applanim.2006.10.010.

Newberry, R. C. and Shackleton, D. M. (1997), 'Use of cover by domestic fowl: a Venetian blind effect?', *Anim. Behav.*, 54, 387–95. http://dx.doi.org/10.1006/anbe.1996.0421.

Nicol, C. J., Bestman, M., Gilani, A. M., De Haas, E. N., De Jong, I. C., Lambton, S., Wagenaar, J. P., Weeks, C. A. and Rodenburg, T. B. (2013), 'The prevention and control of feather pecking in laying hens: application to commercial systems', *World Poult. Sci. J.*, 69, 775–87. http://dx.doi.org/10.1017/S0043933913000809.

Nielsen, B. L., Thomsen, M. G., Sorensen, P. and Young, J. F. (2003), 'Feed and strain effects on the use of the outdoor areas by broilers', *Brit. Poultry Sci.*, 44, 161–9. http://dx.doi.org/10.1080/0007166031000088389.

Norring, M., Kaukonen, E. and Valros, A. (2016), 'The use of perches and platforms by broiler chickens', *Appl. Anim. Behav. Sci.*, 184, 91–6. http://dx.doi.org/10.1016/j.applanim.2016.07.012.

Olsson, I. A. S. and Keeling, L. J. (2005), 'Why in earth? Dustbathing behaviour in jungle and domestic fowl reviewed from a Tinbergian and animal welfare perspective', *Appl. Anim. Behav. Sci.*, 93, 259–82. http://dx.doi.org/10.1016/j.applanim.2004.11.018.

Olsson, P., Lind, O. and Kelber, A. (2015), 'Bird colour vision: behavioural thresholds reveal receptor noise', *J. Exp. Biol.*, 218, 184–93. http://dx.doi.org/10.1242/jeb.111187.

Petherick, J. C. and Duncan, I. J. H. (1989), 'Behaviour of young domestic fowl directed towards different substrates', *Brit. Poult. Sci.*, 30, 229–38. http://dx.doi.org/10.1080/00071668908417143.

Pettit-Riley, R. and Estevez, I. (2001), 'Effects of density on perching behavior of broiler chickens', *Appl. Anim. Behav. Sci.*, 71, 127–40. http://dx.doi.org/10.1016/S0168-1591(00)00174-X.

Pichova, K., Nordgreen, J., Leterrier, C., Kostal, L. and Moe, R. O. (2016), 'The effects of food-related environmental complexity on litter directed behaviour, fear and exploration of novel stimuli in young broiler chickens', *Appl. Anim. Behav. Sci.*, 174, 83–9. http://dx.doi.org/10.1016/j.applanim.2015.11.007.

Pizzari, T. (2016), 'The Wood-Gush legacy: A sociobiology perspective to fertility and the welfare of chickens', *Appl. Anim. Behav. Sci.*, 181, 12–18. http://dx.doi.org/10.1016/j.applanim.2016.01.025.

Prescott, N. B., Wathes, C. M. and Jarvis, J. R. (2003), 'Light, vision and the welfare of poultry', *Anim. Welf.*, 12(2), 269–88.

Reilly, M. W., Koelkebeck, K. W. and Harrison, P. C. (1991), 'Performance evaluation of heat-stressed commercial broilers provided water-cooled floor perches', *Poult. Sci.*, 70, 1699–703. http://dx.doi.org/10.3382/ps.0701699.

Riber, A. B. and Guzman, D. A. (2016), 'Effects of dark brooders on behavior and fearfulness in layers', *Animals*, 6(1), 3. http://dx.doi.org/10.3390/ani6010003.

Richards, G. J., Wilkins, L. J., Knowles, T. G., Booth, F., Toscano, M. J., Nicol, C. J. and Brown, S. N. (2011), 'Continuous monitoring of pop hole usage by commercially housed free-range hens throughout the production cycle', *Vet. Rec.*, 169, 338. http://dx.doi.org/10.1136/vr.d4603.

Richards, G. J., Wilkins, L. J., Knowles, T. G., Booth, F., Toscano, M. J., Nicol, C. J. and Brown, S. N. (2012), 'Pop hole use by hens with different keel fracture status monitored throughout the laying period', *Vet. Rec.*, 170, 494. http://dx.doi.org/10.1136/vr.100489.

Rodriguez Aurrekoetxea, A. (2016), The behaviour and use of space of the domestic fowl in alternative meat and egg production systems, PhD Thesis, University of the Basque Country, Bilbao, Spain. http://dx.doi.org/10810/21173.

Rodríguez-Aurrekoetxea, A. and Estevez, I. (2014), 'Aggressiveness in the domestic fowl: Distance versus 'attitude''. *Appl. Anim. Behav. Sci.*, 153, 68–74. https://doi.org/10.1016/j.applanim.2014.01.007.

Rodríguez-Aurrekoetxea, A. and Estevez, I. (2016), 'Use of space and its impact on the welfare of laying hens in a commercial free-range system', *Poult. Sci.*, 95, 2503–13. http://dx.doi.org/10.3382/ps/pew238.

Rodríguez-Aurrekoetxea, A., Leone E. H. and Estevez, I. (2014), 'Environmental complexity and use of space in slow growing free range chickens', *Appl. Anim. Behav. Sci.*, 161, 86–94. http://dx.doi.org/10.1016/j.applanim.2014.09.014.

Rodríguez-Aurrekoetxea, A., Leone E. H. and Estevez, I. (2015), 'Effects of panels and perches on the behaviour of commercial slow-growing free-range chickens', *Appl. Anim. Behav. Sci.*, 165, 103–11. https://doi.org/10.1016/j.applanim.2015.02.004.

Roper, T. J. and Marples, N. M. (1997), 'Odour and behaviour as cues for taste-avoidance learning in domestic chicks', *Anim. Behav.*, 53, 1241–50. http://dx.doi.org/10.1006/anbe.1996.0384.

Rosenzweig, M. R. and Bennett, E. L. (1996), 'Psychobiology of plasticity: effects of training and experience on brain and behaviour', *Behav. Brain. Res.*, 78, 57–65. http://dx.doi.org/10.1016/0166-4328(95)00216-2.

Ruiz-Feria, C. A., Arroyo-Villegas, J. J., Pro-Martinez, A., Bautista-Ortega, J., Cortes-Cuevas, A., Narciso-Gaytan, C., Hernandez-Cazares, A. and Gallegos-Sanchez, J. (2014), 'Effects of distance and barriers between resources on bone and tendon strength and productive performance of broiler chickens', *Poult. Sci.*, 93, 1–10. http://dx.doi.org/10.3382/ps.2013-03421.

Schenk, A., Porter, A. L., Alenciks, E., Frazier, K., Best, A. A., Fraley, S. M. and Fraley, G. S. (2016), 'Increased water contamination and grow-out Pekin duck mortality when raised with water troughs compared to pin-metered water lines using a United States management system', *Poult. Sci.*, 95, 736–48. http://dx.doi.org/10.3382/ps/pev381.

Sherwin, C. M., Lewis, P. D. and Perry, G. C. (1999), 'Effects of environmental enrichment, fluorescent and intermittent lighting on injurious pecking amongst male turkey poults', *Appl. Anim. Behav. Sci.*, 62, 319–33. http://dx.doi.org/10.1080/00071669986954.

Shields, S., Garner, J. P. and Mench, J. A. (2004), 'Dustbathing by broiler chickens: a comparison of preference for four different substrates', *Appl. Anim. Behav. Sci.*, 86, 291–8. http://dx.doi.org/10.1016/j.applanim.2004.01.003.

Stadig, L. M., Rodenburg, T. B., Ampe, B., Reubens, B. and Tuyttens, F. A. M. (2016), 'Effect of free-range access, shelter type and weather conditions on free-range use and welfare of slow-growing broiler chickens', *Appl. Anim. Behav. Sci.*, 192, 15–23. http://dx.doi.org/10.1016/j.applanim.2016.11.008.

Su, G., Sorensen, P. and Kestin, S. C. (2000), 'A note on the effects of perches and litter substrate on leg weakness in broiler chickens', *Poult. Sci.*, 79, 1259–63. http://dx.doi.org/10.1093/ps/79.9.1259.

Tahamtani, F. M., Brantsæter, M., Nordgreen, J., Sandberg, E., Hansen, T. B., Nødtvedt, A., Rodenburg, T. B., Moe, R. O. and Janczak, A. M. (2016), 'Effects of litter provision during early rearing and environmental enrichment during the production phase on feather pecking and feather damage in laying hens', *Poult. Sci.*, 95(12), 2747–56. http://dx.doi.org/10.3382/ps/pew265.

Tauson, R. (1984), 'Effects of a perch in conventional cages for laying hens', *Acta Agric. Scand.*, 34, 193–209. http://dx.doi.org/10.1080/00015128409435389.

Taylor, P. E., Scott, G. B. and Rose, P. (2003), 'The ability of domestic hens to jump between horizontal perches: effects of light intensity and perch colour', *Appl. Anim. Behav. Sci.*, 83, 99–108. http://dx.doi.org/10.1016/S0168-1591(03)00127-8.

Toghyani, M., Gheisari, A., Modaresi, M., Tabeidian, S. A. and Toghyani, M. (2010), 'Effect of different litter material on performance and behavior of broiler chickens', *Appl. Anim. Behav. Sci.*, 122(1), 48–52. http://dx.doi.org/10.1016/j.applanim.2009.11.008.

Vallortigara, G. (1996), 'Learning of colour and position cues in domestic chicks: males are better at position, females at colour', *Behav. Proc.*, 36, 289–96. http://dx.doi.org/10.1016/0376-6357(95)00063-1.

van Liere, D. W. (1992), 'The significance of fowls' bathing in dust', *Anim. Welf.*, 1, 187–202.

van Liere, D. W., Kooijman, J. and Wiepkema, P. R. (1990), 'Dustbathing behaviour of laying hens as related to quality of dustbathing material', *Appl. Anim. Behav. Sci.*, 26, 127–41. http://dx.doi.org/10.1016/0168-1591(90)90093-S.

Ventura B. A., Siewerdt, F. and Estevez I. (2010), 'Effects of barrier perches and density on broiler leg health, fear and performance', *Poult. Sci.*, 89, 1574–83. http://dx.doi.org/10.3382/ps.2009-00576.

Ventura, B. A., Siewerdt, F. and Estevez, I. (2012), 'Access to barrier perches improves behavior repertoire in broilers', *PLoS One*, 7(1), e29826. https://doi.org/10.1371/journal.pone.0029826.

Vestergaard, K. S. and Sanotra, G. S. (1999), 'Relationships between leg disorders and changes in the behaviour of broiler chickens', *Vet. Rec.*, 144, 205–9. http://dx.doi.org/10.1136/vr.144.8.205.

Vestergaard, K. S., Skadhauge, E. and Lawson, L. G. (1997), 'The stress of not being able to perform dustbathing in laying hens', *Physiol. Behav.*, 62, 413–19. http://doi.org/10.1016/S0031-9384(97)00041-3.

Vezzoli, G., Mullens, B. A. and Mench A. (2015), 'Dustbathing behavior: Do ectoparasites matter?', *Appl. Anim. Behav. Sci.*, 169, 93–9. http://dx.doi.org/10.1016/j.applanim.2015.06.001.

Villagrá, A., Olivas, I., Althaus, R. L., Gómez, E. A., Lainez, M. and Torres, A. G. (2014), 'Behavior of broiler chickens in four different substrates: a choice test', *Rev. Bras. Cienc. Avic.*, 16, 67–76. http://dx.doi.org/10.1590/S1516-635X2014000100010.

Weeks, C. A., Danbury, T. D., Davies, H. C. Hunt, P. and Kestin. S. C. (2000), 'The behaviour of chickens and its modification by lameness', *Appl. Anim. Behav. Sci.*, 67, 111–25. http://dx.doi.org/10.1016/S0168-1591(99)00102-1.

Whiteside, M. A., Sage, R. and Madden, J. R. (2016), 'Multiple behavioural, morphological and cognitive developmental changes arise from a single alteration to early life spatial environment, resulting in fitness consequences for released pheasants', *R. Soc. Open Sci.* 3, 160008. http://dx.doi.org/10.1098/rsos.160008.

Widowski, T. M. and Duncan, I. J. H. (2000), 'Working for a dustbath: are hens increasing pleasure rather than reducing suffering?', *Appl. Anim. Behav. Sci.*, 68, 39–53. http://dx.doi.org/10.1016/S0168-1591(00)00088-5.

Wood-Gush D. G. M., Duncan, I. J. H. and Savory, C. J. (1978), 'Observations on the social behaviour of domestic fowl in the wild', *Biol. Behav.*, 3, 193–205. http://dx.doi.org/10.1016/0304-3762(78)90091-3.

Wright, M. W. and Bowmaker, J. K. (2001), 'Retinal photoreceptors of paleognathous birds: the ostrich (Struthio camelus) and rhea (Rhea americana)', Vision Res., 41(1), 1–12. http://dx.doi.org/10.1016/S0042-6989(00)00227-3.

Yngvesson, J., Gustafson, J., Berg, C., Larsson, I., Gunnarsson, S. and Oden, K. (2011),'A field study of access to day light, ammonia, plumage condition and mortality in loose housed laying hens in south east Sweden', p. 53. In *Proceedings of 30th Poultry Sci. Symp.*, World Poultry Sci.

Assoc. http://www.cabi.org/Uploads/animal-science/worlds-poultry-science-association/WPSA-glasgow-scotland-2011/25_Yngvesson_short_oral.pdf.

Zeltner, E. and Hirt, H. (2008), 'A note on fear reaction of three different genetic strains of laying hens to a simulated hawk attack in the hen run of a free-range system', *Appl. Anim. Behav. Sci.*, 113, 69–73. http://dx.doi.org/10.1016/j.applanim.2007.12.003.

Chapter 14

Hot weather management of poultry

Brian Fairchild, University of Georgia, USA

1 Introduction

Hot weather management of poultry has been a topic of research and discussion for centuries. There is discussion of poultry husbandry practices with regard to managing poultry dating back to writings of the Egyptians and Romans. In terms of managing poultry in hot weather, many studies have been conducted to evaluate the impacts of genetics, biology, chemistry, nutrition and management in keeping birds productive during torrid days of the year. While genetics and nutrition are a part of the process of getting optimal performance from poultry during hot weather, the focus of this chapter will be on the management practices implemented in buildings on the farm. In poultry management, a primary objective is to reduce environmental stress so that the energy derived from the feed consumed can be used for body growth and development. Housing contributes greatly to this objective. Poultry housing will protect birds from predators and contribute to biosecurity by preventing exposure to biological vectors that may carry viruses and bacterial that are harmful to birds and food safety. When properly constructed and operated, a high level of environmental control can be achieved. This means that instead of the bird being exposed to outside temperature swings in excess of 15°C, in well-managed poultry housing conditions the bird could only experience variations of 5°C or less in a given day.

The ventilation system is a primary factor in keeping poultry cool during hot weather, but depending on the type of house other attributes contribute as well. For example, naturally ventilated poultry houses are typically oriented east-west to minimize the amount of direct sunlight entering the house. Direct sunlight on birds can create heat stress even if air temperature would not normally be an issue. The house roof overhang over the side wall will depend on the height of the side wall and the size of the curtain opening. More overhang is required as the side wall height and curtain opening increases. The design of roof will affect how much radiant heat is collected. Steep roof slopes of 45° will collect less radiant heat than a flat roof. Ideally, all roofs would be insulated in such

http://dx.doi.org/10.19103/AS.2016.0011.29

a way to minimize the heat collected. There are multiple methods of insulating roofs (dropped ceiling, polystyrene or polyurethane rigid board insulation, polyurethane foam and reflective insulation) but each has its advantages and disadvantages. The important aspect to remember is that the insulation needs to be done in a way that does not leave cracks or gaps for heat to transfer into the house. A dropped ceiling is popular in many countries creating an attic that can be insulated to prevent heat collected by the tin roof from entering the house into the bird environment. In houses that have tight construction to minimize air leakage these attics can be a source of warmer air during cooler months of the year as well as helping with brooding temperature during all seasons.

2 Sources of heat and heat loss mechanisms

2.1 Sources of heat in poultry housing

In order to control temperature in poultry houses, one has to understand what sources contribute to the heat gain. Many people often think that heat comes from outside the poultry house from high temperatures and radiant energy from the sun. However, with properly constructed poultry housing using modern building specifications, they are insulated to reduce the impact that the outside environment has on the inside conditions in the building. If one calculated all of the heat gain from potential heat sources within a poultry house they would see that the heat gained from insulated surfaces is relatively small compared to what the birds are producing themselves. When it comes to hot weather management the biggest source of heat in insulated poultry houses is from the birds themselves (Fig. 1). A broiler weighing 2.3 kg will produce 60 BTU/hr of sensible heat (60 BTU heat is comparable to 60 lit matchsticks). This heat needs to be removed from the poultry house otherwise house temperature will increase as would bird body temperature.

2.2 Mechanisms of heat loss in birds

All animals have to maintain body temperature by dissipating the metabolic heat that is produced constantly. Poultry lose heat through four primary mechanisms: radiation, conduction, evaporation and convection.

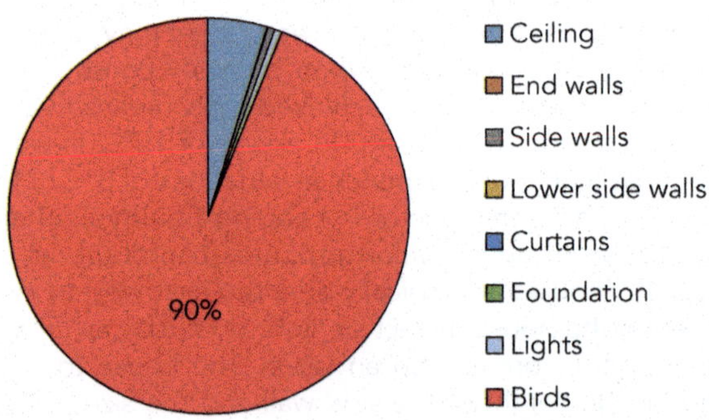

Figure 1 An illustration of the heat sources within a properly constructed and insulated poultry house raising a 2.72 kg broiler with a ceiling insulation $r = 19$ and the walls insulated at $r = 12$.

Radiation

The transfer of heat from the bird to the air. As the environmental air temperature in a poultry house increases, the amount of heat lost from the bird through radiation decreases.

Conduction

The transfer of heat to a solid object that is cooler than the bird body temperature. Typically, this method does not represent a large removal of heat from the bird. In poultry houses floors are covered with bedding to absorb moisture and dilute faecal material. Bedding materials have insulation like properties and they are slow to take on or release heat.

Evaporation

Poultry do not possess sweat glands, but they can evaporate moisture from the surfaces of the mouth and respiratory passages through panting and this is referred to as latent heat loss. Latent heat loss is a primary means of cooling for poultry. Under thermoneutral conditions poultry remove 60% of their heat using latent heat loss/evaporation (Simmons et al. 1997). As the environmental relative humidity (RH) increases, the ability to evaporate moisture through panting decreases. Furthermore, panting can result in respiratory alkalosis. As birds pant, the loss of carbon dioxide is increased, which results in a reduction of carbon dioxide partial pressure and thus bicarbonate in blood plasma. In turn the removal of bicarbonate results in fewer hydrogen ions in the blood which results in an increase in pH of the plasma that can lead to a condition referred to as alkalosis. Dehydration is another issue that can occur with panting if proper volumes of water are not provided to the birds. Poultry producers strive to minimize and prevent conditions that lead to panting. While there are management options to counteract the negative side effects of panting (dehydration, respiratory alkalosis) such as providing ample volumes of water and supplementing electrolytes in the diet or water, it is preferable to keep the birds from panting so that the energy associated with the process continues to go towards growth and development.

Convection

The transfer of heat through a moving fluid. In poultry houses, the air movement leads to a transfer of heat from the birds to the air.

2.3 Sensible and latent heat loss

While poultry may lose heat through these four mechanisms, the two primary ones are evaporation and convection. Heat loss is divided into sensible or latent heat loss. Sensible heat results in a rise in temperature when it is absorbed by a substance. The sensible heat produced by the birds causes a rise in house air temperature. Latent heat is related to a substance that changes from a solid to liquid or liquid to a gas. In poultry production, bird latent heat loss is the evaporation of water from its body that does not increase house temperature but does increase house humidity. Poultry will lose 60% of their heat through latent heat loss and 40% of their heat through sensible heat loss. This partitioning of heat loss will exist even under thermoneutral conditions. It is important to remember that poultry lose more heat through evaporation than people do. Therefore, poultry are much more sensitive to RH than humans. This means conditions that 'feel good' to people may not be the optimum environment for the birds and that poultry producers should monitor RH as well as ambient temperature in their poultry houses.

3 Poultry house ventilation systems

Poultry houses and their ventilation systems are dynamic and have to provide an environment as close to thermoneutral temperature as possible for chicks on the coldest day of the year and market-age or adult birds on the hottest day of the year. The environmental temperature target of poultry is lower as bird age and body mass increases (Fig. 2).

The ventilation system of modern poultry houses can be classified into two modes: minimum ventilation and tunnel ventilation. The minimum ventilation system will utilize 40–50% of the house fan power and air will enter through the side wall or ceiling inlets. The purpose of the minimum ventilation system is to maintain air quality (carbon dioxide, ammonia, carbon monoxide, RH). The primary factor that drives minimum ventilation is moisture levels inside the house. Minimum ventilation removes moisture from the house while maintaining an optimum temperature and air quality environment for bird growth, health and welfare. RH in the range of 40–60% will help remove enough moisture to maintain dry floors. Minimum ventilation can also remove a limited amount of heat from the house and is used in some houses as cross-ventilation. However, this type of ventilation during hot weather is only suited for narrow poultry houses that are 10–12 m wide. In wider houses of 16–20 m, cross-ventilation can result in larger differences in bird environment across the width of the house. In poultry management not only do producers primarily try to reduce bird stress but they are also trying to provide uniform conditions from wall to wall and from front to back of the poultry house. That way every bird in the house is exposed to similar environmental conditions and will have the same opportunity to reach their full genetic potential. In houses of 16–20 m in width with cross-ventilation, the air movement can differ from one side to the other exposing birds to different microenvironments within the same house. This could lead to poor bird uniformity in weight gain, egg production and egg fertility.

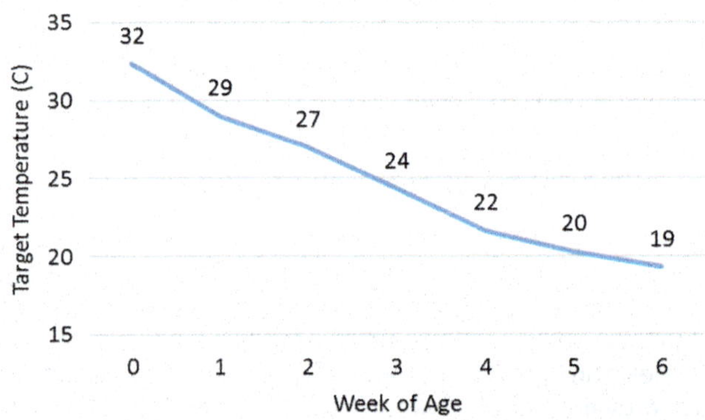

Figure 2 Target temperatures in broiler houses based on age using a radiant heat source. These temperature recommendations are an average from the broiler production manuals of three primary broiler breeder companies (Aviagen, Cobb-Vantress, Hubbard).

Tunnel ventilation was introduced to the poultry industry in the 1980s and can provide a consistent air movement over the birds across the width of the house. When house temperature gets approximately 5°C above the set temperature then the house will switch from minimum ventilation to tunnel ventilation (Fig. 3 and 4). Some poultry houses employ the sidewall inlets used during minimum ventilation to ventilate the houses during hot weather. In tunnel ventilation mode, the side wall inlets are closed, large inlets in front of the house are opened and the fans in the rear of the house will move air out of the building creating a

Figure 3 The front/inlet end of a typical tunnel ventilated broiler house 15 m x 152 m with 7 week old broilers.

Figure 4 The fan/back end of a typical tunnel ventilation broiler house 15 m x 152 m with 7 week old broilers. The picture also shows a 25 cm migration barrier used to keep birds evenly distributed down the length of the house.

negative pressure. The air outside will want to rush in through large inlets at the front of the house. As the air moves down the house at a speed of 3 m/s (the target for average broiler house airspeed is typically 3 m/s, but poultry companies and farmers raising broilers larger than 4 kg may target higher airspeeds as high as 4.5 m/s), not only does this continue to remove heat from the house but a wind chill is created that reduces the effective temperature even more. The effective temperature can be defined as the combination of air temperature, RH and airspeed on the sensation of warmth and cold felt by an animal.

3.1 Tunnel ventilation and bird performance

Using air velocity to remove heat and cool poultry during hot weather has been shown to be beneficial in many ways that include reduced panting (Fig. 5), increased livability, increased weight gain (Fig. 6), better feed conversion (Fig. 7) and less stress on the birds.

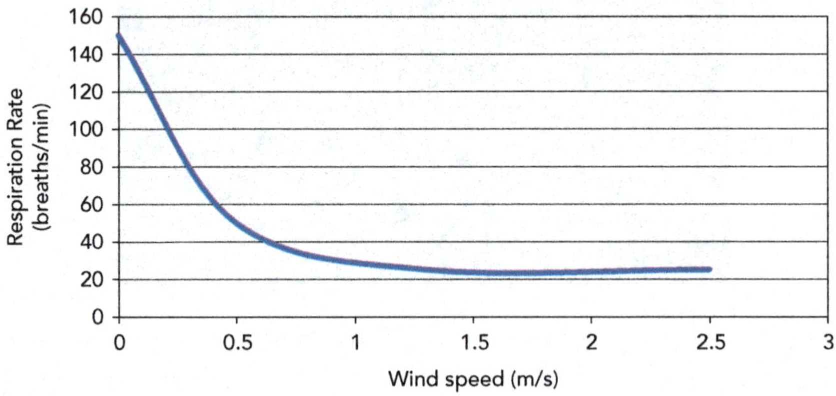

Figure 5 Respiration rate decreased as airspeed increased (Drury and Siegal 1966).

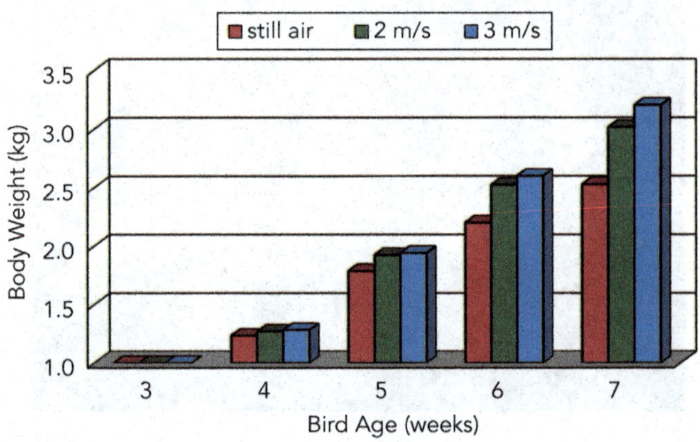

Figure 6 Broiler weights were heavier in treatments with higher airspeed. Older birds benefited from 600 ft/min more than younger birds (Simmons et al. 2003).

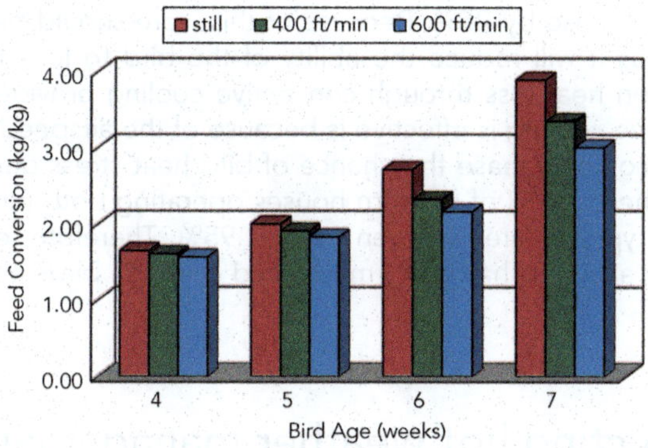

Figure 7 Feed conversion improved as airspeed increased (Dozier et al. 2005).

As mentioned earlier, the houses are designed to cool market-age birds on the hottest day of the year. Therefore, the maximum airspeed of a house will not be required at all times. There are a number of factors that will influence the amount of airspeed required and include bird age, stocking density, body weight, air temperature, RH and feed intake. As a result, younger birds do not need as much airspeed to keep their body temperature normal as older birds because their surface area to mass ratio is greater, they are less feathered, their mass is less and they have more floor space per bird.

Pullet and broiler breeder houses usually have less airspeed requirements than broiler grow out houses for a number of reasons. Pullets grow slower than broilers and weigh less. As a result they produce less heat at 3.5 BTU/lb/h compared to the 6.0 BTU/lb/h for broilers. The stocking density is less than broilers resulting in fewer pounds of live weight per house. These birds are not full fed as pullet growth rate is regulated through feeding programmes. On breeder farms where hatching eggs are produced, there is less heat generated in the house by the birds because there are fewer birds in the house. In addition these chickens are mature and are no longer growing meaning they are producing less heat then broiler do.

3.2 Evaporative cooling

Another component of the poultry house hot weather ventilation system is evaporative cooling. While airspeed is the primary means of cooling birds during hot weather, its effectiveness decreases as air temperature increases above 30°C. The evaporative cooling system reduces incoming air temperature which maintains the effectiveness of airspeed cooling on the hottest days of the year. The two main types of evaporative cooling systems are interior fogging or recirculating pads. Fogging systems introduce water to the house using high water pressure through nozzles to create a small water droplet to the point that it is a mist to enhance evaporation. Pad systems are installed at the tunnel inlet and pull air through a pad that is wetted with water. Fogging systems could be used in both naturally ventilated and tunnel ventilated houses while pad systems can only be used in tunnel ventilated houses. The house water supply should be capable of delivering 10 L/min for every metre length of evaporative cooling pad. Air temperature is reduced by evaporating water. In doing so RH is increased. For every degree decrease in temperature, the RH will increase 2.5%. So if the system is reducing air temperature 5°C then the RH in the house would increase by 25%. Because of this, care should be taken when using evaporative cooling as it will reduce the ability of the bird to lose heat via respiration making it rely more on heat loss through convective cooling provided by the airspeed. The reason evaporative cooling is effective is because of the airspeed. Without airspeed, evaporative cooling could increase the chance of bird heat stress due to the increase in RH. During hot weather the RH of modern houses operating both tunnel ventilation and evaporative cooling typically are between 85 and 95%. Therefore, evaporative cooling should be used after airspeed has been maximized over the birds and air temperatures reach or exceed 30°C.

4 Factors affecting hot weather management

There are many components of poultry house management that can influence the bird environment and body temperature during hot weather. The factors that can influence bird heat loss/gain include bird density, feed intake and ventilation system management. Bird density may seem like a simple variable to control, but one needs to understand the entire context of poultry rearing to fully appreciate the issues. In many areas around the world when birds are placed in houses at the beginning of the flock/production cycle only a portion of the house is used. For example, many farms utilized partial house brooding where 50% of the floor space is used for the brooding period during the first 7–14 days of the flock. There are two forms of partial house brooding: half house and centre house. This is done to reduce the area that has to be heated to brooding temperature, thus possibly reducing fuel usage and energy costs. It also maintains a better bird weight to floor space ratio, reduces the amount of feeders that are needed throughout the house and makes vaccine applications needed during the first 14 days easier to administer. After the brooding period is concluded, the birds are given access to the rest of the house. It takes about 5–7 days to get the birds evenly distributed throughout the entire house length. Once that is achieved then farmers typically install migration barrier to prevent large numbers of birds from accumulating in the front, middle or rear of the house. Ideally, the birds would be evenly distributed down the length of the house. If this is not achieved in hot weather, then there could be more bird heat in areas of the house where there are more birds. Keeping birds evenly distributed down the length of the house also results in exposing them to more uniform air movement. The more birds that are located in a specific area of the house the less likely that all birds are getting equal amounts of air moving across them, meaning that instead of moving over and around them, the air may just move over them.

When a bird's body temperature increases various physiological changes occur to dissipate heat, especially when it pants. During panting carbon dioxide is removed at a greater rate. This can change electrolyte blood serum concentrations over time. During periods of hot weather it is not unusual to see farm managers add electrolytes to the water. Multiple studies have been conducted on the effects of adding electrolytes or vitamins, such as vitamin A (Moreng 1980), vitamin C (Pardue et al. 1984, 1985), and vitamin E (Ferket and Qureshi, 1992), to poultry drinking water. These practices can help birds endure hot-weather-related stress. However, efforts should focus on the prevention of heat stress by focusing on management to keep the bird body temperature in the normal range to ensure optimal growth, feed conversion and welfare.

5 Conclusions

- Modern poultry housing systems have a high level of environmental control providing effective temperatures that are near their thermoneutral temperature allowing the birds to maintain a normal body temperature and remain productive during hot weather.
- The airspeed generated during tunnel ventilation is the primary means of keeping commercial poultry cool during hot weather.
- Evaporative cooling can augment convective cooling generated by the airspeed, but it is not a substitution and should be used when house temperatures get above 30°C.
- Current issues that are likely to be encountered now as well as in the future include but are not limited to:

    - Better education of what temperatures evaporative cooling systems should operate. In many cases, the evaporative cooling system is turned on at temperatures that are too low (28°C or lower). This results in increased RH but reduced cooling efficiency.
    - More emphasis on getting bird density evenly distributed throughout the house. Increased bird density in one area of the house leads to higher environmental and bird body temperature and increases competition for feed and drinker space.
    - Proper design of ventilation systems to provide the desired air exchange and airspeed.

6 Where to look for further information

- Daghir, N.J. ed., 2008. *Poultry Production in Hot Climates*. CABI.
- *Poultry Housing Tips*, University of Georgia Cooperative Extension Service www.poultryventilation.com.
- Department for Environment, Food and Rural Affairs, 2005. Heat stress in poultry-solving the problem. https://www.gov.uk/government/uploads/system/uploads/attachment_data/file/69373/pb10543-heat-stress-050330.pdf

7 References

Aviagen, 2014. Ross broiler production manual. http://en.aviagen.com/assets/Tech_Center/Ross_Broiler/Ross-Broiler-Handbook-2014i-EN.pdf.

Chepete, H. J. and H. Xin, 2001. Heat and moisture production of poultry and their housing systems – a literature review. Proceedings of the 6th International Symposium, Louisville, Kentucky. ASAE Publication Number 701P0201, 319–35.

Cobb-Vantress, 2012. COBB broiler management guide. https://cobb-guides.s3.amazonaws.com/07fc99a0-bbd2-11e6-bd5d-55bb08833e29.pdf.

Dozier, W. A., B. D. Lott and S. L. Branton, 2005a. Growth responses of male broilers subjected to increasing air velocities at high ambient temperatures and a high dew point. *Poult. Sci.* 84:962–6.

Dozier, W. A., B. D. Lott and S. L. Branton, 2005b. Live performance of male broilers subjected to constant or increasing air velocities at moderate temperatures with a high dew point. *Poult. Sci.* 84:1328–31.

Drury, L. N. and H. S. Siegel, 1966. Air velocity and heat tolerance of young chickens. *Trans. ASAE* 9:583–5.

Ferket, P. R. and M. A. Qureshi, 1992. Performance and immunity of heat-stressed broilers fed vitamin- and electrolyte-supplemented drinking water. *Poult. Sci.* 71:88–97.

Genç, L. and K. M. Portier, 2005. Sensible and latent heat productions from broilers in laboratory conditions. *Turk. J. Vet. Anim. Sci.* 29:635–43.

Hubbard LLC, 2016. Broiler management manual. http://www.hubbardbreeders.com/media/manual_broiler_management_en___013796700_1441_27062016.pdf.

Lacy, M. P. and M. Czarick, 1992. Tunnel-ventilated broiler houses: broiler performance and operating costs. *J. Appl. Poult. Res.* 1:104–9.

Lott, B. D., J. D. Simmons and J. D. May, 1998. Air velocity and high temperature effects on broiler performance. *Poult. Sci.* 77:391–3.

May, J. D., B. D. Lott and J. D. Simmons, 1997. Water consumption by broilers in high cyclic temperatures: bell versus nipple waterers. *Poult. Sci.* 76:944–7.

Mitchell, M. A., 1985. Effects of air velocity and convective and radiant heat transfer from domestic fowls at environmental temperatures of 20 and 30°C. *Br. Poult. Sci.* 26:413–23.

Moreng, R. E., 1980. Temperature and vitamin requirements of the domestic fowl. *Poult. Sci.* 59:782–5.

Pardue, S. L., J. P. Thaxton and J. Brake, 1984. Effects of dietary ascorbic acid in chicks exposed to high environmental temperature. *J. Appl. Physiol.* 58:1511–16.

Pardue S. L., J. P. Thaxton and J. Brake, 1985. Influence of supplemental ascorbic acid on broiler performance following exposure to high environmental temperature. *Poult. Sci.* 64:1334–8.

Pesti, G. M., S. V. Armato and L. R. Minear, 1985. Water consumption of broiler chickens under commercial conditions. *Poult. Sci.* 64:803–8.

Reece, F. N., J. W. Deaton and L. F. Kubena, 1972. Effects of high temperature and humidity on heat prostration of broiler chickens. *Poult. Sci.* 51:2021–5.

Scott, T. A. and D. Balnave, 1988. Comparison between concentrated complete diets and self-selection for feeding sexually maturing pullets at hot and cold temperatures. *Br. Poult. Sci.* 29:613–25.

Simmons, J. D., B. D. Lott and J. D. May, 1997. Heat loss from broiler chickens subjected to various wind speeds and ambient temperatures. *Appl. Eng. Agric.* 13:665–9.

Simmons, J. D., B. D. Lott and D. M. Miles, 2003. The effects of high air velocity on broiler performance. *Poult. Sci.* 82:232–4.

Sturkie, P. D., 1965. *Avian Physiology*, 2nd edn. Cornstock Press: Ithaca, New York.

Timmons, M. B., 1989. Improving ventilation in open-type poultry housing. Proceedings of the 1989 Poultry Symposium, University of California, pp. 1–8.

Van Der Hel, W., M. W. A. Verstegen, A. M. Henken and H. A. Brandsma, 1991. The upper critical ambient temperature in neonatal chicks. *Poult. Sci.* 70:1882–7.

Van Kampen, M., 1981. Water balance of colostomized and non-colostomized hens at different ambient temperature. *Br. Poult. Sci.* 22:17–23.

Wilson, J. D., 1983. Ventilation air distribution. In *Ventilation of Agricultural Structures. ASAE Monograph*. American Society of Agricultural Engineers: St Joseph, Michigan.

Chapter 15

Transportation and the welfare of poultry

K. Schwean-Lardner and T. G. Crowe, University of Saskatchewan, Canada

1 Introduction

Commercial poultry may be transported for a variety of reasons under a number of different circumstances, including from a hatchery to a farm, pullets from a rearing facility to a lay facility or broilers/turkeys and end-of-lay (EOL) hens to slaughter facilities (Nielsen et al., 2011). Fuelled by increasing consumption of poultry meat across the world (http://www.fao.org/ag/againfo/themes/en/meat/background.html), immense numbers of birds are transported daily, and the numbers are increasing.

Undoubtedly, the production economics can be affected by the transportation procedure if the health of the birds or quality of the carcass is negatively impacted. In addition, transportation affects the welfare of birds, and the relationship is complex. There is a vast array of factors that singly can influence how birds can cope with transport, including the species or type of bird, age of the bird, previous experiences, extreme temperature variations, relative humidity, vibration, noise, change in social structure, density variations, stress from human interaction, trailer design, catching crew skill and experience, unfamiliar environments, and lack of food and water (Bayliss and Hinton, 1990; Mitchell and Kettlewell, 1994; Warriss et al., 1997; EFSA, 2011; Nielsen et al., 2011; Glatz and Rodda, 2013; Jacobs et al., 2016a,b). What increases the complexity though is the ability of these factors to act simultaneously, interact with each other and exacerbate the situation with larger stresses and challenges.

The impact on the birds can range from slight discomfort to death (Knowles and Broom, 1990; Nicol and Scott, 1990; Mitchell and Kettlewell, 1994). The majority of literature reported in this chapter focuses on the biological and physiological impacts of poultry

http://dx.doi.org/10.19103/AS.2016.0011.30

transportation. Examining the psychological impact is more difficult, and few studies have investigated that component.

2 Impacts on poultry welfare before and during transport

2.1 Pre-loading impacts on transportation welfare

The number of birds dead on arrival (DOA) can be significantly affected by the health status of the flock on-farm. Flocks that demonstrate high levels of mortality during placement (Jacobs et al., 2016a), the grow-out period (Chauvin et al., 2011) or close to slaughter age (Kittelsen et al., 2015; Jacobs et al., 2016a) tend to have higher DOA levels during transport. The relationship, however, is not always consistent, and good culling practices on farm can result in lower DOA levels during transport (Jacobs et al., 2016a), suggesting that DOA levels can be reduced if birds being transported are healthy and fit for transport. Relative humidity in barns can also impact bird welfare, and any birds that are loaded in cold environments while feathers are wet face a significant physiological challenge, as maintaining core body temperature is even more difficult under wet conditions (Hunter et al., 1999). Care in catching, and possibly catching type (manual vs mechanical) can impact DOA levels, with higher levels of injuries during the catching process showing a positive relationship with DOA numbers (Chauvin et al., 2011). Poor catching techniques result in increases in wing fractures (Jacobs et al., 2016b), and it is thought that 40% of bruising occurs during the catching process (Griffiths, 1985).

2.2 Physiological, behavioural or psychological responses to transport

The bird can undergo a number of changes during transport (physiological, behavioural or psychological). These changes may signify a reduction in welfare, but the point at which the stressor becomes a detriment is not clear. However, an understanding of the responses and what they might mean is important in a discussion of transportation welfare. The responses may include

Behavioural

- Panting, an insensible heat loss mechanism, will occur in hot weather. However, the efficiency of this process is limited if humidity levels are high (Brown-Brandl et al., 1997), and death may result if evaporative heat loss is ineffective;
- Physical damage, as demonstrated by injuries and bruising (Griffiths, 1985), and possibly pain may be caused by the loading and transportation processes. Bruising, particularly of the wings and breast area, and broken wings can occur (Jacobs et al., 2016b), which are partially dependent on stocking density. Cold conditions within the transport vehicle can increase the incidence of huddling within the crates, with birds moving over and under each other to move away from colder areas (Strawford et al., 2011). This movement within the crate may also cause injuries or scratches on other birds;

- Tonic immobility, an assessment used to measure fear and likely related to corticosterone production (Jones et al., 1987), may be higher in transported birds (as reviewed by Broom, 1990).

Physiological

- During cold or hot weather transport, birds must attempt to control core body temperature, and will do so through a number of physiological mechanisms. Adjustments will occur in the bird's cardiovascular system (Yahav, 2002). Blood flow patterns adjust (Richards, 1971) when exposed to cold or hot conditions, and changes to haematocrit, haemoglobin concentration and blood volume may occur (Yahav, 2002);
- As exposure temperature declines, energy requirements increase (Yahav, 2002). If core body temperature declines, birds may use more energy in an effort to replace the heat that is lost (Strawford et al., 2011), which could further weaken the birds;
- Both heart rate and respiration rate increase if birds are startled or alarmed (Knowles and Broom, 1990);
- Immune function impairment by alterations to immune cellular structures can occur under transport stress (Hangalapura et al., 2006; Glatz and Rodda, 2013), although most birds may not be transported long enough for this issue to be of importance;
- Stress can result in tissue damage in the bird, resulting in elevated levels of intracellular enzymes such as creatine kinase that are indicative of muscle damage (Mitchell and Kettlewell, 1994; Glatz and Rodda, 2013);
- Stress results in a reaction of the adrenal cortex, which will in turn increase levels of plasma corticosterone (Freeman, 1984; Knowles and Broom, 1990; Glatz and Rodda, 2013);
- Transport stress can have a direct impact on lymphoid tissue in birds, reducing the capacity to develop an immune challenge, and altering the ratio of heterophils and lymphocytes in the body (Knowles and Broom, 1990);
- Transportation causes alarm responses related to stress, resulting in increases in adrenaline and noradrenaline (Knowles and Broom, 1990; Glatz and Rodda, 2013). Increases in these catecholamines can impact body homeostatic and cardiac functioning;
- Birds undergoing the transport procedure have increased packed cell volume and increased concentrations of plasma total proteins (Warriss et al., 1993; Knowles et al., 1995). These are likely related to dehydration;
- While 'hunger' itself has not been studied in transported birds, Warriss et al. (1988) related declines in liver glycogen to the possibility of increasing the perception of hunger.

The remaining sections of this chapter discuss the science related to the impact of transportation on bird welfare. This is not an exhaustive list, and information in some areas is limited. For example, genetic selection may have altered the ways by which birds cope with transport stressors. The metabolic rates of currently reared turkeys is significantly higher in today's bird as compared to random-bred birds from the mid-1960s (Havenstein et al., 2007), which may make it more difficult for them to thermoregulate when exposed to hot conditions. Comparing commercial turkeys to a slow-growing unselected turkey, Mills et al. (1999) found that core body temperature was higher in the selected lines, suggesting that current turkeys may struggle to expel body heat when exposed to warm conditions and may prefer lower rearing and transportation temperatures.

3 Containers and trailers used for poultry transport

3.1 Physical structure of the crates/modules

The structures used to contain birds during transport differ from company to company, with two primary systems used (modules vs crates). Modules (see Fig. 1) involve a metal frame with perforated plastic drawers or containers that are removed from the transport truck with a forklift and placed near the door of or inside the barn for loading. This reduces the distance the catching crews must carry birds and could potentially reduce injury to the birds. Furthermore, the drawers/containers may be positioned to open at least half of the floor area, allowing more space to place birds inside and reducing the chance of injuring birds. Crates (see Fig. 2) may be filled while they remain on the truck (birds are carried from the barn to the truck) or they may also be removed from the truck and located near the entrance or inside the barn, often with the use of conveyors. In order to fill the crates, birds must be forced through a smaller opening on the top of the crate, increasing the risk of injury as compared to modules. Throughout the following text, the word 'crates' will be used when discussing the containers used to transport birds (crates or modules), unless specifically specified.

During transport, the height of space allowed to birds can alter their welfare, but there is no clear consensus as to what should be used. While especially congested spaces (little headspace within crates) can alter the birds' ability to move, excessive movement during transport may result in painful scratches to other birds and loss of balance, leading to bruising or injury. Bird behaviour may vary with different crate heights, and Wichman et al. (2010) reported that turkeys stand (43% of time), step, turn and preen for a larger percentage of their time in 90-cm high crates than in 50-cm high crates (43%). The authors concluded that while the birds in taller crates had more freedom of movement, the opportunity for

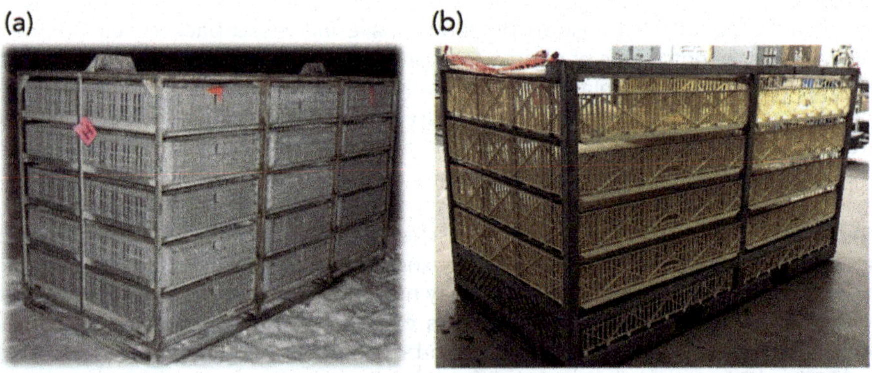

Figure 1 Two examples of broiler transport modules manufactured by (a) Anglia Autoflow Ltd. and (b) Marel Poultry.

Figure 2 Example of a poultry transport crate.

injury and scratching also increased. Vinco et al. (2016) conducted a trial using broilers that compared commercial crates to those modified to double in height. They noted that once the truck initiated movement, birds squatted. No difference in corticosterone levels were found between the birds transported in either crate, and the heterophil:lymphocyte ratio was actually higher in birds transported in the modified crates, suggesting that the greater height increased stress. In addition, more injuries were noted in birds transported in the modified crates. That may be related to vehicle acceleration, as birds attempted to stand in both types of crates. Those in the standard-sized crates, however, returned to squatting before those in the higher crates, and those in the higher crates demonstrated more wing flapping and attempting to climb on other birds. DOA levels were not different in the two groups of birds.

In addition to height, the design of the crates is also important. The containers should be free of any component that could cause injury to the birds, allow air to flow to and from the birds, and allow observations for confirmation of welfare status (Kettlewell and Mitchell, 1994).

3.2 Trailers: design and microclimate

Trailers can either be passively ventilated, using open sides or strategically positioned openings to take advantage of different pressures created around the periphery of a moving vehicle, or actively ventilated, with fans and possibly heating units to ventilate the space where poultry are held during transport. Figure 3 shows examples of passively ventilated transport trailers with (a) side tarps lowered to protect birds during colder ambient conditions and (b) tarps raised for transport during warmer temperatures. Ventilation within the trailers is important to maintain acceptable temperatures and to reduce the risk of high humidity in localized areas in the truck (Mitchell and Kettlewell, 1998). Airflow speed is important, and ventilation rates impact the ability of birds to use conventional heat loss mechanisms (Yahav et al., 2005, 2008). Air speed, however, interacts with ambient temperature to once again make these management decisions complex.

Inside a passively ventilated trailer, the thermal environment can vary significantly (Mitchell and Kettlewell, 1994; Richards et al., 2012). Knezacek et al. (2010) reported that temperatures inside a passively ventilated trailer of a broiler transport truck at an ambient temperature of −27.1°C ranged from 8.9°C to 28.1°C. Similarly, Hui et al. (2003) reported that temperatures inside a passively ventilated trailer while transporting broilers at an ambient temperature of −20.7°C ranged from −14°C to over +26°C.

Figure 4 shows an example of the temperature distribution within trailers transporting broiler chickens. Colder conditions within the trailer, only a few degrees Celsius warmer than ambient temperatures, typically occur near the outside of the load and near air inlets. The warmest conditions occur at the core of the load, closer to the end of the trailer where ventilated air is exhausted (front of the front trailer and rear of the rear trailer). Paradoxically, birds transported in the same vehicle during cold ambient conditions may experience heat stress or cold stress, depending on the location within the load.

Regardless of the trailer type, birds must be protected in cold weather with some type of curtain, and ventilation becomes especially important to remove excess heat, moisture and gases (Mitchell and Kettlewell, 1994).

Figure 3 Passively ventilated trailers with (a) tarps covering the sides of the trailer for transport in colder conditions and (b) tarps stored on the roof of the trailer for transport during warmer ambient conditions.

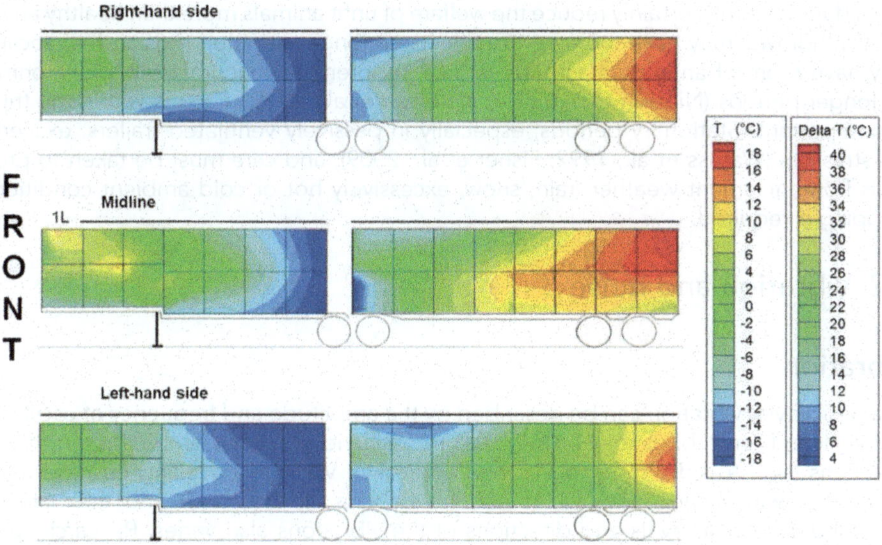

Figure 4 Temperature distribution within a B-train trailer, as presented by Burlinguette et al. (2012). Copied with permission of the *Canadian Journal of Animal Science*.

Trailers featuring forced ventilation have been described in the literature (Dadgar et al., 2010; Spurio et al., 2016; Kettlewell and Mitchell, 2001), and the goal of an actively ventilated trailer is to be able to direct air evenly throughout all areas of the trailer (Kettlewell et al., 2001). They are used in many areas of the world for transportation of poultry, but initial cost and greater maintenance requirements have inhibited wide-scale adoption. Management of conditions within the truck is important, and curtains must be kept closed to allow fans to move air through the load to all parts of the trailer (Kettlewell et al., 2000). If operated correctly, welfare of transported birds is improved (Kettlewell et al., 2001).

4 Other factors influencing poultry welfare during transport

4.1 Trip duration or distance

Distance, or duration, of travel appears to impact bird welfare, with longer distances (or particularly longer times in transit) reducing welfare more significantly than shorter distances. Trips of longer duration result in higher fear levels as compared to shorter trips (Cashman, 1987). Many research studies have demonstrated the link between longer travel durations and DOAs (Vecerek et al., 2006; Voslarova et al., 2007a,b; Machovcova et al., 2016; Vecerek et al., 2016) and bruising (Scholtyssek and Ehinger, 1976).

The procedures or events associated with the trip may be as important or more influential in affecting bird welfare as the travel duration (Nielsen et al., 2011). For example, longer travel durations will certainly reduce the welfare of unfit animals more than healthy animals, food withdrawal may have more severe effects on longer-duration trips and microclimate may have more of an impact as birds remain in potentially uncomfortable environments for longer periods (Nielsen et al., 2011; Kittelsen et al., 2015). Factors within the trip are also important. Stationary periods, especially in passively ventilated trailers, exacerbate the stressors (Warriss et al., 1992; Fisher et al., 2009), and care must be taken to protect birds from inclement weather (rain, snow, excessively hot or cold ambient conditions) if stopping is required.

4.2 Vibration and noise

Vibration

At a basic level, vibration can be described by the amplitude and frequency of oscillations, and while in transit, birds may be exposed to oscillations in each of the three orthogonal directions (side-to-side, front-to-rear and vertically). Vehicles travelling on roads impart vibrations, with varying amplitudes and frequencies, to the birds and their ability to withstand particular levels and directions of the vibrations also varies. Research specific to vibrational impacts on welfare of poultry is limited, but in humans, for example, vehicle vibration can result in discomfort, motion sickness, inability to maintain postural stability, muscle fatigue, blurred vision and reductions in the respiratory volume. In broilers, vibration has been shown to reduce muscle pH (Warriss et al., 1997), and increases in plasma corticosterone and reductions in blood glucose (Carlisle, 1998) are indicative of stress and fatigue when compared to birds exposed to no vibration.

Regardless of species or age, transportation involves an abrupt change to environmental cues, and without doubt, vibration of the truck or trailer, combined with noise from many sources, can affect the welfare of the birds. With regard to affective states, using choice testing or operant conditioning methods, broilers have demonstrated an aversion to vibrational forces (Duggan and Randall, 1995; Randall et al., 1997; Abeyesinghe et al., 2001). The specifics of the vibrational force are important in determining the aversiveness. For example, vertical motion resulting from vibration is more aversive than horizontal, especially at higher frequencies (>1 Hz) (Randall et al., 1997). Vibration also forces birds to attempt to orient or balance themselves, and this may increase fear or distress (Randall et al., 1993; Scott, 1994) and result in an increased occurrence of bruising or injury (Gebresenbet et al., 2011). Muscles must be used in an attempt to stabilize the bird, and over the duration of the trip, this effort could lead to muscle fatigue and exhaustion (Mitchell et al., 1992; Randall et al., 1993; Scott, 1994).

The physiological effects of vibration have been studied in much more depth. When exposed to vibration, the birds must adjust to or counteract these forces via one of a number of mechanisms in addition to stability control. The force appears to initiate an autonomic nervous system reaction (Santurtun and Phillips, 2015), which manifests in a number of physiological processes. Levels of liver and muscle glycogen may be reduced to similar levels as would be seen with feed deprivation, but this appears to only occur when exposed to vibration for longer periods (10 h as compared to 1 h) (Warriss et al., 1993), but Warriss et al. (1997) later found that this may not have been directly related to vibration alone. This latter point may be important when determining the effects of transport in specific vehicles, or over various types of road surfaces. The vibration may cause the birds' internal organs to go into a state of resonance, which, in severe cases, could lead to death (Edwards et al., 1972; Scott, 1994). Vibration affects cardiac functioning in birds (Edwards et al., 1972). Blood pressure and heart rate can change in many animals, resulting in an increase in arterial blood pressure (as reviewed by Scott, 1994). Respiration rates change, with higher oxygen demands to maintain abdominal pressure, which in turn may inhibit the lungs to expand to normal levels (Scott et al., 1994).

While not clearly defined in commercial poultry species, the potential exists that vibrational forces during transport could cause motion sickness (Scott, 1994; Santurtun and Phillips, 2015). Mechanisms may include the sensory rearrangement theory or loss of control of body orientation or equilibrium. Motion sickness has been demonstrated to occur in some avian species, including quail (Ossenkopp and Tu, 1984).

Noise

The impact of noise itself on birds is not clear, although likely the particular volume and/or frequency, and in combination with other stressors, may have further influences. Rutter et al. (1993) found mechanical noise to be aversive to poultry, while Scott and Moran (1993a,b), testing a mechanical conveyor system for moving birds, found no increase in fear levels when compared to birds not exposed to the noise.

4.3 Stocking density

The density of birds placed in crates should be specific, depending on the bird age and type and individual load transport conditions. Research results are not explicit, regarding the ideal stocking density, but suggest that the stocking density should depend on ambient and other specific farm conditions such as temperature, humidity and weather conditions; gender, age and live weight of the bird (Bayliss and Hinton, 1990). The result of incorrect crate density can include an increase in the level of DOAs (Mitchell and Kettlewell, 1994; Vieira et al., 2011). Limited space may not allow birds to use behavioural techniques for thermoregulation (Strawford et al., 2011), but too much space may leave room for injury among the birds.

A major consideration when determining crate density is the ambient environment, in particular temperature and relative humidity. In hot weather, birds rely on a number of heat loss mechanisms, and evaporative cooling from internal organs via panting is an important physiological process that allows this to occur. Similarly, ventilation or airflow over the birds is necessary to help dissipate heat. To allow these functions to occur, crate stocking density should be reduced in hot/humid weather (Bayliss and Hinton, 1990). Similarly, in cold weather, birds adjust their behaviour to keep warm. Strawford et al. (2011) found that if birds were able, they altered their space usage to avoid cold areas and tunnelled under other birds to aid in maintaining appropriate core body temperatures. Densities that would not allow any movement may therefore reduce bird welfare by restricting the use of this behavioural mechanism. These birds were taking advantage of a thermal core that was being created within the crate, and a minimum number of birds would be required to support these conditions. Insufficient bird numbers (excessively low stocking density) would be counterproductive in cold conditions.

In general, high densities in crates can increase the number of DOAs (Nijdam et al., 2004; Whiting et al., 2007; Delezie et al., 2007; Chauvin et al., 2011). Chauvin et al. (2011) hypothesised that reasons for this could be multifactorial, including an increase in humidity from crowded birds leading to higher levels of hypo- or hyperthermia or thermal stress.

4.4 Seasonal factors

DOA levels tend to be higher in some seasons than others (Bayliss and Hinton, 1990), but once again the situation is very complex and will also depend on many of the other factors discussed in this chapter. These include ambient temperature and relative humidity conditions, truck, trailer and lairage details including microclimate, stocking density, bird type, age and prior flock health (Vieira et al., 2011; Bayliss and Hinton, 1990).

Machovcova et al. (2016) found that variations in ambient temperature resulted in differences in DOA levels, with highest DOA levels occurring during periods of warm ambient temperatures (summer, 14–21°C) and lowest levels during colder weather. In their work, there was a tendency to experience higher DOA levels when temperatures dropped below −2°C. Vecerek et al. (2016) reported that highest levels of DOAs occurred during the colder winter period. Localized differences, including climate, equipment and management practices may account for some of the variations in reported data, which may or may not be reported in research articles. For example, geographic locations with high ambient relative humidity coupled with high temperatures make it difficult for poultry to dissipate heat when in transit (Vieira et al., 2011). Alternatively where birds may be transported at −30°C or lower, the dominant challenge becomes maintaining core body temperatures (Dadgar et al., 2010, 2011, 2012).

In addition to the geographic location complexities, there also appears to be an interaction between many transport variables in their impact on levels of DOA (Bayliss and Hinton, 1990; Hunter et al., 1997).

5 Lairage

Once poultry transport trucks reach the slaughter plant, the birds are often maintained in their crates for a period prior to slaughter. This time is referred to as lairage. The space where the birds are held and the duration they are held there can vary significantly from plant to plant. The specific microclimate conditions can have an impact on the level of DOAs on the truck (Bayliss and Hinton, 1990; Vieira et al., 2011). Lairage facilities should be designed for the local environment and well equipped, keeping bird welfare in mind (Bayliss and Hinton, 1990).

Once again, the conditions surrounding lairage likely interact with each other, and care must be taken when interpreting research results based on these different conditions. In hot weather, and again depending on the specific lairage conditions, the distribution of temperatures in the trailer can change, compared to the distribution during transit. In the work done by Richards et al. (2012), lairage resulted in cooler temperatures in the bottom crates, but temperatures actually increased in the top crates. The interaction effects between ambient temperature and lairage time on DOA levels is inconclusive in the literature. While some studies indicate that the number of DOAs increases with longer lairage times (Nijdam et al., 2004), especially in hot environments (Mitchell and Kettlewell, 1994), other works (Vieira et al., 2011) found that DOA levels are reduced during the hot weather within an actively controlled environment. Once again, this indicates the interactive nature of the factors that impact transportation and welfare in poultry. Following transport during cold ambient conditions, Dadgar et al. (2011) showed that longer lairage times negatively affected meat quality factors which could suggest physiological changes occurring in the bird, and these authors suggested limiting the amount of time spent in lairage.

6 Transporting chicks and EOL hens

6.1 Transport of day-olds from hatchery to farm

The transport of chicks has not received the same attention as has the transport of older birds to slaughter. The transport of chicks is done much more carefully and involves trucks with the capacity to provide supplemental heat with active ventilation and computer-controlled actuators. In most cases, food is not supplied to the chicks during the transport process, because the yolk sac and other body stores are capable of supplying adequate nutrients for up to three days post-hatch. Indeed, in the few research programmes noted in the literature focusing on transport of day-old chicks, no differences in weight gain early in life were found when transported with or without feed for either 4–14 h (Valros et al., 2008) or up to 20 h (Pisarski et al., 2005). This does not specifically address affective states and bird welfare, however, and chick hunger has not been studied. Valros et al. (2008) did examine fear levels in their study, but found no effect of the food deprivation period.

6.2 EOL hen transport

The majority of research programmes have studied the effects of transport on broilers, as they represent the greatest number of birds transported (Kettlewell and Mitchell, 1994). However, other birds may represent different challenges. Pre-transport stress can have significant impacts on the welfare of turkeys prior to slaughter, for example (Marchewka et al., 2013), but EOL hens may be subject to stressors unlike other birds (Hester, 2005). Transport of EOL hens is more challenging as these birds often have to travel considerable distances because of a reduced number of processing plants. Many factors from the production period will affect how transport impacts these birds, including the housing system the birds are removed from, age of the bird and methods of catching (Knowles and Broom, 1990; Knowles, 1994).

A major challenge for these birds is the ability to maintain their body temperature during the transport trip. Birds are homeothermic and have the ability to control their body temperature within a narrow range (Yahav, 2009). To do so, however, birds must use a number of physiological and behavioural processes. EOL hens have significant challenges that may make some of these thermoregulation mechanisms more difficult compared to meat birds. In particular, EOL hens are often poorly feathered (Freeman, 1984; Richards et al., 2012), reducing the ability to trap warm air next to the body in cold weather and increasing the rate of heat loss (Knowles and Broom, 1990).

The incidence of EOL hens with broken bones loaded onto the trucks may be higher than in other birds (Gregory and Wilkins, 1989), and there is a possibility of more bone fractures related to the poor skeletal integrity of the birds (Schreiweis et al., 2003). This could mean that longer-duration trips may be even more detrimental to bird welfare. Their high metabolic rate could lead to physiological fatigue (Richards et al., 2012) for EOL hens, meaning that management tools such as duration of feed withdrawal affect these birds more so than others, particularly during cold (Webster et al., 1993) or hot weather (Yahav, 2009).

Mortality levels for EOL hens during transport can be variable (Knowles and Broom, 1990), but are generally higher than would be noted with other bird types or species (Knowles and Broom, 1990; Voslarova et al., 2007a). Work by Voslarova et al. (2007a) also showed an interaction between mortality level and the distance travelled by EOL hens.

An elevated level of care is required during the transport of EOL hens. Keeping trips as short as possible could help reduce mortality levels (Knowles and Broom, 1990). However, with a smaller number of processing facilities to process EOL hens, this would reduce the feasibility of shorter-duration transport. Hot weather management such as keeping birds cool via use of appropriate crate densities, or fans and misters, to aid in thermoregulation during loading and lairage may be helpful, while use of specific truck types or at minimum tarps to help keep birds warm during cold ambient conditions may improve bird welfare (Newberry et al., 1999).

7 Conclusions

Transportation of poultry is complex, and an interaction exists between many variables that will ultimately determine the welfare outcome of a trip. Studying one single factor, whether it be duration, microclimate or lairage, will not result in an accurate understanding of how the biological system of a bird responds. Stressors are additive, meaning that multiple stressors magnify the impact on birds, hence eliminating or at least reducing stressors will be beneficial. Likely, further studies focusing on both singular and combinatorial effects of the aspects of transport stress would add to the understanding of these impacts on poultry species. However, these factors must be within the realistic norms of what the birds will likely experience in commercial practices. These may include combinations such as crate densities along with various air velocities and temperature/humidity combinations for various bird types, at different ages, feather covers and so on.

Regardless, there are specific tools that can be used to reduce or eliminate welfare challenges of transportation. Management practices intended to reduce on-farm mortality levels coupled with good culling practices can improve the welfare implications of transportation. Training for catchers on proper bird-handling methodologies will reduce injury. Reducing the duration of a trip may have significant beneficial effects on DOA levels, but with fewer slaughter plants this may not be realistic. Proper trailer design and management of airflow, possibly with the use of active ventilation systems, may help to produce improved microclimates for birds onboard.

8 Where to look for further information

The transportation of farm animals is an area of active research, and scientists employed within universities and government-sponsored labs continue to investigate the effects of various aspects of transportation on farm animals, including poultry. Results from this work are regularly reported in scientific journal articles (similar to those listed at the end of this chapter) and conference proceedings, and these results are often prospective in nature.

As noted previously, animals may be transported at different stages of the life cycle, for various purposes, and the act of transporting poultry is carried out by various parties. Hatcheries are often involved in the transport of chicks, while processors and third-party haulers typically move birds destined for slaughter. Each of these groups, to varying degrees, are able to provide additional details about the practices and equipment involved in poultry transportation. In addition, producer groups, often in collaboration with researchers, provide extension-type materials, intended to inform people about the processes to produce food. These materials describe the current state of the art in lay terms.

Regulations, informing the safe limits for the transport of farm animals, are developed through consultations with numerous stakeholders, including animal rights activists, consumers, producers, animal health specialists and processors. These regulations are typically readily available from government websites. The effort required to develop and maintain transport regulations is immense; a version of the regulations can be in force for 15 years or more before it is updated. Thus, depending on the jurisdiction, research results and the operational procedures employed by industry can be technologically advanced, beyond the minimum requirements of the regulations. While these regulations outline the requirements of transport, they can often be out of date.

9 References

Abeyesinghe, S. M., C. M. Wathes, C. J. Nicol and J. M. Randall. 2001. The aversion of broiler chickens to concurrent vibrational and thermal stressors. *Applied Animal Behaviour Science* 73:199–215.

Bayliss, P. A., and M. H. Hinton. 1990. Transportation of broilers with special reference to mortality rates. *Applied Animal Behaviour Science* 28:93–118.

Broom, D. M. 1990. Effects of handling and transport on laying hens. *World's Poultry Science Journal* 46:48–50.

Brown-Brandl, T. M., M. M. Beck, D. D. Schulte, A. M. Parkhurst and J. A. DeShazer. 1997. Temperature humidity index for growing tom turkeys. *Transactions of the ASAE* 40:203–9.

Burlinguette, N. A., M. L. Strawford, J. M. Watts, H. L. Classen, P. J. Shand and T. G. Crowe. 2012. Broiler trailer thermal conditions during cold climate transport. *Canadian Journal of Animal Science* 92:109–22.

Carlisle, A. J. 1998. Physiological responses of broiler chickens to the vibrations experienced during road transportation. *British Poultry Science* 39:S48–S60.

Cashman, P. J. 1987. An assessment of the fear levels of broilers during transit. M. Sc. Thesis, University of Bristol.

Chauvin, C., S. Hillion, L. Balaine, V. Michel, J. Peraste, I. Petetin, C. Lupo and S. Le Bouquin. 2011. Factors associated with mortality of broilers during transport. *Animal* 5:287–93.

Dadgar, S., E. S. Lee, T. L. V. Leer, N. Burlinguette, H. L. Classen, T. G. Crowe and P. J. Shand. 2010. Effect of microclimate temperature during transportation of broiler chickens on quality of the pectoralis major muscle. *Poultry Science* 89:1033–41.

Dadgar, S., E. S. Lee, T. L. V. Leer, T. G. Crowe, H. L. Classen and P. J. Shand. 2011. Effect of acute cold exposure, age, sex, and lairage on broiler breast meat quality. *Poultry Science* 90:444–57.

Dadgar, S., T. G. Crowe, H. L. Classen, J. M. Watts and P. J. Shand. 2012. Broiler chicken thigh and breast muscle responses to cold stress during simulated transport before slaughter. *Poultry Science* 91:1454–64.

Delezie, E., Q. Swennen, J. Buyse and E. Decuypere. 2007. The effect of feed withdrawal and crating density in transit on metabolism and meat quality of broilers at slaughter weight. *Poultry Science* 86:1414–23.

Duggan, J. A., and J. M. Randall. 1995. Aversion of broiler chickens to whole-body vibration using a passive avoidance technique. In *Proceedings of the 29th International Congress for the International Society of Applied Ethology*, Exeter, UFAW, pp. 67–9.

Edwards, R. G., E. P. McCutcheon and C. F. Knapp. 1972. Cardiovascular changes produced by brief whole-body vibration of animals. *Journal of Applied Physiology* 32:386–90.

EFSA. 2011. European Food Safety Authority. Scientific opinion concerning the welfare of animals during transport. *EFSA Journal* 9:1966–2091.

Fisher, A. D., I. G. Colditz, C. Lee and D. M. Ferguson. 2009. The influence of land transport on animal welfare in extensive farming systems. *Journal of Veterinary Behavior* 4:157–62.

Freeman, B. M. 1984. Transportation of poultry. World's *Poultry Science Journal* 40:19–30.

Gebresenbet, G., S. Aradom, F. S. Bulitta and E. Hjerpe. 2011. Vibration levels and frequencies on vehicle and animals during transport. *Biosystems Engineering* 110:10–19.

Glatz, P., and B. Rodda. 2013. Turkey farming: Welfare and husbandry issues. *African Journal of Agricultural Research* 8:6149–63.

Gregory, N. G., and L. J. Wilkins, 1989. Handling and processing damage in end-of-lay hens. British *Poultry Science* 30:555–62.

Griffiths, G. L. 1985. Ageing bruises on chicken legs. In *Proceedings of the 6th Australasian Poultry and Stockfeed Convention*, September 23–27, 1985, Melbourne, Vic, pp. 269–99.

Hangalapura, B. N., M. G. Kaiser, J. J. Vanderpoel, H. K. Parmentier and S. J. Lamont. 2006. Cold stress equally enhances in vivo pro-inflammatory cytokine gene expression in chicken line divergently selected for antibody responses. *Developmental and Comparative Immunology* 30:503–11.

Havenstein, G. B., P. R. Ferket, J. L. Grimes, M. A. Qureshi and K. E. Nestor. 2007. Comparison of the performance of 1966- versus 2003-type turkeys when fed representative 1966 and 2003 turkey diets: growth rate, livability, and feed conversion. *Poultry Science* 86:232–40.

Hester, P. Y. 2005. Impact of science and management on the welfare of egg laying strains of hens. *Poultry Science* 84:687–96.

Hui, K. P. C., T. G. Crowe, H. L. Classen, E. M. Barber, T. D. Knezacek and M. R. L. Bantle. 2003. Poultry transportation under Canadian prairie conditions. Presented at the Annual Meeting of the Canadian Society of Animal Science, Saskatoon, Canada.

Hunter, R. R., M. A. Mitchell and C. Matheu. 1997. Distribution of 'dead on arrivals' within the bio-load on commercial broiler transporters: correlation with climatic conditions and ventilation regimes. *British Poultry Science* 38: (Suppl. 1):7–9.

Hunter, R. R., M. A. Mitchell and A. J. Carlisle. 1999. Wetting of broilers during cold weather transport: A major source of physiological stress. *British Poultry Science* 40 (Suppl. 1):S48–S49.

Jacobs, L., E. Delezie, L. Duchateau, K. Goethals and F. A. M. Tuyttens. 2016a. Broiler chickens dead on arrival: associated risk factors and welfare indicators. *Poultry Science*. http://dx.doi.org/10.3382/ps/per353.

Jacobs, L., E. Delezie, L. Duchateau, K. Goethals and F. A. M. Tuyttens. 2016b. Impact of the separate pre-slaughter stages on broiler chicken welfare. *Poultry Science*. http://dx.doi.org/10.3382/ps/per361.

Jones, R. B., G. Beuving and H. J. Blokhuis. 1987. Tonic immobility and heterophil/lymphocyte responses of the domestic fowl to corticosterone infusion. *Physiology and Behaviour* 42:249–53.

Knezacek, T. D., A. A. Olkowski, P. J. Kettlewell, M. A. Mitchell and H. L. Classen. 2010. Temperature gradients in trailers and changes in broiler rectal and core body temperature during winter transportation in Saskatchewan. *Canadian Journal of Animal Science* 90:321–30.

Kettlewell, P. J., and M. A. Mitchell. 1994. Catching, handling and loading of poultry for road transportation. *World's Poultry Science Journal* 50:54–6.

Kettlewell, P. J., and M. A. Mitchell. 2001. Comfortable ride: Concept 2000 provides climate control during poultry transport. *Resource magazine, Engineering and Technology for a Sustainable World* 8:13–14

Kettlewell, P. J., R. P. Hoxey and M. A. Mitchell. 2000. Heat produced by broiler chickens in a commercial transport vehicle. *Journal of Agricultural Engineering Research* 75:315–26.

Kettlewell, P. J., R. P. Hoxey, C. J. Hampson, N. R. Green, B. M. Veale and M. A. Mitchell. 2001. Design and operation of a prototype mechanical ventilation system for livestock transport vehicles. *Journal of Agricultural Engineering Research* 79:429–39.

Kittelsen, K. E., E. G. Granquist, Ø. Kolbjørnsen, O. Nafstad and R. O. Moe. 2015. A comparison of post-mortem findings in broilers dead-on-farm and broilers dead-on-arrival at the abattoir. *Poultry Science* 94:2622–9.

Knowles, T. G. 1994. Handling and transport of spent hens. World's *Poultry Science Journal* 50:60–1.

Knowles, T. G. and D. M. Broom. 1990. The handling and transport of broilers and spent hens. *Applied Animal Behaviour Science* 28:75–91.

Knowles, T. G., P. D. Warriss, S. N. Brown, J. E. Edwards and M. A. Mitchell. 1995. Response of broilers to deprivation of food and water for 24 hours. *British Veterinary Journal* 151:197–202.

Machovcova, Z., V. Vecerek, E. Voslarova, M. Malena, F. Conte, I. Bedanova and L. Vecerkova. 2016. Pre-slaughter mortality among turkeys related to their transport. *Animal Science Journal* doi:10.1111/asj.12692.

Marchewka, J., T. T. N. Watanabe, V. Ferrante and I. Estevez. 2013. Review of the social and environmental factors affecting the behavior and welfare of turkeys (*Meleagris gallopavo*). *Poultry Science* 92:1467–73.

Mills, L. J., M. A. Mitchell and M. Mahon. 1999. Comparison of thermoregulatory ability in fast and slow growing strains of turkey during acute heat stress. *British Poultry Science* 40: (Suppl. 1) S51–S52.

Mitchell, M. A., P. J. Kettlewell and M. H. Maxwell. 1992. Indicators of physiological stress in broiler chickens during road transportation. *Animal Welfare* 1:91–103.

Mitchell, M. A., and P. J. Kettlewell. 1994. Road transportation of broiler chickens: induction of physiological stress. *World's Poultry Science Journal* 50:57–9.

Mitchell, M. A., and P. J. Kettlewell. 1998. Physiological stress and welfare of broiler chickens in transit: Solutions not problems! *Poultry Science* 77:1803–14.

Newberry, R. C., A. B. Webster, N. J. Lewis and C. van Arnam. 1999. Management of spent hens. *Journal of Applied Animal Welfare Science* 2:13–29.

Nicol, C. J., and G. B. Scott. 1990. Pre-slaughter handling and transport of broiler chickens. *Applied Animal Behaviour Science* 28:57–73.

Nielsen, B. L., L. Dybkjaer and M. S. Herskin. 2011. Road transport of farm animals: effects of journey duration on animal welfare. *Animal* 5:415–27.

Nijdam, E., P. Arens, E. Lambooij, E. Decuypere and J. A. Stegeman. 2004. Factors influencing bruises and mortality of broilers during catching, transport, and lairage. *Poultry Science* 83:1610–15.

Ossenkopp, K. P., and G. S. Tu. 1984. Motion sickness in quail: Body-rotation-induced conditioned fluid aversions in C. coturnix japonica. *Journal of Comparative Psychology* 98:189–93.

Pisarski, R. K., I. Pijarska and H. Malec. 2005. The effect of feeding broiler chickens during transportation on yolk sac, body weight and anti-IBVD level in blood. *Journal of Animal and Feed Sciences* 14 (Suppl.1):479–82.

Randall, J. M., W. V. Streader and A. M. Meehan. 1993. Vibration on poultry transporters. *World's Poultry Science Journal* 50:64–5.

Randall, J. M., J. A. Duggan, M. A. Alami and R. P. White. 1997. Frequency weightings for the aversion of broiler chickens to horizontal and vertical vibration. *Journal of Agricultural Engineering Research* 68:2387–397.

Richards, S. A. 1971. The significance of changes in the temperature of the skin and body core of the chicken in the regulation of heat loss. *Journal of Physiology* 216:1–10.

Richards, G. J., L. J. Wilkins, C. A. Weeks, T. G. Knowles and S. N. Brown. 2012. Evaluation of the microclimate in poultry transport module drawers during the marketing process of end-of-lay hens from farm to slaughter. *Veterinary Record* 171. doi:10.1136/vr.100844.

Rutter, S. M., G. B. Scott and P. Moran. 1993. Aversiveness of mechanical conveying to laying hens. *British Poultry Science* 34:279–85.

Santurtun, E., and C. J. C. Phillips. 2015. The impact of vehicle motion during transport on animal welfare. *Research in Veterinary Science* 100:303–8.

Scholtyssek, S., and F. Ehinger. 1976. Transporteinflüsse auf broiler und deren Schlachtkoerper. *Archiv für Geflügelkunde* 40:27–35.

Schreiweis, M. A., J. I. Orban, M. C. Ledur, D. E. Moody and P. Y. Hester. 2003. A comparison of bone mineral density in live broilers and White Leghorns. *Poultry Science* 82 (Suppl. 1):99 (Abstr.).

Scott, G. B. 1994. Effects of short-term whole body vibration on animals with particular reference to poultry. *World's Poultry Science Journal* 50:25–38.

Scott, G. B., and P. Moran. 1993a. Fear levels in laying hens carried by hand and by mechanical conveyors. *Applied Animal Behaviour Science* 36:337–45.

Scott, G. B., and P. Moran. 1993b. Effects of visual stimuli and noise on fear levels in laying hens. *Applied Animal Behaviour Science* 37:321–9.

Spurio, R. S., A. L. Soares, R. H. Carvalho, V. Silveira Junior, M. Grespan, A. Oba and M. Shimokomaki. 2016. Improving transport container design to reduce broiler chicken PSE (pale, soft, exudative) meat in Brazil. *Animal Science Journal* 87:277–83.

Strawford, M. L., J. M. Watts, T. G. Crowe, H. L. Classen and P. J. Shand. 2011. The effect of simulated cold weather transport on core body temperature and behavior of broilers. *Poultry Science* 90:2415–24.

Valros, A., R. Vuorenmaa and A. M. Janczak. 2008. Effect of simulated long transport on behavioural characteristics in two strains of laying chicks. *Applied Animal Behaviour Science* 109:58–67.

Vecerek, V., S. Grbalova, E. Voslorova, B. Janackova and M. Malena. 2006. Effects of travel distance and the season of the year on death rates of broilers transported to poultry processing plants. *Poultry Science* 85:1881–4.

Vecerek, V., E. Voslarova, F. Conte, L. Vecerkova and I. Bedanova. 2016. Negative trends in transport-related mortality rates in broiler chickens. *Asian-Australasian Journal of Animal Science* 29:1796–804.

Vieira, F. M. C., I. J. O. Silva, J. A. D. Barbosa Filho, A. M. C. Vieira and D. M. Broom. 2011. Preslaughter and mortality of broilers in relation to lairage and season in a subtropical climate. *Poultry Science* 90:2127–33.

Vinco, L. J., I. L. Archetti and S. Giacomelli. 2016. Influence of crate height on the welfare of broilers during transport. *Journal of Veterinary Behavior* 14:28–33.

Voslarova, E., B. Janackova, L. Rubesova, A. Kozak, I. Bedanova, L. Steinhauser and V. Vecerek. 2007a. Mortality rates in poultry species and categories during transport for slaughter. *Acta Veterinaria Brno* 76:101–8.

Voslarova, E., B. Janackova, F. Vitula, A. Kozak and V. Vecerek. 2007b. Effects of the transport distance and the season of the year on death rates among hens and roosters in transport to poultry processing plants in the Czech Republic in the period from 1997 to 2004. *Veterinami Medicina* 52:262–6.

Warriss, P. D., S. C. Kestin, S. N. Brown and E. A. Bevis. 1988. Depletion of glycogen reserves in fasting broiler chickens. *British Poultry Science* 29:149–54.

Warriss, P. D., E. A. Bevis, S. N. Brown and J. E. Edwards. 1992. Longer journeys to processing plants are associated with higher mortality rates in broiler chickens. *British Poultry Science* 33:201–6.

Warriss, P. D., S. C. Kestin, S. N. Brown, T. G. Knowles, L. J. Wilkins, J. E. Edwards, S. D. Austin and C. J. Nicol. 1993. The depletion of glycogen stores and indices of dehydration in transported broilers. *British Veterinary Journal* 149:391–8.

Warriss, P. D., S. N. Brown, T. G. Knowles, J. E. Edwards and J. A. Duggan. 1997. Potential effect of vibration during transport on glycogen reserves in broiler chickens. *The Veterinary Journal* 153:215–19.

Webster, A. J. F., A. Tuddenham, C. A. Saville and G. B. Scott. 1993. Thermal stress on chickens in transit. *British Poultry Science* 34:267–77.

Whiting, T. L., M. E. Drain and D. P. Rasali. 2007. Warm weather transport of broiler chickens in Manitoba. II. Truck management factors associated with death loss in transit to slaughter. *The Canadian Veterinary Journal* 48:148–54.

Wichman, A., M. Norring, M. Pastell, B. Algers, R. Pösö, A. Valros, H. Saloniemi and L. Hänninen. 2010. Effect of crate height during short-term confinement on the welfare and behaviour of turkeys. *Applied Animal Behaviour Science* 126:134–9.

Yahav, S. 2002. Limitations in energy intake affect the ability of young turkeys to cope with low ambient temperatures. *Journal of Thermal Biology* 27:103–8.

Yahav, S. 2009. Alleviating heat stress in domestic fowl – different strategies. *World's Poultry Science Journal* 65:719–32.

Yahav, S., D. Shiner, J. Tanny and S Cohen. 2005. Sensible heat loss – the broiler's paradox. *World's Poultry Science Journal* 61:419–33.

Yahav, S., M. Rusal and D. Shinder. 2008. The effect of ventilation on performance body and surface temperature of young turkeys. *Poultry Science* 87:133–7.

Chapter 16

Developments in humane slaughtering techniques for poultry

Andy Butterworth, University of Bristol, UK

1 Introduction

Stunning of animals prior to slaughter is a practice that has developed almost global acceptance as the understanding of the need to provide 'immediate insensibility' during the slaughter process has become widely accepted. Some countries and territories do not permit stunning before slaughter for religious or cultural reasons, but in many parts of the world it is now assumed, and in many countries, demanded by law, that stunning will take place as the first of the steps required to kill a commercially slaughtered animal (Berg and Raj, 2015).

The requirement for stunning has as its basis that animals are sentient beings (EC, 2009) and that the welfare of the animals can be best protected, according to current understanding of the cognitive and physiological basis of suffering, by induction of unconsciousness as the first and enduring stage of the process of killing. Council Regulation (EC) No 1099/2009 (EC, 2009) on the protection of animals at the time of killing defines 'stunning' in Article 2 (f) as *'any intentionally induced process which causes loss of consciousness and sensibility without pain including any process resulting in instantaneous death'*.

For most methods subsequent to stunning, there will be the necessity to induce a non-recoverable state (death) by exsanguination, cessation of the function of the heart or circulation, or irreversible destruction of brain and spinal cord tissue (by pithing or decapitation) so that recovery of consciousness is not possible – that is, that 'unconsciousness

http://dx.doi.org/10.19103/AS.2016.0011.31

lasts until death' occurs. For these reasons, most permitted killing methods comprise a set of stunning and killing methods, which are known to be effective, for the species in question. European Food Safety Authority (EFSA) has recently published guidance on the assessment criteria for studies evaluating the effectiveness of stunning methods regarding animal protection at the time of killing (EFSA, 2013). The eligibility criteria that must be fulfilled to achieve the required definition of 'effective stunning' in the EU (EFSA, 2013) are

a. Immediate onset of unconsciousness and insensibility;

OR

b. Absence of avoidable pain, distress and suffering until the loss of consciousness and sensibility
c. Duration of the unconsciousness and insensibility (until death).

In Europe, the welfare of poultry at the time of killing is protected by Regulation (EC) 1099/2009 (Anonymous, 2009). Article 4 on stunning interventions regulates that *'animals shall only be killed after stunning in accordance with the methods and specific requirements related to the application of those methods set out in Annex I of the Regulation'* and *'that the loss of consciousness and sensibility shall be maintained until the death of the animal'*.

Furthermore, the methods referred to in Annex I of Regulation EC 1099/2009, which do not result in instantaneous death shall be followed as quickly as possible by a procedure ensuring death, such as bleeding, pithing, electrocution or prolonged exposure to anoxia. The methods for stunning and killing poultry permitted by this regulation are (in some estimated order of commonality of use in the EU) (EC, 2012);

- Electrical water bath stunning (81% of broilers [EC, 2012]) – the birds move through a water bath suspended by the legs on a shackle, and their heads are immersed in water which carries the stunning current from the head through the body to the electrical contact in the shackle.
- Gas stunning (19% of broilers) – usually progressing to gas killing through hypoxia – use of gases, commonly carbon dioxide, or carbon dioxide in association with inert gases such as nitrogen, argon, or inert gases only.
- Head-only electrical stunning – the use of manually operated 'tongs' which produce current flow across the head, and are most commonly used in turkeys, ducks and breeder poultry.
- Penetrating and non-penetrating captive bolt, firearms – used for emergency on farm or casualty killing of birds on farm, and sometimes used for killing of larger birds including turkeys and ostrich.
- Head-to-body electrical stunning – theoretically possible for use on poultry, applying the current from the head to electrodes on the body and inducing both a stunned state and cardiac arrest – predominately used in sheep, cattle and pigs.

2 The physiological basis of stunning

In general, 'stunning' or the 'stunned state' has been assessed by methods which examine the degree and extent of unconsciousness. This has been assessed in the laboratory and in the slaughterhouse by using electroencephalograms (EEGs) to ascertain the depth and duration of unconsciousness induced by a stunning method. Such EEG measures include assessment of patterns of electrical activity associated with, and characteristic of, epileptiform activity – and the detection of profoundly suppressed or quiescent EEGs, and abolition of evoked electrical activity in the brain including the suppression of somatosensory evoked potentials. For a stunning method to be considered effective, some or all of these requirements are generally required to have been met, and this forms the basis of the EFSA guidance on the assessment criteria for evaluating the effectiveness of stunning interventions (EFSA, 2004).

How quickly does a bird have to be induced into unconsciousness for a stunning process to be considered humane? This question has troubled animal welfare scientists and legislators over many years, and the pragmatic response of the law, based on scientific findings, has been to consider the concept of 'immediate' – and EFSA (2004) has described a time frame of '*within a second*' as a measurable time limit under 'practical conditions' which may be considered as 'immediate'. In practice, much research indicates that electrical stunning systems can induce an insensible or unconscious state in a shorter time than this, perhaps of the order 100 ms (Gregory and Wotton, 1989, 1990). Modified atmosphere systems (briefly described below) do not achieve unconsciousness in these short time frames.

However, the practical implementation of the electrical water bath stunning system across the world is complicated by the diversity of waveforms, frequencies and currents, and the variability in the physical apparatus seen in slaughterhouses across the globe. When it is considered that broiler chicken alone are slaughtered at a rate of approximately 45 000 000 000/365 days/24 hours/60 minutes = 85 000 birds a minute in countries across the world, it is clear that the 'effectiveness' of stunning systems for poultry may have a major role in determining how humanely mankind may be considered to be 'treating the most common farmed animal on the terrestrial surface'.

3 Electrical stunning methods

3.1 Waterbath – Shackle

In many commercial stunning systems, birds are suspended live by their legs in metal shackles to enable automated electrical stunning (Sparrey et al., 1992; Sparrey and Kettlewell, 1994) and the extent of discomfort, pain or injury caused by the shackle can be affected by shackle design, the care with which the birds are hung in the shackle and the duration of hanging before stunning.

Effective electrical stunning involves stimulation of the brain with an electrical current and frequency capable of inducing generalized epilepsy, characterized more specifically by the occurrence of synchronized 8–13 Hz EEG activity. This characteristic EEG finding is known to produce persistent unconsciousness and insensibility (Lambooij, 1981). If the electrical stunning also induces cardiac ventricular fibrillation, then this will lead to death, and no potential for return to consciousness after stunning – and some stunning systems are designed with cardiac arrest as a specific aim.

For the majority of broiler chicken, effective electrical stunning can be achieved by delivering 120 mA per bird (not divided between multiple birds) in a water bath stunning unit, supplied with a 50 Hz sine wave alternating current (AC). Where multiple birds are concurrently stunned in the same water bath, as is the case in many stunning systems, all of the birds will not receive the same effective stunning current, as the current will be determined (at a constant voltage) by the impedance in the pathway of each bird (Sparrey et al., 1992; Wooley et al., 1986). In practice, this means that birds of different size, weight or having different degrees of 'electrical contact with the shackle' may receive significantly different currents (Hindle et al., 2010). EFSA (EFSA, 2014) has recommended that alternatives to water bath stunning should be developed, as the complexity of multiple bird waterbath stunning systems used in poultry slaughterhouses are not, in all cases, capable of ensuring humane killing and hence providing good animal welfare. In some circumstances, electric water bath stunning may also result in poor meat quality due to the profound muscle contractions that can be generated in the bird during the application of the stunning current, particularly at higher stunning currents. These electrically induced contractions may result in fractures and in internal muscle and other tissue haemorrhaging, resulting in blood splash (Kranen et al., 2000). Another potential animal welfare problem, which is directly associated with meat quality issues stemming from electrical stunning, is the temptation of slaughterhouse staff to alter the current settings on the stunning unit to reduce haemorrhagic meat quality issues, resulting potentially in improvement in meat quality issues, but simultaneously risking reduced effectiveness of stunning.

Induction of cardiac arrest alone may be possible by placement of the electrodes so as to avoid conduction through the head, but to cause flow through the thorax. This may result in death through cardiac fibrillation, but cannot be considered humane, as the bird will not achieve unconsciousness during the process as no direct stimulation of the brain will occur, as is required to cause epileptiform unconsciousness. Work by Raj and co-workers (Raj and O'Callaghan, 2004a,b; Raj et al., 2006a,b,c) has shown that induction of epileptiform activity in the brain of chickens is dependent on the waveform and frequency of the current used in the water baths, and this forms the basis of the rationale for the minimum currents stipulated for different frequency ranges in the EU Regulation 1099/2009.

3.2 Head-to-cloaca

In the Netherlands a method has been devised to stun birds by using electrodes acting on the head and the cloaca – the head is immersed in a water bath and a second electrode placed on the cloaca. This system stuns the bird and also induces cardiac arrest (Lambooij et al., 2008, 2012). This method has the potential to overcome the major deficiency of water bath to shackle stunning systems – which is the sharing of current between the birds in the water bath, which risks variation in stunning current between birds. However, the system has not been adopted in large-scale commercial practice as yet, and the birds do still require to be shackled and automatically positioned – so this does not offer any advantages in these respects over 'conventional' water bath stunning.

3.3 'Dry electrode' head-only electrical stunning

Small-scale poultry operations may use a head-only electrical stunning system – for instance for stunning of geese, ducks or turkeys during on-farm killing. The advantages of individually applied stunning to single birds is that a known and regulated current can be provided to each bird, eliminating the risk associated with current variability seen in water bath systems in which multiple birds are stunned simultaneously (Lambooij et al., 2014). This 'dry electrode' individually stunned concept has been adapted to commercial systems, where birds are placed in moving stunning cones to restrain them and to prevent wing flapping, and electrodes make contact with either side of the head. In these commercial-scale systems, the stunned birds are then transferred from the moving cone line by the machine to a shackle line (Berg and Raj, 2015).

4 Gas stunning systems

It has been known for many centuries that it is possible to kill animals by using a range of gases – gases to exclude oxygen, or gases to cause direct depression of the CNS. However, the use of gas or 'modified atmosphere' systems have only become common for killing of poultry and pigs in the last quarter century, as, at a commercial scale, the use of gas offers some potential advantages to the slaughter house operator (and some potential risks as well). The advantages – to the slaughterhouse and potentially to the welfare of the bird – are that the birds are stunned in the crate they were transported in, or whilst sitting on a moving belt, and do not need to be suspended alive in the shackles, but are shackled when unconscious. Unlike the case for electrical stunning systems, the size of the bird does not affect the efficacy of the stunning method, and the hanging of unconscious birds is perceived to be less stressful to human workers (Gerritzen et al., 2013). Modified atmosphere stunning is also perceived to offer meat quality advantages through reduced muscle haemorrhage and flapping damage, which can be associated with electrical stunning.

Gas stunning systems, however, do require high investment (Berg and Raj, 2015), and the potential welfare disadvantages of gas stunning revolve around the fact that modified atmosphere stunning is not, and never can be, immediate – as described above, in *'within a second'* (EFSA, 2004). The physical replacement of atmospheric air (atmospheric air has approximately 78.09% nitrogen, 20.95% oxygen, 0.93% argon, 0.039% carbon dioxide and small amounts of other gases) with other gases in any gas-based system will take a finite amount of time. Because of this, and because it takes a period of time for the gas to either displace sufficient oxygen or cause depressive effects on the CNS, gas stunning will not lead to immediate unconsciousness. Unconsciousness is instead gradually achieved, and it is the time taken to achieve unconsciousness and the potential for aversion or distress in the bird during this time of induction that is the principal cause of concern in some observers regarding gas stunning systems.

A number of gases, and gas mixtures, are in use commercially for modified atmosphere stunning – including carbon dioxide, where the birds are first exposed to low concentrations of < 40% by volume in air, and then, once the birds have become unconscious, exposed to higher concentrations of up to 80–90% by volume in air (Gerritzen et al., 2013). Carbon dioxide reacts with moisture on mucous membranes to form carbonic acid, and this itself is aversive (Anton et al., 1992), and additionally, birds have chemoreceptors sensitive to increased circulating carbon dioxide, and birds will react to increased dissolved CO_2 in the circulation with signs of responsiveness, headshaking, increased respiratory rate, and sometimes flapping, before unconsciousness is induced (McKeegan et al., 2006). The secondary exposure to high concentrations of CO_2 as described above ensures that unconsciousness lasts through the bleedout process. Carbon dioxide is also commonly used in combination with other gases – argon, nitrogen or both. Exposure to inert gases alone leads to progressive hypoxia by reduction in available oxygen in the stunning chamber. However, hypoxia without sedation can cause very pronounced flapping in birds, and so, inert gases are usually mixed with carbon dioxide which itself has anaesthetic and depressive effects on the CNS which can overcome the flapping induced by the hypoxia (McKeegan et al., 2007).

4.1 Gas killing for emergency purposes or for depopulation in laying hens

In Sweden, laying hens can be killed on-farm with CO_2 in a mobile container or with CO_2 inside the barn and during this type of killing, a veterinarian is required to be present. Over 1.5 million laying hens have been killed by using these methods on 150 separate occasions. Berg et al. (2014) describe the time until birds became unconscious after coming into contact with the gas, based on time until loss of balance, as being from between 3 and 5 minutes.

These authors concluded that, on the whole, 'the CO_2 whole-house gas killing of spent hens was carried out in accordance with the appropriate legislation and that death was achieved reliably'. However, these authors also identified that risks to animal welfare remained as a result of the use of 'on farm' CO_2, and that miscalculations of house volume, improper sealing or premature ventilation turn-off in the house had the potential to cause animal welfare problems.

For purposes of emergency slaughter in the case of notifiable or controlled disease, or physical emergency (flooding, inability to feed the animals due to snow), it may be necessary to kill flocks of birds 'in situ'.

McKeegan et al. (2013) explored the use of high-expansion gas-filled foam as a potentially humane, emergency-killing method. Broiler chickens, laying hens, ducks and turkeys were exposed to foam generated by passing compressed gas through a pressure vessel containing a premixed solution of water and 3% foam concentrate (HTF 1000, Ajax BV, Netherlands). The foam generated contained either air, N_2 or CO_2 and was delivered to a chamber containing birds under standardized conditions. During exposure, measurements of oxygen concentration in the foam were carried out, and the birds were fitted with sensors to measure cardiac and brain activity. Examination of the birds' behaviour when immersed in the foam included headshakes and brief bouts of wing flapping, and both CO_2 and N_2 foam rapidly induced ataxia and loss of posture, and also vigorous wing flapping in all of the bird species, with wing flapping being recognized to be characteristic of anoxic death. The use of air-filled foam did not result in loss of consciousness because when the foam broke down locally around the birds breathable air was released.

The physiological responses of the birds to both N_2 and CO_2 foam were that the birds showed bradyarrhythmia (slowing of the heart) and changes in the measured EEG which indicated an average time-to-loss of consciousness in N_2 foam of 30 seconds in hens and 18 seconds in broilers. For CO_2 foam, the average time to loss of consciousness was 16 seconds in broilers and 15 seconds in turkeys. Euthanasia thus appeared to be achieved quite rapidly, which may be explained by the very low oxygen concentrations (below 1%) available within the foam. Some obvious concerns had been raised as to whether the mode of action was occlusion of the airway by the physical presence of foam, but post-mortem examination showed that this was not the case. These authors (McKeegan et al., 2013) propose that these trials provide the proof of principle that gas-filled foam may provide an effective method of euthanasia, potentially of use in emergency-killing situations for poultry.

McKeegan et al. (2011) also studied the use of 'whole house' CO_2. Liquid CO_2 was passed into a farm building containing 28000 end of lay hens, producing a gas concentration of 45% within about 20 minutes. EEG measures showed signs associated with unconsciousness in an average time of 7.8 minutes from the onset of gas delivery to the building. The birds showed deep breathing whilst 'substantively conscious', and birds died at between 12 and 22.1 minutes after the onset of gas delivery. The authors discuss the possible concerns of aversion (CO_2 reacts with water on mucous membranes to form carbonic acid, and this is regarded as highly aversive in high concentrations) and low temperatures resulting from the application of liquid CO_2 – the authors state that in this study the birds did not die from hypothermia.

5 Other systems: maceration, manual neck dislocation, concussive stunning and low atmosphere pressure systems

5.1 Maceration and gas systems in hatcheries

In the United Kingdom, the Welfare of Animals at the Time of Killing (England) Regulations 2015 No. 1782 states that 44

(1) No person may kill surplus chicks which are less than 72 hours old in hatchery waste except by one of the following methods;
(a) maceration (in accordance with Table 1 of Chapter I and Chapter II of Annex I)
(b) exposure to a gas mixture (in accordance with Table 3 of Chapter I and Chapter II of Annex I and this paragraph)
or
(c) where there is no other method available for killing, cervical dislocation (in accordance with Table 1 of Chapter I and Chapter II of Annex I).
(2) No person may kill surplus chicks which are less than 72 hours old in hatchery waste by exposure to a gas mixture unless the chicks are placed in the gas mixture and remain in the gas mixture until dead.
(3) The killing of surplus chicks which are less than 72 hours old in hatchery waste must be as rapid as possible.

EU, Council Regulation (EC) No 1099/2009 defines maceration as 'immediate crushing of the entire animal', which may be used in chicks up to 72 hours and egg embryos. This regulation then goes on to require that 'this method shall provide instantaneous maceration and immediate death of the animals. The apparatus shall contain rapidly rotating mechanically operated killing blades or expanded polystyrene projections. The capacity of the apparatus shall be sufficient to ensure that all animals are killed instantaneously, even if they are handled in a large number.' Around the world many different designs of macerators are used in hatcheries and places where unhatched eggs (chicks that are entirely developed but that have not yet hatched) or hatched chicks that have been rejected by hatchery staff for health or other reasons must be killed in the hatchery.

The Welfare of Animals at the Time of Killing (England) Regulations 2015 No. 1782 states that chicks must be placed into the gas mixture and remain there until they are dead, and that gas mixtures can be

i. *90% argon (or other inert gas) and a maximum of 2% total oxygen by volume (2% total oxygen by volume is the proportion of oxygen in a 90% argon, 10% air mixture, as the amount of oxygen in air is 20.9%)*
ii. *a maximum of 30% carbon dioxide and a minimum of 60% argon (or other inert gas) with no more than 2% total oxygen by volume*
iii. *the highest possible concentration of carbon dioxide from a source of 100% carbon dioxide (this concentration is not recommended by the HAS as high concentrations of CO_2 are known to be very acidic and aversive)*

The aesthetic aspects of the destruction of young animals by whole maceration or by use of gas have raised many objections to these methods of killing (e.g. see *The Independent* newspaper, 2015).

5.2 Manual neck dislocation

Manual cervical dislocation (MCD) is a widely applied method used to kill individual casualty birds. EU, Council Regulation (EC) No 1099/2009 defines MCD or cervical dislocation as 'manual or mechanical stretching and twisting of the neck provoking cerebral ischemia', for use in poultry up to 5 kg live weight.

A number of techniques and equipment are used across the world for neck dislocation. The mechanism of inducing unconsciousness and death is stretching of the neck to damage the spinal cord, brain stem and the blood vessels in the neck. When applied efficiently, MCD destroys the neural connection between brain and body and causes rapid unconsciousness due to a drop in local brain blood pressure. Pronounced wing flapping and body movements are a common sequel to MCD and can indicate that the inhibitory reflexes in the spinal cord have been lost, indicating effective application of the MCD. Historically, some mechanisms have been used to 'crush' the neck of birds, but this is considered not to have the rapid onset of MCD and therefore is no longer recommended by organizations that provide guidance on poultry slaughter (HSA, 2016).

There are widely voiced concerns over the humaneness of the MCD method, with concern over the time to loss of consciousness and death (Gregory and Wotton, 1990; Sandilands et al., 2012) and issues such as training of people who carry out MCD on the farm and of assessment of the effectiveness of the method by the operator. The effects of fatigue on the effectiveness of this method if it has to be applied in anything other than small numbers of animals have also been raised, and in the EU, Council Regulation (EC) No 1099/2009 states:

These methods (MCD) shall not be used as routine methods but only where there are no other methods available for stunning.

These methods shall not be used in slaughterhouses except as a back-up method for stunning.

No person shall kill by MCD or percussive blow to the head more than seventy animals per day.

MCD shall not be used on animals of more than three kg live weight.

5.3 Concussive stunning/killing

Concussive stunning requires the application of highly concentrated energy to the skull, to cause accelerative changes in the brain, which, if effectively applied, will result in unconsciousness. EU, Council Regulation (EC) No 1099/2009 describes this method as a 'percussive blow to the head, a firm and accurate blow to the head provoking severe damage to the brain, for use in poultry up to 5 kg live weight'. The critical elements of effective application are the 'force and location of the blow'.

In practice, this is achieved either by the application of a mechanical device that delivers very rapid acceleration to the skull or by the use of manual hand-operated devices such as a 'priest' (a heavy rod).

The Humane Slaughter Association (HAS, 2016) guide to practical slaughter of poultry outlines methods of 'best practice' in the use of concussive stunning equipment, saying 'it requires complete commitment in order to destroy the brain in one fast blow. If not applied accurately and with sufficient force, it will cause extreme distress and suffering' (HSA, 2016). Examples of commercial concussive stunning devices include the TED stunner (Hulet, 2016), which uses the energy derived from gas combustion to power a portable a concussive stunner, with different models (different energies) for chickens and for turkey. Accles and Shelvoke produce a poultry killing device (CASH killer) which is commonly used on chicken, turkeys, ducks and geese (Accles and Shelvoke, 2016).

5.4 Low atmosphere pressure system (LAPS)

A well-known painting, '*An Experiment on a Bird in the Air Pump*', an oil-on-canvas by Joseph Wright of Derby painted in 1768, and now in the National Gallery, London (http://www.nationalgallery.org.uk), illustrates the principle of low atmospheric pressure stunning. In this picture, a cockatiel is seen expiring as air is withdrawn from a glass container by a pump. Currently, there is significant international interest in low atmosphere pressure stunning, as it offers the potential for batch stunning/killing of birds without use of imported gases (CO_2/nitrogen/argon) and the birds can be rendered unconscious in their transport module.

As with gas stunning systems, the gradual removal of air by pumps and induction or progressive hypoxia resulting from very low oxygen tension take a period of time (McKeegan et al., 2013). In commercial systems, the time taken to create the low pressure by pumping out air, and once the chamber is sealed, is of the order of 30 seconds. Concerns have been voiced by animal welfare organizations that the reduction in pressure (to an equivalent of 10000 m altitude) within a short time could result in pain associated with pressure changes in sinuses, viscera and other closed body spaces that contain gas – but currently no published work in detail assesses the effects of pressure changes on body tissues in birds as a result of LAP. Studies have indicated that time to loss of posture once low pressure is achieved was from 20 to 69 seconds (McKeegan et al., 2013), the average time from first movement to loss of posture was 64.9 seconds (Vizzier-Thaxton et al., 2010) and that the chickens were inside the LAPS for 2 minutes after loss of posture and for a total of 2.5 minutes. McKeegan et al. (2013) also reported that the onset of wing flapping was from 96 to 159 seconds, and it is proposed by some authors that the time to onset of wing flapping may be used as a behavioural indicator of the time to onset of unconsciousness. Therefore, the time to loss of consciousness could be as long as 159 seconds after exposure to LAPS (McKeegan et al., 2013).

LAPS is approved by the USDA for commercial use in the United States (Vizzier Thaxton, 2015) and is gaining commercial application for large commercial poultry slaughterhouses in that country. Following analysis of existing published data, EFSA (EFSA, 2014) concluded that it was unclear from existing data whether the rate of decompression used in the LAPS system induces unconsciousness and death without causing avoidable pain and suffering, and concluded that LAPS could not currently be permitted as an approved stunning and killing method in the EU.

6 Future trends

Globally, a variety of methods are used to kill poultry on-farm, as individual animals for small-scale culling, for emergency purposes and for commercial purposes in slaughterhouses and hatcheries. There are welfare and aesthetic concerns with many of these methods, as the killing of animals will always induce concern in observers as to the speed, effectiveness and 'foolproofness' of the methods used.

Two potential 'future' methods for commercial-scale stunning have been proposed – transcranial magnetic stimulation (TMS) and the use of microwave energy.

TMS creates an intense magnetic field around the skull and can result in behavioural signs and an appearance of the EEG characteristic of unconsciousness; however, TMS is currently only at an experimental stage (Lambooij et al., 2011).

Microwave energy at between 300 MHz and 300 GHz (Small et al., 2013a,b) has been shown to lead to an increased temperature of the brain, and an increase to between 43°C and 50°C (Small et al., 2013b) can induce reversible unconsciousness, without tissue damage, but again, no commercial stunning application has yet been created from this discovery.

Further research in the area of welfare science at the time of killing could provide more confidence and surety that the methods used can produce 'as good a death as is practical' (*euthanatos – Greek* εὐθανασία; 'good death': εὖ, *eu*; 'well' or 'good' – θάνατος, *thanatos*; 'death'). And if societal concerns or science indicate that what is being done should be altered, then there should be a responsiveness to 'change' to accommodate improvements driven by increases in knowledge, or as a result of societies changing the understanding of what constitutes a 'good death'.

7 Where to look for further information

Humane Slaughter Association (HSA) (2016), Practical Slaughter of Poultry A guide for the smallholder and small-scale producer. http://www.hsa.org.uk/downloads/publications/hsa-practical-slaughter-of-poultry.pdf (accessed on 15 March 2016).

European Commission (EC) (2009), *Official Journal of the European Union* Council Regulation (EC) No. 1099/2009 of 24 September 2009 on the protection of animals at the time of killing.

8 References

Accles and Shelvoke (2016), http://www.acclesandshelvoke.co.uk (accessed on 16 March 2016).

Anonymous (2009), Council Regulation (EC) No 1099/2009 of 24 September 2009 on the protection of animals at the time of killing.

Anton, F., Euchner, I. and Handwerker, H. O. (1992), Psychophysical examination of pain induced by defined CO_2 pulses applied to the nasal mucosa. *Pain* 49: 53–60.

Berg, C. and Raj, A. B. M. (2015), A review of different stunning methods for poultry – animal welfare aspects. *Animals* 5: 1207–19. doi:10.3390/ani5040407.

Berg, C., Yngvesson, J., Nimmermark, S., Sandström, V. and Algers, B. (November 2014), Killing of spent laying hens using CO_2 in poultry barns. *Animal Welfare* 23(4): 445–57.

Bock (2016), TED Stunner. http://www.tedstunner.com (accessed on 16 March 2016).

EFSA (2004), Welfare aspects of the main systems of stunning and killing the main commercial species of animals. *EFSA J.* 45: 1–29.

EFSA (2013), Guidance on the assessment criteria for studies evaluating the effectiveness of stunning interventions regarding animal protection at the time of killing. *EFSA J.* 11(12): 3486.

EFSA (2014), Scientific opinion on electrical requirements for poultry waterbath stunning equipment. *EFSA J.* 12(7): 3745.

EFSA (2014), Scientific opinion on the use of low atmosphere pressure system (LAPS) for stunning poultry. *EFSA J.* 12(1): 3488.

European Commission (EC) (2009), Council Conclusions on a Universal Declaration on Animal Welfare 2934th Agriculture and Fisheries Council meeting Brussels, 23 March 2009 Council of the European Union.

European Commission (EC) (2009), Official Journal of the European Union Council Regulation (EC) No 1099/2009 of 24 September 2009 on the protection of animals at the time of killing.

European Commission (EC) (2012), Study on various methods of stunning for poultry. Framework Contract for evaluation and evaluation related services – Lot 3: Food Chain. Final Report, p. 123.

Gerritzen, M. A., Reimert, H. G. M., Hindle, V. A., Verhoeven, M. T. W. and Veerkamp, W. B. (2013), Multistage carbon dioxide gas stunning of broilers. *Poult. Sci.* 92: 41–50.

Gregory, N. G. and Wotton, S. B. (1989), Effect of electrical stunning on somatosensory evoked potentials in chickens. *Br. Vet. J.* 145: 159–64.

Gregory, N. G. and Wotton, S. B. (1990), Effect of stunning on spontaneous physical activity and evoked activity in the brain. *Br. Poult. Sci.* 31: 215–20.

Gregory, N. G. and Wotton, S. B. (1990), Comparison of neck dislocation and percussion of the head on visual evoked responses in the chicken's brain. *Vet. Rec.* 126: 570–2.

Hindle, V. A., Lambooij, E., Reimert, H. G. M., Workel, L. D. and Gerritzen, M. A. (2010), Animal welfare concerns during the use of the water bath for stunning broilers, hens, and ducks. *Poult. Sci.* 89: 401–12.

Hulet, R. M., Cravener T. L. and Bock R. (2013), *Evaluation of Captive Bolt Method of Turkey Euthanasia Device (TED) for Humane Euthanasia of Poultry Department of Animal Science*. Penn State University, University Park, PA 16801, Bock Industries, Inc., 156 Bock Lane, Philipsburg, PA 16866. http://www.tedstunner.com/publications1.html (accessed on 16 March 2016).

Humane Slaughter Association (HSA) (2016), Practical Slaughter of Poultry A guide for the smallholder and small-scale producer. http://www.hsa.org.uk/downloads/publications/hsa-practical-slaughter-of-poultry.pdf (accessed on 15 March 2016).

Independent Newspaper (2015), http://www.independent.co.uk/life-style/food-and-drink/hatched-discarded-gassed-what-happens-to-male-chicks-in-the-uk-10088509.html (accessed on 16 March 2016).

Kranen, R. W., Lambooij, E., Veerkamp, C. H., van Kuppevelt, T. H. and Veerkamp, J. H. (2000), Haemorrhages in muscles of broiler chickens. *Worlds Poult. Sci. J.* 56(2): 93–126.

Lambooij, E., Reimert, H. G. M., Verhoeven, M. T. W. and Hindle, V. A. (2014), Cone restraining and head-only electrical stunning in broilers: effects on physiological responses and meat quality. *Poult. Sci.* 93: 512–18.

Lambooij, E., Reimert, H. G. M., van de Vis, J. W. and Gerritzen, M. A. (2008), Head-to-cloaca electrical stunning of broilers. *Poult. Sci.* 87: 2160–5.

Lambooij, E., Reimert, H. G. M., Workel, L. D. and Hindle, V. A. (2012), Head-cloaca controlled current stunning: assessment of brain and heart activity and meat quality. *Br. Poult. Sci.* 53(2): 168–74.

Lambooij, E. (1981), Some neural and physiological aspects of electrical and mechanical stunning in ruminants. PhD. Thesis, University of Utrecht.

Lambooij, E., Anil, H., Butler, S. R., Reimert, H., Workel, L. and Hindle, V. (2011), Transcranial magnetic stunning of broilers: a preliminary trial to induce unconsciousness. *Animal Welfare* 20: 407–12.

McKeegan, D. E. F., McIntyre, J. A., Demmers, T. G. M., Lowe, J. C., Wathes, C. M., van den Broek, P. L. C., Coenen, A. M. L. and Gentle, M. (2007), Physiological and behavioural responses of broilers to controlled atmosphere stunning: implications for welfare. *Animal Welfare* 16: 409–26.

McKeegan, D. E. F., Sandercock, D. A. and Gerritzen, M. A. (2013), Physiological responses to low atmospheric pressure stunning and the implications for welfare. *Poult. Sci.* 92: 858–68.

McKeegan, D. E. F., McIntyre, J., Demmers, T. G. M., Wathes, C. M. and Jones, R. B. (2006), Behavioural responses of broiler chickens during acute exposure to gaseous stimulation. *Appl. Anim. Behav. Sci.* 99: 271–86.

McKeegan, D. E. F., Wathes, C. M., Reimert, T., Hindle, V. A., Boulcott, P., Demmers, J. G. M., Gerritzen, M. A. and Sparrey, M. (2013), Physiological and behavioral responses of poultry exposed to gas-filled high expansion foam. *Poult. Sci.* 92: 1145–54. http://dx.doi.org/10.3382/ps.2012-02587.

McKeegan, D. E. F., Sparks, N. H. C., Sandilands, V., Demmers, T. G. M., Boulcott, P. and Wathes, C. M. (December 2011), Physiological responses of laying hens during whole-house killing with carbon dioxide. *Br. Poult. Sci.* 52(6): 645–57.

Raj, A. B. M. and O'Callaghan, M. (2004a), Effect of amount and frequency of head-only stunning currents on the electroencephalogram and somatosensory evoked potentials in broilers. *Animal Welfare* 13: 159–70.

Raj, A. B. M. and O'Callaghan, M. (2004b), Effects of electrical water bath stunning current frequencies on the spontaneous electroencephalogram and somatosensory evoked potentials in hens. *Br. Poult. Sci.* 45: 230–6.

Raj, A. B. M., O'Callaghan, M. and Knowles, T. G. (2006a), The effect of amount and frequency of alternating current used in water bath stunning and neck cutting methods on spontaneous electroencephalograms in broilers. *Animal Welfare* 15: 7–18.

Raj, A. B. M., O'Callaghan, M. and Hughes, S. I. (2006b), The effect of amount and frequency of pulsed direct current used in water bath stunning and neck cutting methods on spontaneous electroencephalograms in broilers. *Animal Welfare* 15: 19–24.

Raj, A. B. M., O'Callaghan, M. and Hughes, S. I. (2006c), The effects of pulse width of a pulsed direct current used in water bath stunning and neck cutting methods on spontaneous electroencephalograms in broilers. *Animal Welfare* 15: 25–30.

Sandilands, V., Sandercock, D., Sparrey, J., McKeegan, D. E. F., Hopkins, J., Raj, A. B. M. and Sparks, N. H. C. (August 2012), Approaches to culling poultry humanely on farm. *Br. Poult. Abstr.* 8(1): 1–55.

Small, A., McLean, D., Owen, J. S. and Ralph, J. (2013a), Electromagnetic induction of insensibility in animals: a review. *Animal Welfare* 22: 287–90.

Small, A., McLean, D., Keates, H., Owen, J. S. and Ralph, J. (2013b), Preliminary investigations into the use of microwave energy for reversible stunning of sheep. *Animal Welfare* 22: 291–6.

Sparrey, J. M., Kettlewell, P. J. and Paice, M. E. A. (1992), Model of current pathways in electrical water bath stunners used for poultry. *Br. Poult. Sci.* 33: 907–16.

Sparrey, J. M. and Kettlewell, P. J. (1994), Shackling of poultry: is it a welfare problem. *World's Poult. Sci. J.* 50: 167–76.

Wooley, S. A., Brothwick, F. J. W. and Gentle, M. J. (1986), Tissue resistivities and current pathways and their importance in pre-slaughter stunning of chickens. *Br. Poult. Sci.* 27: 301–6.

Vizzier-Thaxton, Y. (2015), From idea to reality: the development of the low atmosphere stunning system. Proceedings 'Recent Advances II' – HSA International Symposium, 16 to 17th July, Zagreb, Croatia, p. 36.

Vizzier-Thaxton, Y., Christensen, K. D., Schilling, M. W., Buhr, R. J. and Thaxton, J. P. (2010), A new humane method of stunning broilers using low atmospheric pressure. *J. Appl. Poult. Res.* 19: 341–8. doi:10.3382/japr.2010-00184.

Index

Lightning Source UK Ltd.
Milton Keynes UK
UKOW06n2148300817
308252UK00001B/1/P

9 781786 760722